D0861197

INTERSECTION

HOW ENTERPRISE DESIGN BRIDGES THE GAP BETWEEN BUSINESS, TECHNOLOGY AND PEOPLE

INTERSECTION

HOW ENTERPRISE DESIGN BRIDGES THE GAP BETWEEN BUSINESS, TECHNOLOGY AND PEOPLE

MILAN GUENTHER

AMSTERDAM • BOSTON • HEIDELBERG • LONDON
NEW YORK • OXFORD • PARIS • SAN DIEGO
SAN FRANCISCO • SINGAPORE • SYDNEY • TOKYO
Morgan Kaufmann is an imprint of Elsevier

Intersection is an aptly named book, both in its content and the evolving world of leading innovation that it describes. This book explains why new programs like ours are changing the form, content, and purposes for business education. Businesses need to challenge and rethink nearly everything about their processes, procedures, strategies, and cultures in order to successfully lead into the future.

Nathan Shedroff, Program Chair, MBA in Design Strategy, California College of Art

Intersection is a landmark book on how to design enterprises in today's hyper-connected world. Founded on the relationships between identity, architecture and experiences, Milan Guenther's work explains both the change in mindset and practical approaches that are needed for success. If you are still designing your enterprise from the inside outwards, *Intersection* will convince you to think again. If you are already designing your enterprise outside-in, use this wonderful book to validate your work.

Chris Potts, author of *FruITion* and *RecrEAtion*

If you think that words like cross-channel, ecosystem, or bridge experience have no place in the world of large-scale corporate strategies, this is the book for you.
Intersection reframes enterprise-level design challenges through the lens of complexity, and provides us with a thorough interdisciplinary framework and a holistic approach to Enterprise Information Architecture that bridges business, technology and people into one seamless, dynamic design process.
A must read.

Andrea Resmini & Luca Rosati, authors of *Pervasive Information Architecture*

This innovative book will give you a clear vision of organizations as interaction-driven businesses. Reading *Intersection* will enlighten you with deep insights into how to manage design in our contemporary world.
First, you will understand that all large-scale design projects have an impact on organizational design and need to be managed as such: as Enterprise Design projects.
Second, this book conveys—what most business people avoid to see—that any enterprise starts as an abstract intangible concept, but will only come to life through tangible signs, things, and places.
Third, you will understand better how designers think in systems, what is holistic thinking, and how design skills are useful for managing the interaction between all stakeholders by aligning identity, architecture, and experience of your enterprise.

Dr. Brigitte Borja de Mozota, Design Management Institute Life Fellow / Université Paris I / Ecole Parsons

Sustainable enterprises increasingly depend on a process incorporating inter-disciplinary knowledge on both strategic and operative levels. This book merges valuable insights from various perspectives into a holistic framework.

Marc Stickdorn & Jakob Schneider, Editors of *This is Service Design Thinking*

Acquiring Editor: Andrea Dierna
Development Editor: Robyn Day
Project Manager: Mohanambal Natarajan
Editorial Design: Benjamin Falke, Sarah Matzke, Dennis Middeke, Eva Pika, Andrea Santema
Copyeditor: Philip Hellyer

Morgan Kaufmann is an imprint of Elsevier
225 Wyman Street, Waltham, MA 02451, USA

© 2013 Elsevier, Inc. All rights reserved.

No part of this publication may be reproduced or transmitted in any form or by any means, electronic or mechanical, including photocopying, recording, or any information storage and retrieval system, without permission in writing from the publisher. Details on how to seek permission, further information about the Publisher's permissions policies and our arrangements with organizations such as the Copyright Clearance Center and the Copyright Licensing Agency, can be found at our website: *www.elsevier.com/permissions*.

This book and the individual contributions contained in it are protected under copyright by the Publisher (other than as may be noted herein).

Notices
Knowledge and best practice in this field are constantly changing. As new research and experience broaden our understanding, changes in research methods or professional practices, may become necessary. Practitioners and researchers must always rely on their own experience and knowledge in evaluating and using any information or methods described herein. In using such information or methods they should be mindful of their own safety and the safety of others, including parties for whom they have a professional responsibility.

To the fullest extent of the law, neither the Publisher nor the authors, contributors, or editors, assume any liability for any injury and/or damage to persons or property as a matter of products liability, negligence or otherwise, or from any use or operation of any methods, products, instructions, or ideas contained in the material herein.

Library of Congress Cataloging-in-Publication Data
Application Submitted

British Library Cataloguing-in-Publication Data
A catalogue record for this book is available from the British Library.

ISBN: 978-0-12-388435-0

For information on all MK publications
visit our website at *www.mkp.com*

Printed in China

13 14 15 16 17 10 9 8 7 6 5 4 3 2 1

Working together to grow
libraries in developing countries

www.elsevier.com | www.bookaid.org | www.sabre.org

ELSEVIER BOOK AID International Sabre Foundation

CONTENTS

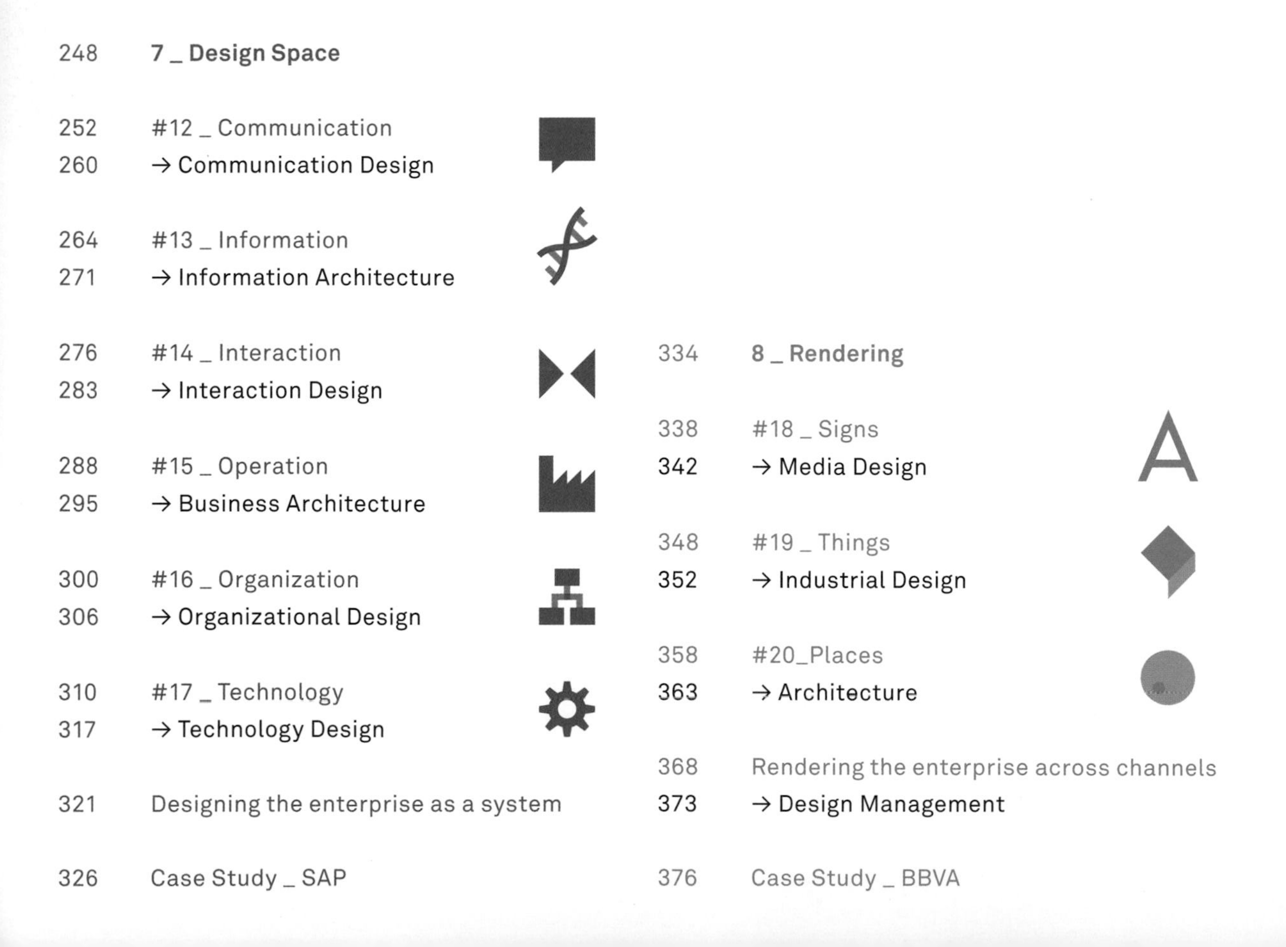

INTRODUCTION

_ WHY DID WE CALL THIS BOOK INTERSECTION?

When Jenifer Niles, then my editor at Morgan Kaufmann, proposed that name, I was intrigued.

This is a book about design, clearly. But it touches many adjacent or related areas, by approaching problems from different perspectives, aiming to bridge viewpoints and concerns, and connecting design to today's complex social ecosystems.

Therefore, the title expresses very well one of the key themes of this publication: looking beyond the immediate task, beyond your own comfort zone and background, beyond a briefing or project scope, and embracing viewpoints and practices other than your own.

Intersection gives you a thinking model, a methodological framework, and a vocabulary to do just that. It is a resource to apply design thinking and practice to challenges you consider relevant and important to tackle. It promotes both a mindset and an approach that enables you to take a step back, and look at the big picture of everything that matters when approaching a difficult design challenge. The face of companies, organizations, public services, and other types of enterprise is changing. Formerly clear lines are fading away—between online and offline, internal and external, owned and shared, customer and user, social and business, branding and operations. When thinking holistically about a complex challenge, such distinctions just don't seem to make much sense anymore.

© 2013 Elsevier Inc. All rights reserved.

One of the immediate consequences of these shifts is the need to align, bridge and connect; more and more professionals are calling themselves architects, designers, or consultants. If you are among those, you inevitably face the challenge of transforming ecosystems, regardless of your particular background, focus area, or level in an organization. Such systemic challenges go well beyond the problems of designing products, web sites, or services in isolation.

This is not a hands-on book promoting definite methods or tools to be used in such a setting. Instead, it is about the interrelationships and dependencies between the various concerns you will meet, and how to align different conceptual decisions on common course.

Every design process in such a setting involves acting in a space of great uncertainty, making a series of conceptual decisions, and producing real outcomes. It means taking risks, embarking on a journey without a defined end, and chasing opportunities as they emerge. It requires working closely with your peers, partners, stakeholders, users, and customers, as well as developing a clear vision of where you want to be. I wrote this book to help you with that task. And, more importantl, to have some fun in tackling your particular design challenges at the intersection of business, technology, and people.

Milan Guenther
May 2012

_ THE STORY BEHIND THIS BOOK

In 2007, I attended a talk at the Ecole Nationale Supérieure d'Art in Nancy, France, where I was spending a year as a student. The presentation was about a new campus that would bring together three independent schools, planning their joint future to educate the next generation of graduates. They should benefit from a vivid exchange between a school for art and design, a business school, and a technology institute, with joint classes and projects that crossed the boundaries of disciplines.

The team around Parisian architect Nicolas Michelin presented their idea of the new campus, with models and renderings of the buildings and their surroundings. They began their presentation with a thinking model, a system of interconnected concerns that drove their decisions—light, space, materials, social life, ecology, wayfinding... all aspects that have to be brought together in one coherent vision.

I have been involved in many different projects between User Experience, Information Systems, and Communications in the enterprise. The complexity of different concerns to be addressed, and the interplay of viewpoints always struck me as the most difficult challenge in design.

So I began drawing a model of what matters to my work, what has to be aligned and brought together, and I have revised and refined it over the past few years based on experience and many great conversations. This book is about that model.

_ DESIGN, STRATEGY, AND THE ENTERPRISE

When I attended design school, the term *strategy* was not used very often. To some design students, using design and strategy in one sentence seemed like a contradiction, mixing artistic merit with the quest for profits.

In practice, and regardless whether for profit, the relevance of strategy to design work cannot be undervalued. We have seen design projects produce results that were great by themselves, but that fail completely to deliver on the intentions of the clients commissioning them. Such projects are usually doomed even before they begin, by pre-determining the outcomes that they are expected to deliver.

Designers are used to being asked to deliver web sites, mobile apps, logos, or other things, working with a long list of ready-made requirements, or beautifying something existing, even when the problem to be solved actually lies completely elsewhere. Concerns that were considered to be outside the scope of the design project lead to random decisions in neglected areas, ultimately producing misaligned concepts and leading to failure and overall disappointing results.

At the same time, the scope of design work grows rapidly, from visuals to interactive systems and services. To step up to this challenge, designers today look beyond individual artifacts to the entire experience with a brand, a service, or an organization. Key to this is a dialogue about the strategy behind a design initiative: understanding, questioning, rephrasing, and clarifying the goals to be achieved with a task is what makes a design initiative relevant to the problem to be solved.

At eda.c, we have experienced both the failure of projects due to predetermined results and a constant expansion of the scope of our work. One concept that we found particularly relevant to exploring the actual problem behind a briefing, and setting the true scope of a project, is the notion of enterprise.

What exactly is an enterprise? Although there is no agreement on the definition and meaning of the term, it is used widely in the areas of business and IT, and has also been adopted in the world of Marketing and Branding. While the individual definitions vary, all uses share a common basic premise: that key challenges companies and other organizations face are best tackled by addressing them in a holistic and coherent fashion.

In *Intersection*, the enterprise can be seen as the space of market players, people, and stakeholders across the ecosystem that an organization is embedded in. It comprises the structures put in place to facilitate exchanges and transactions, such as services, channels, systems, processes, and decision rules. It provides the setting for the tools, systems, artifacts, or media we produce to address this audience.

And, finally, it also encompasses the variety of motivations, meanings, experiences, and personal contexts we are designing for in the end.

_ WHO INTERSECTION IS FOR

Looking at the enterprise level means understanding one's work in terms of the overall system. This book is for everyone involved with designing and transforming enterprises at that level and scale:

- _ *Executives and strategists* looking to apply strategic design in their organizations, developing products, services, models, structures, and systems as part of a bigger whole, driving performance and competitiveness
- _ *Designers and architects* working on design challenges that require expanding their view on the enterprise as a playing field, looking beyond particular domains, projects ,or intended outcomes
- _ *Consultants and technologists* being caught between the views, concerns, and interests of their clients and stakeholders, and looking to employ strategic design to generate a way to move forward
- _ *Entrepreneurs and visionaries* faced with the challenge of creating their enterprise from scratch, making the right decisions with regard to all relevant concerns and making their vision tangible

Intersection is especially for you if you are not exactly clear how to describe what you do, if you are always struggling with your official job title, or always creating your own roles. For some of us, this ambiguity is a part of our professional lives, and serves us well when navigating the complexity of an enterprise environment. This book is your guide on that journey.

_ HOW INTERSECTION IS STRUCTURED

This book includes 10 Chapters and is organized into three parts.

- _ *Part 1*, comprising Chapters 1 to 3, describes the thoughts behind the messages of this book, and provides the basic thinking to understand enterprise-people relationships, interdisciplinary work, and a design approach to strategic challenges.
- _ *Part 2* consists of Chapters 4 to 8 and describes the Enterprise Design Framework, the main part of *Intersection*. It takes you on a journey across the 20 aspects we found relevant in strategic design work on the enterprise level, starting at from a set of Big Picture questions to develop a conceptual Design Space, to finally come to a Rendering of results.

_ *Part 3* is about the practical side of strategic design work, with Chapter 9 mapping the framework to a typical design process and Chapter 10 describing challenges of organizational design practice.
_ The *Case Studies* at the end of Chapter 1-9 tie the abstract concepts and ideas to real-world illustrations of design work and the aspects applied. Some are taken from our work at eda.c; others are not.
_ Finally, a total of 24 *approaches to strategic design challenges* are portrayed throughout the book. They describe professional practices, knowledge areas and disciplines that pertain to the concerns being discussed in this book.

HOW TO USE INTERSECTION _

_ As an *introduction* into strategic design, and the various domains, disciplines, and approaches you might encounter when tackling design challenges on the enterprise level—this works by reading in a linear way or a rather chaotic fashion, however you prefer
_ As a *link* between otherwise disparate concerns, and a way to align them, making professionals and stakeholders work together, see commonalities and differences, and enable cross-disciplinary collaboration—starting with an aspect close to your own field and going on from there
_ As a *guide* in challenging projects, helping you to ask the right questions and showing up ways to make conceptual decisions in a complex space of intertwined aspects—looking at the aspects you find the most relevant in your context and exploring the references to others
_ As a *reference* for aspects and concerns that you find important in your projects, pointing you to related thinking, professional disciplines, approaches, and methods—using *Intersection* like a dictionary
_ As a *thinking aid* to come to an overall vision of where you want to be, making the journey from exploring the environment of your challenge, constraints and opportunities, to finally decide on outcomes and their introduction into the enterprise—mapping the framework to your context and going through the aspects to be addressed

PART 1

Thoughts on Enterprise Design

The first part of this book discusses the nature of enterprise-people relationships, as well as some of the challenges with which an enterprise is confronted in considering and managing these relationships. Their strategic significance is what sparks the need for a holistic design approach, to reshape the enterprise and make it meaningful to people.

To benefit from those relationships, enterprises must take into account insights from multiple disciplines to envision a way to integrate them into their long-term visions, daily operations and communications.

They must leverage the contributions and interrelationships of different sciences, disciplines and practices from engineering and technology, the human condition and society, and bridge these silos to tackle complex challenges.

This part portrays strategic design as a way to synthesize new solutions in a space of constraints and rapidly evolving opportunities, and to foster innovation starting with human relationships.

1 _ THE RELATIONSHIP CHALLENGE

Enterprises are everywhere, playing a vital role in our lives. Basically all human endeavors have reached a level of scale and complexity that makes them depend on an ecosystem of interrelated organizations and technology. Most of us rely upon enterprises several times a day, even for the most fundamental tasks of daily life. They are ubiquitous in our world of consumer brands and services, and visible in the mass of organizations of all sizes we are in touch with as consumers, employees, investors, or in other roles.

When dealing with enterprises, we are used to strange and often quite frustrating experiences. They seem to make even simple transactions awkward and complex. Straightforward activities such as booking tickets for a journey, paying your taxes, subscribing to health insurance, or resolving a problem with your energy supplier require customers to embark on a laborious journey, jumping between call centers, online forms, and missing information. They make us shift between different contacts, tools, and communication channels. They lose track of the conversation, get stuck in inflexible procedures, and often fail to deliver what they promised. Most of us have had a lot of such experiences, be it with companies, government institutions, or other types of enterprises, making them appear slow, rude, and inhumane.

When looking at the big picture of human-enterprise interaction, there are numerous examples of failed relationships. While most of them are simply annoying and just make you go somewhere else, some examples of failed relationships have a profound impact on people's lives. They result in lost customers, demotivated employees, or even scandals being echoed in mass media.

© 2013 Elsevier Inc. All rights reserved.

_ EXAMPLE

As an example, think of the advent of digital media and the fundamental shift it caused in the music industry. With the virtualization of information, long established business models did not work anymore. This development required companies to reinvent their business models from scratch, to incorporate new technologies and distribution channels. The line between content producers and consumers has become blurred. New market players reflect the changed behaviors and preferences of their customers by delivering a wide range of niche content purely online.

This sparks the need to consider people relationships much more profoundly, making them central to the enterprise. At its heart, every enterprise is about its relationships to people. This includes customers addressed with its activities, employees carrying them out, and other stakeholders involved in or impacted by their execution. By approaching stakeholders with a strategically relevant relationship to the enterprise as what they are—as human beings with needs and demands to be met—businesses can uncover tremendous opportunities by intentionally addressing their concerns.

This book is about employing the approach and the discipline of design to do just that, turning relationships into viable business assets. By taking a creative and deeply humanist approach to problem solving, design can be used to tackle strategic relationship challenges and make visible a desired future state of the enterprise.

_ DESIGNING ENTERPRISES

Today's enterprises are deeply embedded in hyperconnected and constantly shifting economic and social environments. This confronts them with numerous challenges in their management, planning, and daily business execution. Some of these challenges are related to large-scale transformation processes such as digitization, globalization, or shorter production and consumption cycles, while others are the results of shifts in specific industries, regions, or markets.

The transformation of the music industry shows how a shift in an industry not only means new technology, but also changed business models that disrupt and transform the existing order. Every change also alters the mindsets of people involved, consumers and market players alike. It is their new interests and goals, interactions and exchanges, values and beliefs that drive the transformation process. Even if triggered by some external influence, like the invention of a new compression algorithm for audio data in the example above, it results in new or evolved relationships between people, and between people and enterprises.

For any enterprise, this translates to concrete business challenges to be addressed with a coherent strategy, in order to survive and succeed in the marketplace. It involves achieving and maintaining competitive advantage, establishing new operating models, transforming organizational structures, and turning new technologies into useful business differentiators.

While established management approaches offer a large variety of tools to tackle these challenges, they seem to do so thinking in silos. They address topics such as improved operational efficiency, better product quality, or reaching new target groups with Marketing efforts. Seen from a human perspective however, they fall short of addressing relationships as what they really are—the key driver behind any business strategy. This section outlines some of the fundamental challenges enterprises face in creating and sustaining their relationships to people, those being the foundation of a strategic design initiative.

Designing to create value _

To survive in the marketplace, enterprises constantly seek to differentiate themselves against other market players. There are different ways to approach this challenge. Businesses make use of research to envision innovative products and leverage technologies, they examine and optimize work methods and processes to increase operational efficiency, or outsource parts of their business in order to excel at their core activity. On the customer side, they aim for new and better products and services, faster delivery, or lower prices. The success of these approaches has made them a common practice for many business executives. But while in theory every customer would prefer to buy from the best and the cheapest, reality shows a different picture: that we are different, enterprises and people alike. People make their choices on where to buy, who to work for, or what to invest in based on a wide range of influencing factors like brand identity, values, trust, and always in a unique individual context.

Therefore, it is crucial to any business to differentiate itself by shaping and sustaining its relationships with people. This thinking has resulted in initiatives like Customer Experience management, Investor Relations or Talent Management, applying an outside-in view on the enterprise through the eyes of its key stakeholders—customers, investors, and staff members. Instead of just differentiating on product features, performance, or abstract notions of quality, they aim at expressing the value an enterprise provides to the people it is related to, in a way that makes it unique *just for them.*

A design approach as an integral part of the business strategy enables enterprises to systematically create value propositions for the people they address. It allows us to integrate stakeholder-specific approaches in an overarching vision, and to design artifacts and systems that are useful and meaningful for everyone in touch with the enterprise. It fosters relationships by exploring, meeting, and exceeding real human needs and turning them into business initiatives, products, and services.

Designing to connect and exchange _

Today's dynamic business environment requires enterprises to connect to a fluctuating, highly interconnected group of people, reaching beyond organizational boundaries. They need to establish strong links to customers and to collaborate with business partners in loosely bound networks along the value chain. They have to enable their employees to make the decisions needed for business development and daily operations, and exchange knowledge and information with colleagues and external contacts.

In order to gain trust and thrive in their ecosystem, they must drive an open exchange with employees and business partners, and conduct authentic conversations with customers. They have to be visibly engaged, communicate consistently and behave appropriately in virtual spaces, where they can exercise only limited control over dialogues between individuals. The way a company interacts with and through people such as employees, clients, applicants, or investors shapes its identity and image, regardless of whether an exchange takes place in person, an online portal, or on Facebook. Even the voice of a single customer or employee can have a big impact on an entire enterprise, being amplified by the dynamics of people connected in global social networks.

The availability and exchange of information is a crucial success factor for successfully mastering this dialogue. Enterprises must establish strong links to all of the people they are connected to, and provide the right information and messages, at the right time and in the right format, adapted to the individual role and context of use.

Traditional approaches to information management and communication, such as business intelligence or advertising, seem to be too narrowly focused on their respective silos to provide suitable solutions for this challenge. Despite new technology and communication channels becoming available, we still have to deal with a mass of disconnected tools and media. We a hard time filtering relevant information, making sense of it, and using it for our individual purposes.

A strategic design approach is a means to tailor information and messages to all people being addressed, and to establish sincere and honest relationships. Starting with people helps organizations find out about individual information needs, and to use communication channels to provide information in a way that suits their respective audience. Those communication channels form the foundation for successful stakeholder relationships, being the basis for defining and conveying an organizational identity.

The Internet and related advances in information and communication technology (ICT) are transforming economic activity, much as the steam engine, railways and electricity did in the past.
OECD

Designing to adapt and grow _

Enterprises have to constantly adapt to changes in their environment, to deal with new situations, or to seize opportunities as they emerge. Transformation is required for them to keep up with new market developments, perhaps with an adapted business model, a new product, an acquisition, or simply to become compliant with changed legal requirements.

An agile enterprise needs to be structured and organized in a way that allows for transformation processes to take place on all levels. Instead of static structures encoded in organizational silos with fixed processes, it has to be designed with agility in mind. This new perspective is reflected in management paradigms that strive to turn organizations into lean and responsive entities. On the technology side, this has resulted in new approaches: the fixed, monolithic assets used to drive business processes, such as enterprise resource planning software used for operations, are replaced by modular network structures made of individual components available on demand, which can be reused and recombined to meet new needs.

To be able to react and adapt quickly to new conditions, enterprises depend largely on the agility of people involved in their activities. Actual work gets done by people solving real tasks, embedded in larger business processes. On the level of teams and individuals, business agility is often hampered by the technical and cultural difficulties involved in transforming how work actually gets done. Individual agility depends on the ability of people to adapt their work methods, collaborate and make decisions, across organizational boundaries and in a timely manner. It involves transforming the activities performed and introducing new tools to staff, customers, and other stakeholders interacting with the enterprise.

To achieve such a level of agility, transformation processes need to take people into account. Enterprises must enable the capabilities needed to accomplish tasks and carry out work in a flexible way, and to adapt collaborative workflows, roles, and responsibilities to changed situations. This also involves allowing for deviation from the way an activity was originally planned, and to customize and tailor tools to reflect new requirements.

By looking in detail at the interactions taking place, tasks being done, and contexts of use, a strategic design initiative can come up with solutions that fit. This in turn helps the enterprise to significantly increase productivity and agility. It enables businesses to empower people with a choice of tools and services as composite structures, allowing for customization and adaptation. It allows us to design explicitly for the transitions between different elements and media, fostering flexibility, and turning technology into valuable tools for its users.

_ HUMANIZING TECHNOLOGY

Many seem to believe that the key to doing this, and tackling all the challenges described before, is the better use of technology. The increasing influence of digital systems on our private and professional lives are hard to deny, as are the possibilities they offer. The ongoing digitization of our economy is making information systems the essential backbone of all exchange in society, with elements like emails, databases, and algorithmic automation supporting all sorts of human activities. For enterprises, these systems are the enablers to automate operations, reach more customers and markets, and to manage their conversations, business processes, and cases. This gives IT and practitioners of the related discipline of Information Systems, a key role in designing the modern enterprise.

INFORMATION SYSTEMS

Today, the improvement of organizations and the information systems in them is not a matter of making more information available, but of conserving scarce human attention so that it can focus on the information that is most important and most relevant to the decisions that have to be made.

Herbert A. Simon

Information Systems as a discipline and practice has emerged to respond to the increasing need of organizations to improve their capabilities to process, store, manage and make sense of data. Although in theory, an information system does not require using a computer and usually not all of its parts are automated, in practice those systems are almost always widely based on digital technology. Today, a large part of the business processes in a company are IT-dependent, being at least partly automated and relying on digitally stored data.

The information systems that drive the processes of an enterprise are complex structures, composed of technology, people, and process components facilitating a dynamic flow of data. Common practice in IT departments is to treat these systems as though they were invisible, working in the background to support business operations—the concerns of efficiency, reliability, quality, continuity, and security are prevalent on the IT executive's agenda.

While this perspective on information systems will remain important in fulfilling their organizational role, new technical developments and new usage paradigms are forcing IT executives to think beyond invisible operations. They need to support the users of their systems by providing valuable information and better, more usable functionality.

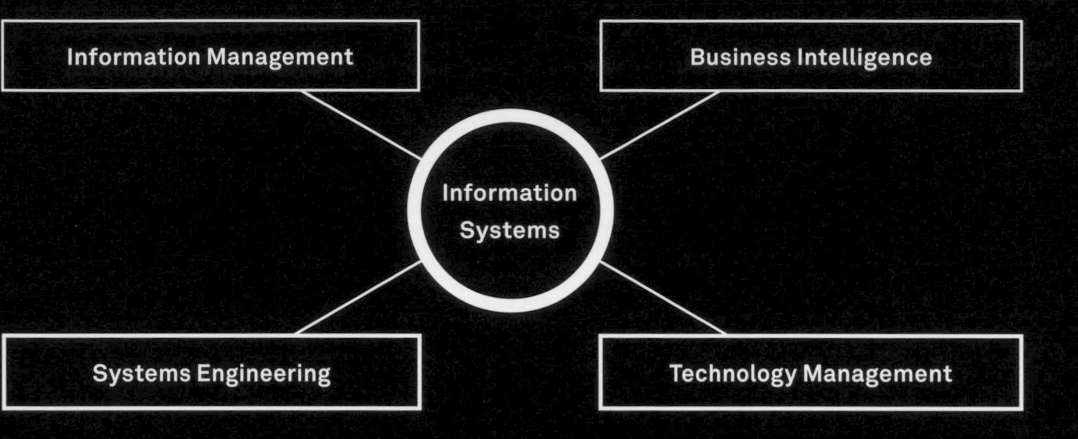

Information Systems is a field of knowledge exploring the nature of these systems and the way they are structured, their usage by people and organizations, and their impact on economy and society. In most organizations, there is some sort of IT department applying the practice of information systems for developing, maintaining, and evolving the systems used for its operations.

Information Systems is applied in enterprises in various sub-practices:

_ INFORMATION MANAGEMENT

the way an organization deals with information relevant for it to function and progress, makes it available as a resource, and uses it in its business processes and activities.

_ BUSINESS INTELLIGENCE

supporting decision making by identifying, gathering, and analyzing business data and using it in visualizations, reports, dashboards, or other forms of information media.

_ SYSTEMS ENGINEERING

the practice of developing, customizing, and using information systems based on requirements elicitation, data modeling, and implementation in IT systems and infrastructure.

_ TECHNOLOGY MANAGEMENT

the use of technology infrastructure and applications in organizations, including planning, implementation, operational support, and lifecycle management.

A new role for IT _

Sometimes, modern enterprises manage to do the impossible. They delight customers with great offerings, establish trustful relations to their staff, and engage the interest of a community of stakeholders with a powerful idea. They employ technology for streamlined operations and information exchange, while at the same time listening to people and adapting to their needs. While information technology plays a vital part in such settings, it seems to have significantly shifted its role.

With digital tools getting more important for society and the economy every day, organizational IT departments find themselves under immense pressure from various stakeholders to reinvent their function. No longer just providing a necessary commodity to manage data at low cost, they must become an enabler of technology-driven innovation. This is visible in the employees using their private phones and tablets for work, customers interacting with enterprises on Facebook, or products embedded in Internet-based service platforms.

In order to meet this demand, they must recognize that digital systems used in enterprises are no longer just the background data storage and processing units they used to be. Customers expect to interact with an enterprise using digital channels available at all times, and employees integrate their personal device landscape into their work practice. While the need to make systems work smoothly to run the business efficiently is still there, it is not enough to reflect the new role of information systems: they are the channels for interaction and communication, both inside and outside the organization.

IT must either start leading the business models and evolution of the organization, or become a commoditized utility while the business figures out the moves on their own.

Dion Hinchcliffe, Dachis Group

They are what people see and use when interacting with other people or businesses. Information systems are a part of a company's products and services, in the form of websites and applications. They have evolved from being background components to providing the stage for people interactions, acting as the communication channel for every business process and every transaction, as websites, applications, or kiosks. Even when invisible to customers in the case of a direct exchange with representatives, information systems support customer-facing staff by making available information in context for the service to be provided.

Yet most organizations today are still far away from leveraging the potential and opportunities made possible by the larger role of digital systems in people's daily lives. Businesses struggle to address process agility and change, and to overcome the paradigm of purely IT-driven automation. Employees find their needs unaddressed, having to work with isolated tools which are hard to use and inflexible, that bury relevant information deep in a mass of data, and force them to manually bridge gaps between applications or media. Customers are kept outside the operational systems with meaningless digital brochures on the web or with useless apps for their smartphone. Their desired interactions require writing emails or talking to call centers because of a lack of useful online interaction capabilities. A single user has to interact with a wide range of tools, media, and systems which reflect different roles, relationships, tasks, information types, or data sources.

Digitization and shifts in the perception of technology have resulted in new kinds of information systems, that are intelligent, pervasive, embedded, and ubiquitous. These new paradigms are the basis to use design as a technology enabler for the enterprise, bringing together different systems in a way that is meaningful to their users:

_ MODULAR AND CO-CREATED SYSTEMS

instead of the monolithic software solutions or applications of the past, individual pieces of software take the form of distributed and reusable components, being loosely coupled and integrated in larger contexts and individual usage scenarios. Users can compose and co-create their own individual solution on the fly out of flexible components, going beyond packaged solutions.

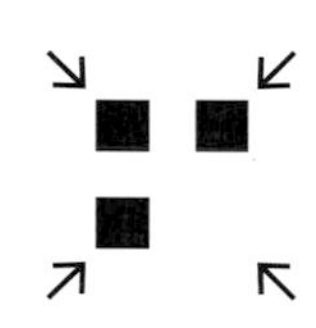

_ OPEN AND INTERCONNECTED SYSTEMS

information systems of all kinds are being opened to the outside world, where people from different organizations and consumers communicate and interact in shared or public environments, exchange information, and collaborate. These portals and communities are owned by a particular organization, shared with business partners, or made available by third parties.

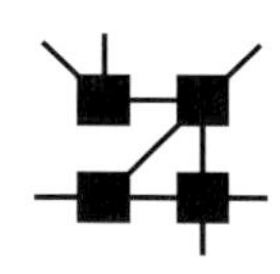

_ UBIQUITOUS AND MOBILE SYSTEMS

resources and capabilities are becoming available at all times in the Cloud, in various use contexts and situations—at work, at home, or in transit. Devices like tablets and smartphones, and a whole set of emergent natural interfaces and systems embedded in everyday things enable ubiquitous interaction, like an invisible information layer covering the world.

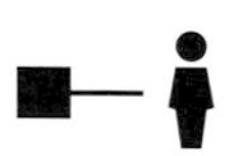

_ INTELLIGENT AND ADAPTIVE SYSTEMS

algorithmic intelligence, rapid data processing and mining, semantic data structures, and knowledge about roles and needs enable systems to proactively adapt themselves. It allows identifying and providing information that is useful for a system's target users in a way that supports their activities.

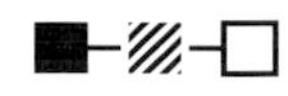

_ Design and Technology

As explored at the beginning of this Chapter, relationships between enterprises and people today suffer deeply from the bad experiences they keep producing. With all the technology we have today, it should be clear that just introducing more systems and more data for its own sake does not solve any problems. Even advanced organizations are not leveraging technology to gain competitive advantage, but are merely using digital tools to facilitate and streamline their existing activities. In order to take on the evolved role of information systems that support people in various contexts and facilitate enterprise-people relationships, enterprises need to redesign and restructure themselves around human needs.

This makes the task at hand that of bridging the gaps between business, technology, and people. Instead of asking what is possible, it is about making sense of it all, and seeking the sweet spot where the technical possibilities, human needs, and business opportunities overlap.

A strategic initiative enables enterprises to leverage new technologies as part of their architectural structure, supporting brand identity and human experiences by connecting with people in a meaningful fashion. In order to work, they must integrate information systems in a way that they provide value to the enterprise and people involved in all its activities. They must take into account business requirements, human needs and aspirations, as well as technical possibilities and constraints to come up with an overall design that makes sense. This is a complex challenge that requires looking at problems from different angles—the theme discussed in the next Chapter.

AT A GLANCE

Designing or redesigning an enterprise means shaping and evolving its relationships to people. The use of information systems has shifted from being only the backbone to providing the stage for people interactions. This makes a meaningful fusion of business goals, human needs, and technology key to any such transformation.

Recommendations

_ Formulate business strategies that create, renew, and strengthen relationships by creating and delivering real value to people

_ Establish links to people that allow an open exchange of relevant, meaningful, and useful information

_ Build dynamic business structures around human needs that flexibly support people achieving their goals

_ Employ an enterprise-wide design approach to leverage technology as a means to connect to people and facilitate business relationships

CASE STUDY _ AEG

AEG'S ENTERPRISE

Making useful electrical appliances and devices for households and industrial usage, *perfekt in Form und Funktion*.

The idea of using design to address strategic relationship questions may seem quite new, emerging with the approaches of design thinking and experience design. To some, it surely feels like taking a huge leap from today's predominant role of design in organizations, often relegated to lower-level execution or external service providers. In such a setting, design is merely an afterthought to choices of business and technology.

Therefore, the first case study we have chosen for Intersection is one from history, taking place in Germany during a period of industrialization shortly after the turn of the century. The roots of the industrial giant AEG go back into the year 1883, and by the time of this case study the Berlin-based company had become one of the world's largest manufacturers of electrical equipment. Commissioned as an Artistic Consultant in 1907, Peter Behrens pioneered the domains of industrial design and architecture, and is one of the inventors of what is today known as corporate brand identity. Being a painter by trade and architecture autodidact, his ability to think beyond the boundaries of any single discipline paved the way for a more holistic use of design.

The work of Peter Behrens still profoundly impacts design practice in Europe and beyond. Artists, architects, and designers such as Walter Gropius, Ludwig Mies van der Rohe, and Le Corbusier worked for his architecture company, incorporating his thinking into the *bauhaus* and the emerging formal design education. Decades later, two world wars and decades of rule by fascist ideologies brought this expanded cultural role of design to an end. As it is practiced today, design seems to have lost some of the ideas pioneered by these early practitioners.

Symptomatic of this view of design is the separation into subdisciplines, and the purely tactical, short-sighted application of design, which leads to incoherence between different designed elements. It is worthwhile to look back again, learning once more to see the whole of a design challenge, not just the individual parts in isolation.

AEG arclamps and fan by Peter Behrens

_ DESIGN CHALLENGES

Originally hired to work on smaller projects and single problems such as advertisements or manuals, Peter Behrens soon convinced his client that the actual design challenge was the relationships among these elements. Therefore, he can be seen today as the first designer or design consultant working in an enterprise context—the company contracted him to look after everything visible they produced, used, and offered to their target markets. Although he was never an employee of AEG, he significantly influenced the company's strategy and endeavors by aligning it with the different disciplines of design.

The challenges of designing enterprises that we explored before can be found in this work. He developed a system of fundamental design principles and universal elements, spanning across the company's product portfolio, brand identity, and corporate architecture long before these terms were invented. These foundations were guiding the design choices, striving to align the design of industrial assets, communications, and products. They enabled the company to clarify its vision and mission, provide valuable propositions to the market, reach out to their target groups, and adapt their portfolio and operations to new developments.

With this toolkit, Peter Behrens designed communication tools and visual brand elements, office buildings and worker homes, factory floors and production facilities, household appliances and industrial devices. His approach enabled AEG to overcome the dual paradigms of practical considerations of function and the arbitrary decisions of style. Instead, they made each designed element part of a larger design, applied to the whole enterprise.

Design is not about decorating functional forms – it is about creating forms that accord with the character of the object and that show new technologies to advantage.

Peter Behrens, Artistic Consultant at AEG

THE ROLE OF TECHNOLOGY _

As a technology company, AEG was clearly driven by a culture of engineering. Therefore, the work of Peter Behrens also holds insights for applying design in today's technology-infused world, where information systems and digital devices support enterprises and people alike. Although the technologies were quite different at the time, the human and business needs still prevail, especially in technology-centric business environments.

In a time when artistic sense was considered irrelevant for commercial and industrial activities, Peter Behrens defined the task of a designer as making technologies and their functions accessible to people, and expressing the character of an object in its form. He applied this philosophy thoroughly, which could be seen today as an early user-centric approach to design. The task shifts from decoration and ornamenting to shaping technology based on its usage and role, and applying this throughout all applications in the shape of products, machines, tools, and work environments.

Based on this approach, also systems considered outside the realm of industrial production were considered a part of the design work. Advertising, catalogues, manuals and other communication channels are examples of pre-digital information systems, informing people in their roles as customers or users of AEG products, as well as staff working on their production. They were facilitating the company's business processes such as production and sales. In retrospect, we find the elements of strategic enterprise-wide design approach, based on a union of humanism, technology, and business brought together in a variety of applied arts, whose results helped the enterprise to adapt to changes in its environment and steer its development. We believe that this kind of overarching thinking is the foundation of using design strategically, in both yesterday's and today's enterprise environments.

Evolution of the AEG Logo

2 _ BLURRING BOUNDARIES

When looking at relationships between enterprises and people, it becomes clear that reshaping and transforming them is no trivial task. Because relationships fail or succeed at the enterprise level, it is also on this level that they need to be considered. In order to address them in a coherent and holistic fashion, design practice must get beyond individual target groups, products, artifacts, or channels, and wholeheartedly embrace the complexity and diversity of those relationships. In such a setting, enterprises are forced to constantly adapt and evolve to new situations.

Just in our private lives, digitization and its consequences change the world around enterprises more rapidly and thoroughly than ever before. Although disruptive technology plays an important role in these shifts, what really matters for the enterprise is their impact on the people it is related to. Changing the way people and businesses interact with each other, these shifts result in faster communication, more transparency, new knowledge distribution, and shifted powers among market players. They have profound effects on the relationships to people, and prove to be quite difficult to deal with in terms of both scale and complexity. They apply not only on the level of enterprises and business ecosystems, but also on entire industries and on society and our economy as a whole.

The new situations that enterprises suddenly find themselves in put them into a position where they are forced to act and adapt, but the nature of the problem makes it difficult to find a suitable approach. In practice, strategists often struggle to define how and where to start the innovation process once such a challenge is recognized.

© 2013 Elsevier Inc. All rights reserved.

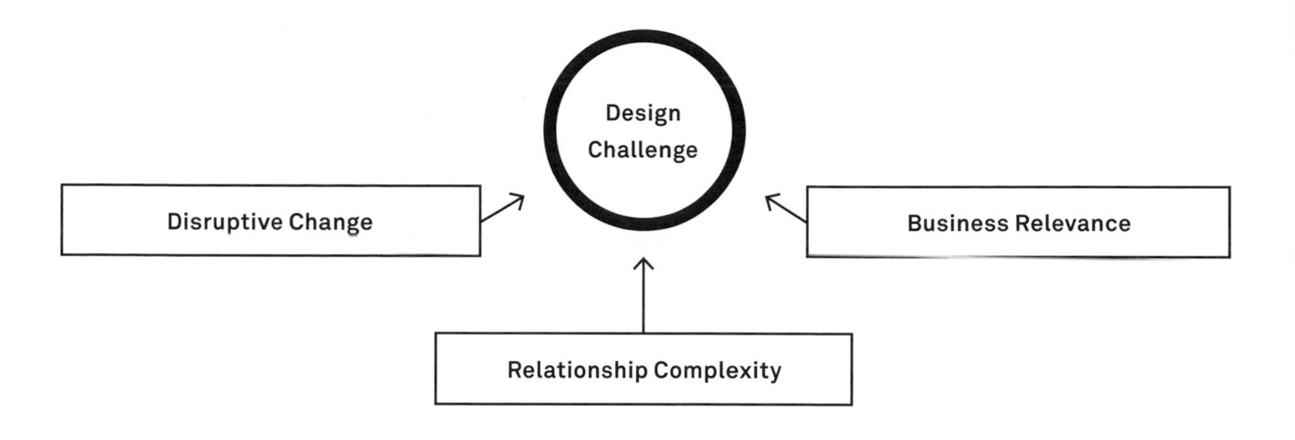

In our consulting work at eda.c, we often find relationship challenges in the enterprise context to be driven by three major characteristics:

_ DISRUPTIVE CHANGE

although the arrival of some kind of disruptive technology or trend often plays a major role when thinking about the problem, its potential impact is hard to predict. The issue is part of a dynamic environment that shows a large diversity of substructures, and affects the dynamic, unpredictable behavior of many of the elements and actors involved.

_ RELATIONSHIP COMPLEXITY

the issue involves or impacts a large variety of people (stakeholders) who have some kind of influence on the strategic direction of the enterprise. Their diverging concerns, values, and interests make it difficult to choose a direction. Addressing this requires taking into account a strong human and social dimension.

_ BUSINESS RELEVANCE

successfully dealing with the problem is critical to survival in the marketplace and business development, because it endangers the current market differentiation and the way things are done in the enterprise, but may also hold a great opportunity. Although the transformation process has already started, there seems to be no obvious answer to cope with the issue.

_ EXAMPLE

Today's level of connectedness makes it possible for people to connect with each other as well as with brands and enterprises in a more dynamic way, causing a major change in the way enterprises have to communicate with their customers, prospects and other stakeholders. The role these new communication devices played in the political revolutions in the Middle East illustrates their potential: people used them to communicate, to coordinate their efforts and to use the power of tribes acting collectively. Organizations attempting to repeat traditional advertising campaigns on Facebook were caught by surprise when critical voices were amplified and their attempts to control the situation failed, resulting in a PR disaster impacting daily business.

People documenting events in Egypt

In their book *Dilemmas in a General Theory of Planning* from 1973, design researcher Horst Rittel and city planner Melvin Webber coined a label for such challenges, calling them *wicked problems*. While not all relationship challenges qualify as wicked, they share a level of uncertainty that makes them hard to tackle. They typically apply to the enterprise as the larger ecosystem around an organization, and require holistic thinking about its particularities as a complex socio-economic and socio-technical system. The problem is literally shifting while we attempt to solve it. Their inherent complexity and ambiguity mean that conventional, rational approaches to resolve them turn out to be ill-suited. Instead of analyzing the problem in detail before making an informed decision about what to do, they require acting with incomplete information about parameters, conditions, and requirements, and in an environment where even the known parameters constantly change.

INNOVATING ACROSS DOMAINS AND DISCIPLINES _

Relationship challenges do not halt at the boundaries of particular disciplines, practices, thinking styles, or knowledge areas. Due to the diversity of domains touched by such a challenge, it is often unclear what kind of competence would be best suited to approach the problem, and who to involve as an expert. This Chapter explores the idea of approaching relationship challenges as enterprise-wide design problems, by bridging different domains and their views. It strives for a shared understanding among experts and stakeholders, to achieve a common idea about the conditions and characteristics of a desirable outcome.

Professional disciplines bring with them their philosophies, approaches, biases, and core competencies. Being experts in their field, professionals view the world in a certain way, and apply known recipes to approach a problem. While being thorough in their own field, they are often unaware of the views and insights from other domains.

Tackling a relationship challenge therefore requires the involvement of many contributors, in varying depth and breadth depending on the individual issue. This section explores the particular contributions of some of the major knowledge areas to the resolution of relationship challenges in the enterprise, reflecting also disciplines, initiatives, and practices that are typically present in modern organizations. As an important side note, the mindsets and approaches described in this section should not be read as descriptions of people, but as divergent attitudes that can apply alone or in combination with individuals, teams, and entire organizations.

_ Business: uncovering opportunities

The effects of a digitized society and economy are widely visible today, especially in the level of competence many business people have in using and understanding new technology. This has caused specialized business disciplines such as management and strategy, Marketing, Finance, or Legal to investigate an ongoing *digital shift* in business, economy, and society. They attempt to understand the transformation processes and to translate this understanding into actionable initiatives for organizations and their leadership.

Professionals in the area of business strategy and operations are confronted with these kinds of relationship challenges across all industries and subdomains. The task of understanding the potential and influence of a change is undertaken by strategists close to the business decision makers rather than by technologists close to its technical root cause—the underlying trigger that made it happen in the first place. These strategists concentrate on the role of technology as tool or medium, exploring risks and opportunities as well as value propositions.

From a business perspective, the interesting part is how technology supports new ways of communicating, interacting, doing business, or even committing crime, thus causing cultural change in society and affecting the way the economy works. This provides the basis for innovative business models, that change the way a market works and how a business positions itself in it.

Breakthrough innovations that thrive in such dynamic environments, regardless of industry or time, are always based on a spirit of entrepreneurship (or enterprise), which is vital to achieving a transformation in a complex setting. It is fundamentally about replacing something existing with something new that provides more value. It requires actively seeking the opportunities that lie sleeping in technical developments or other transformations, developing a vision of how to realize that value. The hard part is then turning such technologies into a powerful enough idea that it provides the basis for an enterprise, a strong brand, and a compelling offering. Today's business thinking, particularly in large organizations, is often not ideal to generate and pursue such a vision, largely due to its roots in a worldview of simplified models, such as those used in financial economics.

The scale and speed at which innovative business models are transforming industry landscapes is unprecedented.

For entrepreneurs, executives, consultants and academics, it is high time to understand the impact of this extraordinary revolution.

Alexander Osterwalder and Yves Pigneur in *Business Model Generation*

Such thinking focuses on detailed planning, predicting, and measuring. It strives to reduce uncertainty and maximize the chances for success. It assumes a world that can be captured in a mathematical model, where the basis for a successful strategy formulation lies in a thorough analysis of the market environment, leading to a positioning that provides competitive advantage. Many prevalent management paradigms taught at business schools are based on the idea that the key challenge is to improve and recombine what already exists, from improving operational efficiency to increasing product quality, and expanding the feature-driven product matrices used in Marketing.

This thinking is apparent in many business-related knowledge areas, and places a large emphasis on gaining empirical insights through market research and acting upon them. Instead of recognizing the people involved in the enterprise as what they are—real human beings with human needs—it is based on generic concepts of consumers, staff, and shareholders. Difficult decisions are simplified by replacing human participants with rational actors, assigning to them the aspiration to maximize quantifiable values such as return on investment or the number of features in a product.

These models are indeed very valuable approaches to developing an understanding about the economic side of a problem, but their inherent simplifications make them fall short when dealing with the complexity of relationship challenges. While people actually do value useful functionality and often attempt to make rational choices, the reality of social and cultural complexity often results in unexpected behaviors. The economic approaches lack the more qualitative and often quite uncertain human element of any business success, although the strategic significance of addressing these aspects is widely visible.

> The appropriate manifestation and use of technological advancements can bring about powerful change with regards to the mind, body and soul. These benefits are made possible by advances in engineering, yet they will not be found by engineering advances alone.
>
> **Jon Kolko in *Thoughts on Interaction Design***

_ Engineering: exploring the possible

Successfully applying any technology first requires an enterprise to understand how it works. Therefore, to rely on expertise apparent in engineering disciplines seems to be an obvious choice for a business when a new technology arrives. In many organizations, especially those that have technology at the heart of their business model, engineers are the major force behind innovation and change. In such a business context, innovations as well as the research activities seeking to generate them are equaled by the disruptive effect of new technologies.

The technology-centric mindset is one of creative exploration of the possible and the feasible, seeking to achieve new benefits by understanding and transforming how things work. This is based on deductive, analytic thinking and attempts to predict and control a system's behavior. While the closely related natural sciences are trying to explain the world as we see it, engineering disciplines use that knowledge to invent new things that did not exist before.

This emphasis on analysis as a thinking mode lets engineering approaches concentrate on the details *inside the box*, attempting to develop understanding about the hidden causalities, invisible to those people who just use a technology. It enables the planning and construction of new realities, dealing with questions about implementation, maintenance, performance, security, and optimization, and relying on empirical results and formal logic to make it actually work.

When approaching relationship challenges that are embedded in enterprise challenges or opportunities, a technology-centric perspective helps to extend the boundaries of what is possible, developing technical solutions that might prove disruptive to the current state. Technology-centric, algorithmic thinking is the basis for any automation in the enterprise, and the key prerequisite for all the advanced interactive systems and infrastructures most people use now every day. Those inventions were made possible by engineering and IT professionals, making faster data processing, larger storage capacities, artificial intelligence, touch interfaces, and other inventions available to people via smaller, more powerful, and more accessible devices. Engineers understand and build things that, to most people, seem to be magical.

Technology-driven innovation has caused major transformations in life and business—especially when looking at larger time scales and key inventions like the computer. However, there is a shift in perception going on even in engineering cycles that just adding more capabilities, more information, and faster processing is not the answer to all challenges. Engineering-driven innovation is primarily about what is possible, not what is desirable. It fundamentally cares about the details, develops an understanding of structures that a single mind cannot possibly tackle, and develops solutions that are complicated in nature in order to solve a complicated problem. With the ongoing digitization of everything, many realize that the engineering mindset, while offering tremendous opportunities for innovation, is ill-suited to converting them to value.

_ EXAMPLE

The arrival of social networks has forced the professional practice of Public Relations to reorient itself, moving away from one-way mass messages toward influencing individual communication streams. They have to explore ways to use these new channels for communication with their target groups, thereby reinventing their tools and methods. On a larger scale, politicians, lawyers and judges attempt to reflect evolving usage patterns and possibilities in legislation and judicial practice. Governments are only beginning to address the copyright implications of file sharing, while with *BitCoin* there is already a virtual, independent currency emerging. On the business side, small start-ups like Skype transformed entire markets by combining new technology with a bold business model, turning phone calls into a free service.

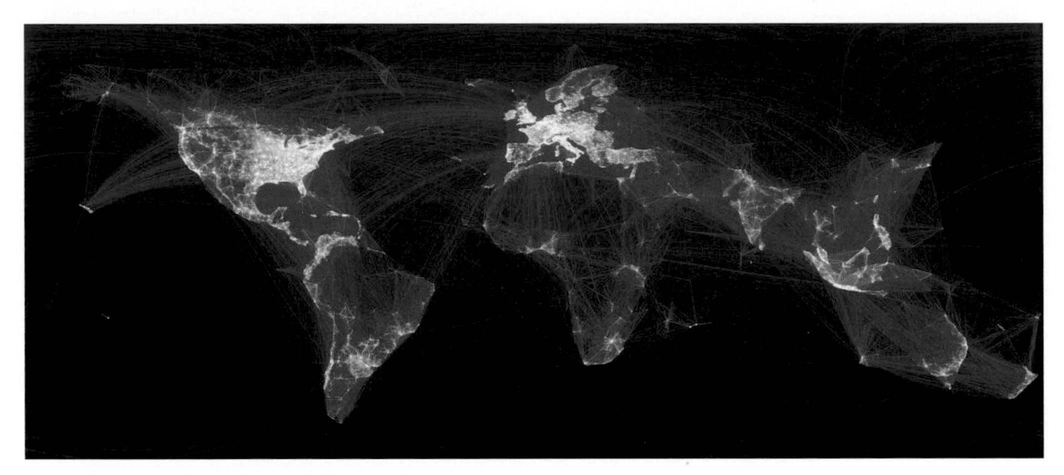

A visualization of Facebook connections

Many engineers seem to prefer to work from defined requirements, assuming that the desirable result of their work is already clear; the challenge is simply how to achieve it. In reality, only a deep inquiry into usage, context, and meaning makes technology fit into people's lives, and enables enterprises to turn it into useful innovations. Technology changes the way we live in both professional and private contexts, and any business needs technologists to benefit from such a change. Determining the individual impact on life and business to make such a change meaningful to people or to turn it into a business opportunity is a different matter.

People: understanding the human condition _

All innovations are fundamentally about people. Any decision made in the areas of technology and business results in perceivable social outcomes. With all the technology there is and all the business models that make its usage viable, what matters in the end is whether it is possible to give it a meaningful role and purpose in our personal or professional lives. Only if that role is anticipated and understood can a change be an opportunity to a business, for example, by turning a new technology into a new product or service offering. Today's entrepreneurs still lack business tools to incorporate human-centric thinking into their strategy, sparking the increasing need for business executives to reconnect with the people side of business. To do so, they have to turn to insights from disciplines exploring the subject of all our endeavors: the human being, the human condition, and society.

Business practice today employs a large variety of knowledge areas around the humanities and social sciences to better understand people and the relationships between them. Disciplines such as psychology, anthropology, sociology, communication studies, linguistics, and cognitive science provide detailed insights into how people are experiencing, thinking, and collaborating in the enterprise. Using varying scientific backgrounds, these fields explore human nature in all its diversity, and provide a basis for grounding a business strategy or initiative in a larger theory of human and social behavior. They help us develop an understanding of fundamental and universal elements of any enterprise as a social system, such as values and cultural norms, routines and habits, motivation and communication. Applied in the right way, they provide explanations and predictions regarding the way people act and react to change. They allow us to learn about people and their behaviors, and to inform our decisions, strategies, and leadership.

In modern organizations there are many practices and professionals with a professional background in humanities and social sciences, such as HR, Marketing, or Organizational Development. Although they share the general subject matter—the study of human beings—their individual focus, philosophies, approaches, and work methods vary significantly. A key challenge in the enterprise context is therefore to identify a human-centered approach that is suitable for contributing valuable insights given a specific challenge. The traditional role of these disciplines in the enterprise suffers from being marginalized as a part operationalizing a given business strategy.

In their book *Subject to Change*, the authors from Adaptive Path take a look at the ways that today's business practices incorporate knowledge about people in their activities. The discipline of human factors enables us to optimize the usability and efficiency of technical systems, but thereby reduces people to being just a part of the system to be engineered, without considering the emotional or subjective aspects of their experience. In a similar fashion, consumer psychology is misused by traditional Marketing and advertising to detach a story from the actual product or service, to simply convey a persuasive message. Just as in the case of applying simplified economic models to relationship challenges, these applications suffer from a too simplified model of people in relation to the enterprise—as robots functioning in a machine, part of an efficiency-driven automation, or as mindless consumers subjected to sales-driven persuasion efforts.

Applied in isolation, these approaches all fall short of addressing relationship challenges. In fact, most people problems can be considered wicked problems by themselves, simply because the human reality is so complex and difficult to grasp. This has caused some business people to believe that their way to success has a lot more in common with fine arts than with applying scientific approaches. Indeed, there are many success stories that seem to follow a rather intuitive and expressive approach, driven by a strong idea and a charismatic leader. In reality such success is usually based on an intuitive, empathetic understanding of people. Even if not consciously pursued, the element of appropriately addressing the intended audience is present in any success story.

_ EXAMPLE

When Apple introduced their first iPod in 2001, it marked the beginning of a radical shift in their business. A few years later, the company has expanded from IT into the entertainment industry, based on giving their existing products and services new roles and meanings. While technology innovations clearly played a role in this development, there were only few really new things in the offering from an engineering point of view. Instead, Apple concentrated on designing a seamless, overarching experience for their customers, taking the tasks of finding, buying, and listening to music seriously. This made them dominate the marketplace, solving the challenges of a complex technology, new competition, unknown territory, and risks related to selling digital content. Although technology was an enabling factor, Apple's success is largely based on a purposeful design for an evolved relationship with their customers, and exploiting the opportunities therein.

Involving expertise from the people disciplines has the goal of making this aspect visible and learning more about it. It provides a starting point for turning a business idea into a success, by exploring the conditions of a human experience that is yet to be defined, including perception, cognition, behavior, emotion, and context. Methods applied in this field, such as ethnographic research and social modeling, enable enterprises to achieve a deep understanding about the people they address, both on a level of empirical analysis and of personal empathy. They allow us to develop a detailed understanding of the mental models and characteristics of the people that drive the enterprise-people relationship, validating ideas in the field and putting innovations to the test.

Therefore, successfully applying this knowledge requires a clear definition of its role in the innovation process. The main challenge in order to get actionable insights from human-centric methods is to ask the right questions, select a suitable approach out of many options, and interpret the insights from research activities in a way that it delivers actionable results. This interpretation needs to be fused into a solution approach that bridges the human-centric viewpoint with others, such as those of engineering and business. Again, the *people people* cannot do it alone; they just add another fundamental piece to the puzzle of dealing with major transformations.

All planning processes are, at their core, vehicles for communication with employees at all levels and between business units. That is particularly true of processes that tackle wicked issues.

John C. Camillus in *Harvard Business Review (May 2008 edition)*

_ CONNECTING THE DOTS

Although in most organizations, many of the disciplines mentioned earlier are somehow present, enterprises struggle to leverage their intellectual capital to tackle complex relationship challenges. It is necessary to take into account many viewpoints, insights, competencies, and influencing factors to envision a possible direction to go. Moreover, because of the large number of stakeholders that need to be involved, any resolution depends on continuous communication and the successful alignment of a wide range of diverging concerns.

The place, domain, or background that sparks the motivation to tackle such a problem can be anywhere in the enterprise. People having great ideas usually spend some time thinking about a problem from their own perspective, then talking to people with a different one, and reflecting on it again. The challenge is to facilitate these dialogues, making people talk to each other and helping them to translate between their individual languages and viewpoints. This is particularly true for any approach to tackling a complex relationship challenge, where different domains need to be blended in order to turn them into a coherent strategy.

This section explores ways to achieve this, by establishing cross-functional teams, and placing creative generalists in connector roles to translate between domains.

We found that the best project teams we worked with share some common traits of team culture, which proved to be critical success factors when seeking a holistic response to a complex relationship challenge in the enterprise:

_ DIVERSITY

team members come from different organizational units and varying professional backgrounds, including those with a business, technology, and people perspective. Also, diversity in terms of age, cultural backgrounds, seniority, and experience is encouraged.

_ OWNERSHIP

people feel responsible and passionate about achieving a good overall outcome, and develop a culture of shared problem and solution ownership beyond their individual contributions.

_ RESPECT

deep expertise in a certain domain is respected by other team members without losing interest, and people are engaged in a continuous open exchange also about the details of a problem.

_ INTEGRATION

team members are connected to the enterprise, share thoughts and results with the rest of the enterprise, and reach out to involve stakeholders and other experts from the outside beyond just obtaining facts and opinions.

_ Enterprise teamwork

Virtually all transformations a business undertakes require specialist expertise and deep domain knowledge. If a transformation is so important that it is on the executive agenda, it must be done the right way, typically by involving expert knowledge. Therefore, most organizations bring together many professional backgrounds, even though the structure of disciplines and practices present in an individual enterprise varies widely depending on the respective industry and market segment.

Any domain comes with its own terminology, approaches, context-specific experiences, and ways of thinking. This leads to the familiar silos around specific domains, bundling specialist expertise and experience, and avoiding exchange and collaboration with other domains. This is widely visible on many organizational charts and the problems are apparent in daily business—specialist groups such as HR, IT, or Communications end up in having a separate department, culture, silo of competence, and agenda.

But capturing and tackling a relationship problem requires connecting different domains, to manage the many challenges of complexity and interaction, to bridge silos established by separate groups, and to overcome the uniform thinking they employ.

Even in a small enterprise, there is no single person able to cover every domain involved in such a challenge in full depth. When the problems to be tackled are merely cross-cutting concerns appearing regardless of competence area or unit, each of those areas has to deal with the issues, but cannot solve the problem in isolation. To approach such a problem, work needs to be partitioned between different people with a deep expertise in their respective domain and also the power and influence in the enterprise to make change happen. Consequently, any resolution depends on some excellent teamwork.

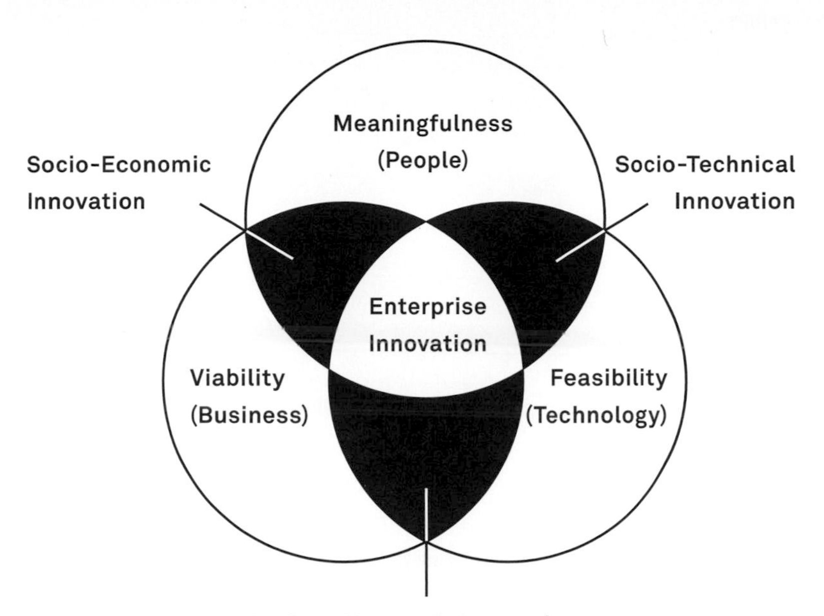

In a famous talk on Design Thinking at *TED*, Tim Brown from IDEO identified different areas of innovation that cross the domains of business, technology, and people, and a need to address these overlaps to seize the opportunities therein. Building on that idea, here is our classification of innovations in the enterprise:

_ SOCIO-ECONOMIC INNOVATION

opportunities to convey the business idea to customers, employees, and other stakeholders on an emotional level, making its purpose obvious and embedding it in culture, and building on human capabilities and characteristics to drive success.

_ TECHNO-ECONOMIC INNOVATION

opportunities to leverage the potential of technology to make the business operate and perform, by improving processes and capabilities, achieving higher value with fewer resources and less effort.

_ SOCIO-TECHNICAL INNOVATION

opportunities to make a technical development serve a new purpose in people's lives, by adapting it to human needs, creating compelling offerings, and facilitating interactions and transactions.

_ ENTERPRISE INNOVATION

opportunities to achieve innovation on a relationship level, by combining new technical options with new business models to generate new meanings for people.

_The role of connectors

All three of the perspectives explored before provide valuable input to get to a meaningful, feasible and viable outcome. Creating and sustaining fruitful enterprise-people relationships facilitated by information systems requires aligning technical, business, and human elements to serve a common purpose. Cross-functional teams are essential for exploring the problem and ways to resolve it holistically, determining the opportunities and constraints, stakeholder concerns and success conditions. A key challenge is the multi-competency required to come to a good result. This makes it necessary to establish a connection between those elements, and among the people advocating for them.

Using our interpretation of Tim Brown's framework, almost any significant innovation can be seen as an enterprise innovation, because it touches all three areas and benefits from all types of innovation. However, people deeply grounded in a specific domain tend to focus on that area alone, especially in the enterprise context where such areas are established groups such as departments or homogeneous teams. Collective thought processes and groupthink in organizational structures focus on the familiar terrain, preventing anyone from stepping out of their silo.

To overcome this separation and consider all types of innovation, many organizations use generalist roles typically called architects, designers, or consultants. They work in cross-domain teams to apply a high-level view, bridge viewpoints, and deliver a holistic perspective. These roles act as connectors between viewpoints, translate into the different expert languages, and align efforts in pursuit of a common goal.

People in connector roles have to balance the different modes of thinking and possess domain knowledge in different areas. They apply formal rigor to design a system that fits, but balance this with a strong emphasis on so-called *soft skills* such as leadership, social dynamics, communication, and creativity. They engage in a social planning process involving all experts, team members, and stakeholders. Such a way of thinking has been labeled *integrative* or *hybrid* thinking, building on the idea of design thinking and emphasizing the idea of connecting different domains, problem spaces, and viewpoints.

HYBRID THINKING

Design Thinking's fundamental emphasis on creating meaningful, human-centered experiences provides the core for Hybrid Thinking, which is an emerging "discipline of disciplines."

Hybrid Thinking goes beyond Design Thinking by integrating other forms of thinking to take on the most ambiguous, contradictory and complex problems.

Nick Gall, Gartner

The term *Hybrid Thinking*, coined by Dev Patnaik and now advocated in the fields of business and IT, refers to the ability to integrate different ways of thinking, jump between them, and align them to work towards a universal goal. Before him, Roger Martin had used the term *Integrative Thinking* to describe the related concept of a single mind capable of integrating different states. It defines a certain mindset and perspective on problems (especially wicked ones), together with a set of basic characteristics and skills that apply to any initiative for transformation.

Hybrid Thinking expands on the ideas of Design and Systems Thinking, both as the starting point for a challenge and as a focal point of any outcome. It is based on empathetic, intuitive thinking to create meaningful human-centric experiences. It follows the idea that the approach and thinking of great designers apply to any problem, even to those outside the traditional realm of design, to things like processes, business models, or enterprise transformation. When combined and integrated with other ways of approaching a problem, especially those applied in Systems Thinking to deal with complexity, it can be used to explore the problem space and generate possible outcomes. The hybrid thinker is able to deal with opposite states of mind and integrate different conceptual worlds.

Expanding on the definition from Nick Gall, in our view hybrid thinkers can come from any background, but share a particular way to approach a challenging problem:

_ VISUAL AND SYSTEMS THINKING

they seek systemic solutions to make disparate elements work together and respond to a challenge coherently. They have the ability to make hidden relations, complicated thoughts, and visionary ideas visible to others, either in sketches and schemes, or using other means of communication and documentation.

_ INTEGRATIVE AND COLLABORATIVE

they are able to understand different views and approaches, to speak to and translate between professionals from various domains, to work with diverse groups and different mindsets. They are deeply interested in ideas and contributions from the outside their own background or environment.

_ CREATIVE AND OPTIMISTIC

they draw inspiration and ideas from various sources, and are passionate about achieving a good outcome on a personal and emotional level. They have an open mind for new ideas and are ready to identify and seize opportunities as they emerge.

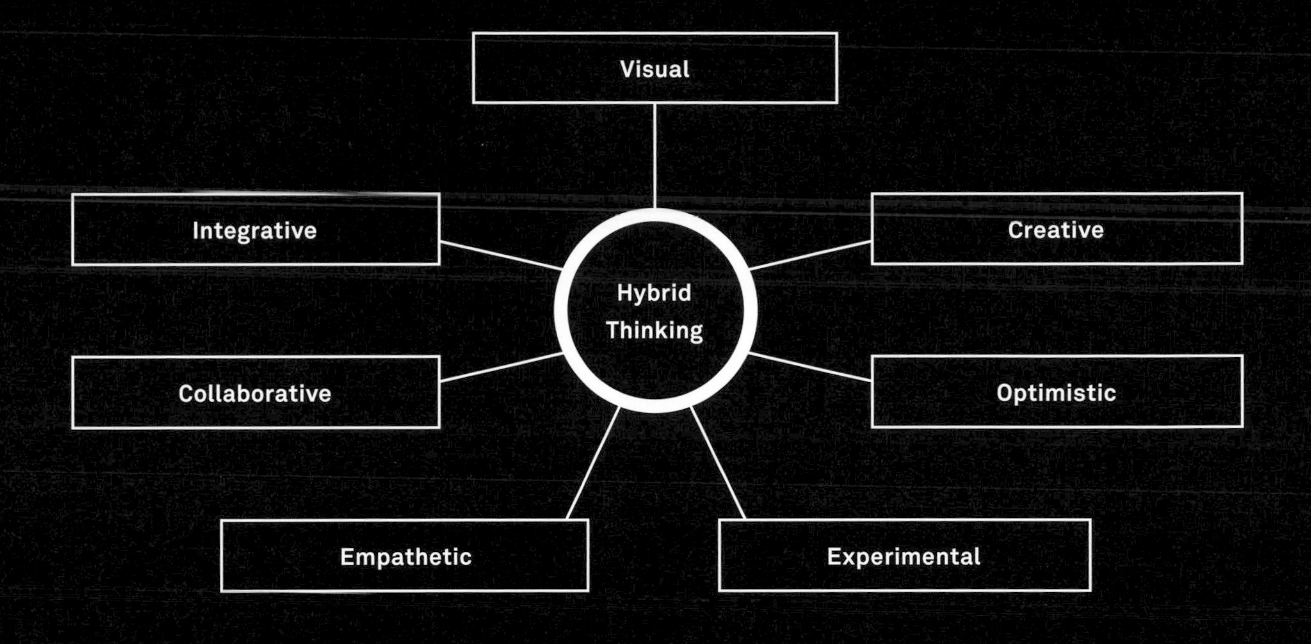

_ EMPATHETIC AND EXPERIMENTAL

they approach a problem by developing empathy for the people involved in or addressed by a transformation, using playful exploration and co-creative approaches in the ecosystem. They try out and validate potential solutions as experiments, are prepared to fail in order to learn something and try again.

_ UNDERSTANDING AMBIGUITY AND RESILIENCE

they understand and deal with paradox or contradicting constraints and requirements, and are comfortable to working with a degree of complexity that makes it impossible to think everything through before deciding what to do. They are conscious of the dynamics of complex relationships, and design flexible and adaptive solutions with resilience in mind.

Innovation, however, demands that you see the world through multiple lenses at the same time, and draw meaning from seemingly disparate points of data.

Dev Patnaik, CEO and Managing Associate, Jump Associates

_ What if...?

Relationship challenges in the enterprise require a holistic answer by achieving a coherent vision of its future state. A fundamental assertion established for wicked problems is that any attempt to think them through completely before acting is doomed to fail. They cannot be *solved*, since their fundamental tradeoffs prevent any solution from addressing them as a whole. They have to be approached differently, by envisioning a positive outcome for a change, ensuring its meaningfulness, viability, and feasibility, and trying to influence the system to get there. Instead of relying too much on detailed analysis, hybrid thinkers rely on holistic synthesis to generate options and enable decision making.

More than just bridging domains and translating between them to address complex relationship challenges, enterprises need to use the linked views of hybrid thinkers to generate visions of the future by asking "What if...?" questions as a starting point for purposeful innovation. This follows an idea that Charles Peirce called *abductive* logic, or as Roger Martin puts it, the *logic of what might be.*

In order to generate such a vision as the basis to tackle complex relationship challenges, enterprises need to develop a design competency. Design can be employed to go beyond what exists today, extending the boundaries of thought, and creating new options for strategy and future development.

AT A GLANCE

Relationship challenges faced by enterprises as complex socio-technical and socio-economic systems come close to wicked problems. They require holistic, cross-domain thinking beyond the conventional analytic and decision-centric mindset.

Recommendations

_ Involve expertise and methods from different professional backgrounds, including those with business-, technology-, and people-centric perspectives

_ Establish innovation teams across organizational boundaries, fostering a culture of shared problem ownership

_ Place hybrid thinkers in connector roles to translate between domains, bridge viewpoints, coordinate efforts, and synthesize holistic approaches

_ Use an enterprise-wide design approach to turn this synthesis into tangible visions of a meaningful, viable, and feasible future

CASE STUDY _ LA 27E RÉGION

LA 27^{E} RÉGION'S ENTERPRISE

Innovating public services, politics and administration in France through interdisciplinary design studios, local experimentation, and cultural transformation programs

Enterprises take on many forms, at least when applying the definition we use for the purposes of this book. When it comes to innovation and design work across different domains and disciplines, some of the most intriguing examples for projects and programs can be found in the public sector. This section is about such an initiative, La 27e Région (French for the 27th region). It has been launched as a non-profit organization based in Paris, supported by the 26 regions of France, the European Union, and a consortium of sponsors. The organization tackles a wide range of challenges that matter for the regions, in areas such as education, public transportation, or urban planning.

La 27e Région is based on the idea that the classic methods to governance and project management in public administration fall short of delivering effective solutions for today's complex relationship challenges. Today, work in that area is based on approaches of top-down commissioning and formal approval procedures, supported by expert studies, financial audits, quantitative analysis and surveys—the same methods applied for decades in European public bodies. They are used with the assumption that the desirable state is more or less clear, and that the sole challenge is to determine how to get there most efficiently.

In order to radically shift the approach to public initiatives, La 27e Région has turned to an interdisciplinary, participative, and design-based approach inspired by methods of design, ethnography, and urban place-making. To rethink administrative challenges with the focus on people relationships instead of on the public bodies dealing with them, the organization facilitates design studio workshops over the timespan of multiple weeks, taking as their model the practice of *artists in residence* common in fine arts and cultural production. In 2012, the organization launched *La Transfo*, another program to embed these practices in the working methods of the regional institutions themselves.

Envisioning a public service landscape

_ DESIGN RESIDENCIES

A design residency is a joint initiative project between La 27e Région and a public institution or entity launching a project in one of the regions of France, generally with a total duration of around 3 months and involving the formation of a cross-disciplinary team. Over the course of 2 to 3 weeks, the team gathers on site to collaboratively engage in a creative process, immersed into the local context they seek to transform.

Residencies are based on the diversity of participants in terms of roles, backgrounds, disciplines, stakeholder relationships, and levels of seniority and experience. Led by a designer, the team of residents includes domain experts, sociologists and anthropologists, architects, and urban planners. This group works with administrative and technical staff from the institution commissioning the project, as well as citizens impacted by a transformation, or the people addressed with a public offering.

Interestingly, the same diversity of actors is also present in today's rather traditional approaches to public initiatives. The key difference is the way they work together, breaking out of domain silos, formal procedures, and predetermined results, and facilitating a creative dialogue to come up with ideas and potential solutions. The 3-week-long stays have the goal to make all participants appreciate the lives of the people they address. They are a prerequisite for developing empathy and building trust, and lead to surprising results not attainable with conventional methods of public program management.

We strongly believe that public administrations need "friendly hackers" to provoke real change from within.

Stéphane Vincent, Director, La 27e Région

GETTING THERE _

To facilitate this process, La 27e Région has developed a framework to support Hybrid Thinking and a results-oriented methodology. Using a staged approach, this diverse group goes through a process of joint exploration of the topic and their environment, of mapping, framing, and creative synthesis, and of prototyping and testing, all deeply immersed into the local context.

A system of collaborative roles enables the team of residents to facilitate the dialogue between people from different backgrounds and with varying concerns, and to co-create potential outcomes with the group. In small workshops and using a set of design tools prepared by La 27e Région, the participants elaborate scenarios and develop ideas to address the topics, and continuously document their work.

Depending on the nature of the problem, the resources available, and client expectations, a residency can have different outcomes. Most projects start in an ideation stage, aiming primarily to generate powerful ideas to address a complex challenge. Others result in stories and sketches of a concept, taking into account technical possibilities or resources that steer research and development efforts. The most elaborate stage of a residence is a prototype development phase, where residents put their concepts into and collaboratively explore how it could work in detail.

An outcome could describe a technical system, a designed environment, abstract process descriptions or plans, or something entirely different. Regardless of the particular outcome, the residency concludes with a handover to the partner, having the goal to ensure its long-term adoption. These results provide the basis for public policies, giving politicians and the administration new strategic choices, and connecting public bodies and services to the people they address.

3 _ THE DESIGN-MINDED ENTERPRISE

In order to address complex relationship challenges in the enterprise and come up with an approach, executives in charge of a strategy for the enterprise need an instrument suited to this kind of problem. Beyond the management paradigms of analyzing the past and optimizing the now, determining a meaningful, viable, and feasible future direction requires facilitating a design process—developing insight, synthesizing an approach, generating a vision, and turning this intent into an actionable plan.

The area of design has gained significant traction in business circles, recognizing that it can drive transformation processes and add significant value. In many organizations however, design is still reduced to a very limited role, also due to the self-perception of many designers as stylists without major influence on the conceptual decisions behind a design. Such a design is seen as being merely an ornamentally addition to things that others have invented, or the arrangement of elements others have conceived. It prefers questions where the answer could be anything, like choices of color or visual patterns. It abdicates the responsibility for real decisions to other disciplines. Typically, it is used only at the lowest level, given the least thought and analysis, and obviously falling short of producing any result that matters on a strategic level.

Fortunately, as the discipline grows and matures, more and more professionals realize that any enterprise that benefits from design at a strategic level has developed an understanding that goes beyond this limited view of a superficial afterthought. This Chapter is about using design competency to link viewpoints and tackle strategic issues.

© 2013 Elsevier Inc. All rights reserved.

ABOUT STRATEGIC DESIGN _

Design can be used by hybrid thinkers in connector roles as a strategic tool, to envision a future for the enterprise, regardless of the particular function or level of the organization. Its unique characteristics and capabilities to deal with ambiguous, complex challenges become apparent even in the more traditional industrial or communication design that focuses on tangible and visible outcomes.

Confronted with strategic issues in today's enterprise context, business leaders essentially face wicked problems when trying to come up with an approach to tackle key relationship challenges. According to research from Gartner, a prerequisite to successfully approaching such a challenge is to be aware that it defies conventional methods of problem solving. There are always parts of a problem setting that are determined by constraints, requirements, and other hard facts, but what makes a pure analytic or inductive approach ill suited is the larger part that is underdetermined or even completely undetermined. The space between problem and solution cannot be bridged by analysis alone.

A design-led approach can be employed to tackle any kind of strategic challenge, by generating a vision of what might be and lead the innovation efforts. However, using such an approach to strategic problems does not exclude other ways of thinking from being successfully applied, such as Process Reengineering or Balanced Scorecards. Those methods to improve quality and efficiency based on measurement and analysis, well suited to optimize the process used to put a vision in place.

In this context, it is critical to identify hybrid thinkers in the enterprise, and to work with them when applying design to a strategic challenge. They provide the basis to steer the conversation, bridge different viewpoints, make sense of the mess, and bring together different approaches.

_ EXAMPLE

Imagine you ask a communication designer to come up with a logo for a new company. The result will probably contain the company name, or a graphic to appear next to it, to convey that name as a message. Asked for his or her choices of what is depicted, what color, shapes and fonts are used, the designer might still be able to state concrete reasons – but this already gets quite difficult, because those decisions involve highly intuitive judgment, driven by talent and experience.

The performance of that logo for that company depends on how well it supports the business in establishing, retaining, and renewing relationships to people—the effects of design decisions are almost impossible to predict. That's why design professionals today consider a logo to be merely a key part of a wider system of visual brand identity, embedded in an individual context of use, perception, and purpose. This enables them to contemplate its function in a system alongside any constraints that apply, and opportunities to stand out. They think about that system being applied in different contexts, instead of just creating an individual artwork in isolation.

Brand. Brand

Brand

In this example, the problem to be solved is largely undetermined. The number of possible solutions to the given problem—visually representing and differentiating a newly created business—seems to be almost infinite. A large number of decisions made by the designer are impossible to derive from the client brief, because large parts of the problem are unknown or ill defined. The outcome will be part of a larger system of communication, and requires making some completely unconstrained choices (some would call this space the artistic part of design). The proposed solutions can of course be measured against the known requirements, for example, technical feasibility in different media, but this validation is incomplete since the actual success factors of the underlying relationship challenge are unknown.

Instead of thinking something through completely before acting, it requires a blend of analytic, intuitive, and systemic thinking, where the emergent result is just one of many possible outcomes—the essence of a design approach. Solving every detail of a problem is impossible due to the fundamental trade-offs between the decisions, so that a good design aims to achieve a significant improvement instead of a comprehensive solution.

In our experience, most professional designers therefore refer to the quality of design work simply as being good or bad design. Though often impossible to state why, they form an opinion whether a result works (solves the most important issues) and fits (integrates well into context and environment). Key to achieving such a design is the constant endeavor to create something new and fantastic.

From artifacts to strategy _

As described earlier, the role of design in most organizations is still far away from realizing its potential as an instrument to make sense of and connect different ideas. This is due to a limited perception of its role, characteristics, and capabilities. Overcoming this state requires involving professional designers who can work beyond the visible, on a conceptual and strategic level.

Most organizations are using design practice to shape signs and objects, as part of brand identity and Marketing, or for Product Design—but they understand its role as one of producing visible outcomes, without any major influence on strategic choices. More conceptually focused design disciplines are rapidly growing both in terms of practitioners and organizations using their approaches. Areas of practice such as User Experience or Service Design redefine behaviors involving interactive systems and services. Only a few organizations seem to successfully employ design practice applied to all sorts of systems, on a level where it informs business strategy formulation in the light of complex challenges.

Richard Buchanan developed a framework at Carnegie Mellon University, differentiating *Four Orders of Design* with varying levels of influence and types of results. There are different variants of the framework cited in design literature, but we found the following definitions useful to describe different aspects of design:

_ THE DESIGN OF SIGNS

or first order design, is about designing the symbols used in communication processes, such as in Graphic and Information Design. It is about conveying messages and persuasive arguments, syntax, and semantics, to enable understanding and facilitate information exchange.

_ THE DESIGN OF OBJECTS

or second order design, is about designing physical objects being used by people for some purpose, such as in Industrial Design and Architecture. It is about selecting and using materials, designing tools, and embodying technology, to support usage and integration in a physical context.

_ THE DESIGN OF INTERACTION

or third order design, is about designing the behavior of systems and considering the actions of people, such as in Interaction and Service Design. It is about designing processes, transitions and activities over time, defining the different states and options to choose from.

_ THE DESIGN OF SYSTEMS

or the emerging fourth order design, is about designing dynamic systems and environments, such as in Organizational Design or Design Thinking. It is about designing the transformation of a system's structures, functions, and flows, taking a hybrid look at the system and its dimensions and constraints.

The four orders can be used to describe four different contexts where design applies, four levels of maturity in thinking about design, or simply four different approaches to address a design problem.

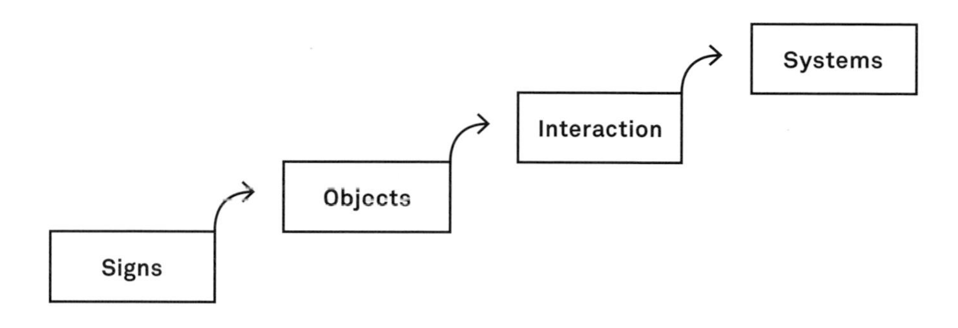

Richard Buchanan's *Four Orders of Design*

While the four orders in Richard Buchanan's framework loosely map to the various fields and subdisciplines in design, it is important to note that any design challenge includes aspects of all four orders, even if hidden from the designer and other parties involved. They are best seen as intertwined layers of design, to be explored in different depths depending on the individual context, as places to generate ideas and foster innovation.

To be employed at a strategic level in the enterprise, design approaches must be inclusive of all four orders, and consciously address the outcomes from a design initiative, from the most abstract to the most tangible. To make that happen, design needs to be embraced and understood in the organization, both to apply design methods to strategic problems and to understand and align the contributions of design professionals.

THE DESIGN COMPETENCY _

In daily life as well as in business, everyone makes design decisions. A design process happens every time separate things get combined, arranged, and aligned to fit a specific purpose, regardless whether those things are kitchen items, project plans, or people in an organizational chart. Design as an approach can be applied by anyone in the enterprise, and certainly anyone can participate in that process and contribute to it. Business professionals unknowingly practice design every day by reinventing parts of the business to address a change, chase an opportunity, or fix a problem. Applying design in a professional manner involves a core set of activities and approaches, and also offers a large variety of methods and tools. This section is an overview of the design approach and profession, and provides a basis for applying the framework and methodology outlined in the following parts to enterprise design challenges.

_ EXAMPLE

A building used as corporate headquarters is a good example of *Second Order* design according to Richard Buchanan's framework, because of its physical manifestation and the dominant role of the architect. Depending on the individual project, however, it might also be seen as a symbol as in *First Order* design, playing a role in the organizational communication or brand identity. Its interiors are used by staff or other people to work, collaborate, or meet with clients, all activities that can be explored using a *Third Order* design approach. Finally, it can also be seen as part of the enterprise as a socio-technical system, a physical context and infrastructure for its work culture, service provisioning, and operational functions, and therefore an example where *Fourth Order* design can be applied.

Headquarters of China Central Television, Beijing

_ Understanding people

Most design philosophies are based on a fundamental understanding of the people you are designing for. Traditional market research as used in Marketing practice or requirements elicitation has a place to inform design, but merely with the goal to develop a thorough understanding of the context, not to derive any direct conclusion or design decision from them. Going beyond quantitative research and collecting stakeholder opinions as requirements, design research seeks to generate deep insight into human reality, including environment, daily routines, concerns, and aspirations.

Using ethnographic methods and modeling techniques, this research strives for immersion into the lives of those being addressed by a design. By undertaking field studies, observation and inquiry in context, and qualitative methods, it dives into the very details of the design problem. Any such research benefits largely from involving professionals with a background in a people-centric discipline and who are familiar with empirical methods and social research. However, a significant goal is to expose any team member involved in design decisions to real people, to view the things from their point of view in the actual context, and to develop both knowledge and empathy.

In the enterprise context, design research expands further to include all stakeholders impacted by or involved in a transformation. Understanding their motivations, goals, and context via immersion and empathetic thinking enables us to address stakeholder concerns on an entirely different level, and to understand the technology and business side of the problem represented by those stakeholders. Besides conducting research work directly with people, designers gain knowledge secondary desk research, drawing insight from structures, culture, environment and physical artifacts, domain knowledge, or other sources to further expand their view of the design space.

Exploration and synthesis _

Based on the research results, the generative activity of design commences with an open exploration phase, by making sense out of a large set of information about the problem space. A thorough design research usually results in a vast amount of seemingly disparate findings and data. This includes transcripts and observations from people research, any quantitative data and analysis available, as well as findings from other methods of research.

It also includes generating ideas and options independent of constraints and limitations, shifting the problem definition to discover hidden spots for improvement and innovation, and challenging known assumptions and beliefs. During the process, the weight of each design decision increases, while new ideas get harder to fit into the design. Simply solving the problem gets more difficult the deeper the exploration dives into the problem space. What seems to be a space of infinite choice at the beginning turns out later to be a wicked mess of related elements and influencing factors to account for, each design decision further constraining the way forward.

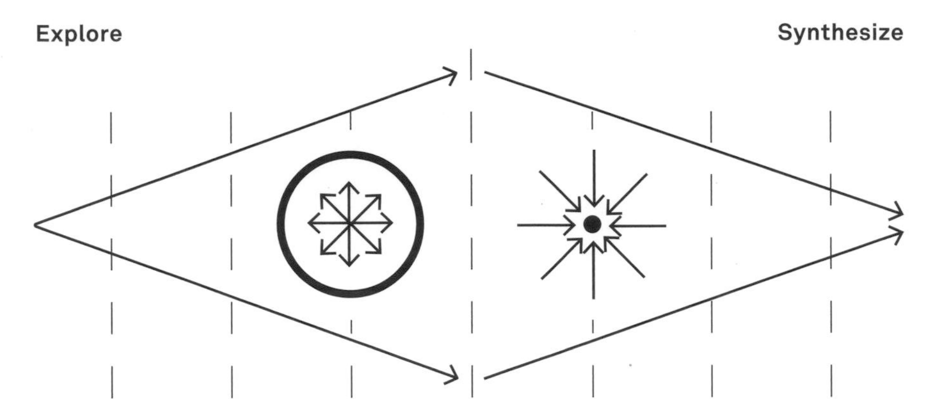

The process of exploration and synthesis

In his book *Exposing the Magic of Design,* Jon Kolko describes the transition from research-based exploration to generating design options as a sensemaking activity, the task of quickly generating meaning out of a lot of data. More than analyzing that data and drawing conclusions, design synthesis introduces new elements into the thinking process by relying on the designer's experience and inspiration, discussion with peers and experts, and thoughts about feasibility and viability.

The synthesis activity is where the *abductive* thinking happens, which in the end enables the design process to produce unexpected and innovative results. Findings from the research are connected, ordered, sorted, and related , in order to create multiple models of the design space. This is often done visually, as affinity, flow, or other conceptual diagrams, sometimes using large walls of sticky notes and drawings. It involves reframing the problem to look at it from various stakeholder perspectives, shifting between different aspects, and mapping mental models onto conceptual models. The goal is to define a mental map of the problem, and generate new knowledge by introducing elements from outside.

This is the basis for what Jon Kolko calls *experience scaffolding,* by providing a framework of elements that characterize the outcome of the design process, and exploring their role in people's experience. This activity involves exploration of possible paths via sketches, developing narratives to place the design in a context of a person's life, and expanding the design space to cover the context where it is applied. In design synthesis, the ability to empathize with those people affected by a design guides the decisions, and helps to ground all activities in a real-life context. This placement of potential results in their context of use is the basis for prototyping and validation.

Designing and validating results _

The conceptual models, insights, and approaches derived from design synthesis are used to envision prototypes, resulting in interactive models of the outcome of a design process. Rather than just proving the feasibility of a concept to prepare for implementation, prototyping is used in design as a tool for creative exploration, as a thinking aid to try out different directions and further ideation prior to choosing a definite path. It serves as a communication tool for aligning a design with stakeholders and target groups, as a source for further questions and possible directions, and can be used to involve people from outside the core design team into the activities of exploration and synthesis. Instead of discussion, opinions, and compromise, the interactivity of prototypes helps to embed the development and evaluation of designs in a realistic setting.

Prototypes mean different things to different people, but share the common goal to put a design to the test and validate it against the problem space. They are usually different from sketches and conceptual models in the sense that they attempt to capture a certain part or aspect of a design holistically and comprehensively, enabling detailed exploration, testing, and iterative redesign. The nature and effort of a prototype therefore vary widely, depending on what exactly is to be validated and the stage in the design process. Design professionals combine multiple very different prototypes to represent the different aspects of their design using a reasonable effort.

Prototyping is a powerful tool to turn the results from framing and sensemaking into visions of a possible future. The visible, tangible, and interactive part can be used to explore different domains of design work, including simulations of flows in the background such as business processes or financial streams. As the indispensable result of all design work, prototypes indicate the direction of the transition, and help to define and refine a solution leading to its implementation, based on the constant and relentless inquiry for the best possible outcome.

_ The design profession as a strategic resource

Just as in other professions such as engineering or management, there is a difference between applying a design approach and working as a design professional. In *Sketching the User Experience*, Bill Buxton asserts that he thinks about design as a profession that is as rich as mathemetics or medicine. It holds room for different philosophies and subdisciplines, a large variety of methods and tools, a lifelong learning and refining of professional skills, and a large and diverse body of knowledge. There are specialist designers working on particular media, subjects, or industries, such as web or interior designers, or even specialists in particular tools.

But design, like other disciplines, is impacted by the convergence of media and technology, causing separate subdisciplines to overlap. Products show some of the qualities of media and behavioral systems, communication media are used as tools, environments are becoming media, and all of them acting as the stage to connect organizations and individuals. A new breed of generalist designers moves towards the third and fourth order design disciplines, and brings together many different design domains to create a coherent response to a problem.

We found that designers working on strategic challenges in the enterprise are not just very experienced, but also tend to have just such a generalist background. They are applying design methods and skills in ambiguous, unfamiliar, and complex environments, to generate visions that are stunning, unexpected, and elegant.

Besides possessing hybrid thinking skills and mixed professional backgrounds allowing them to bring people, domains, and thoughts together into a systemic concept, they possess an eye for outstanding aesthetics, without necessarily doing the visible design part themselves. They find pleasure in elegance both in a mathematical and a visual sense and strive for simplification in terms of structures. They have a preference for clearly arranged systems and unity, but allow for necessary diversity. These values are balanced against a pragmatic idea of the end result.

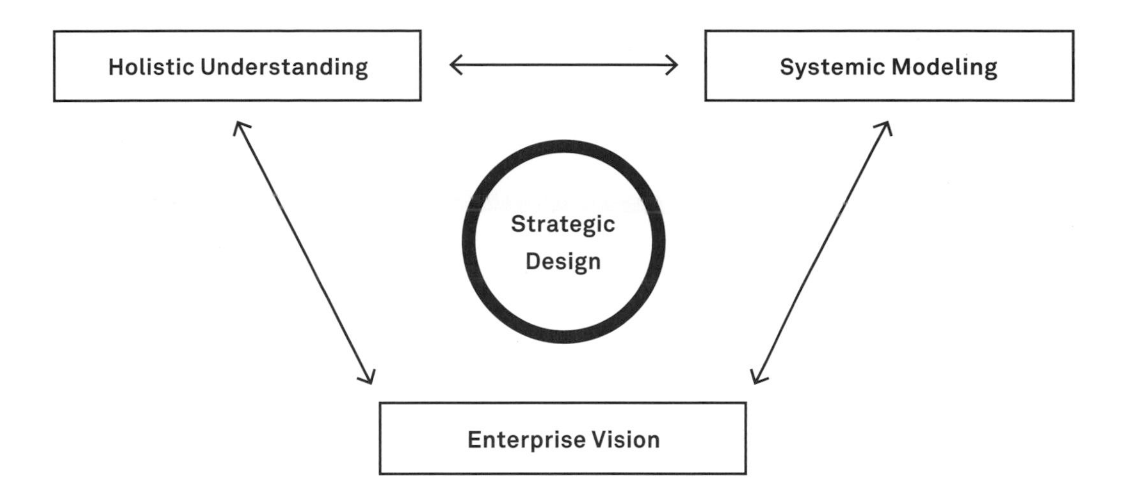

Strategic designers can resort to a set of methods and tools, frequently jumping between different phases of activity:

_ HOLISTIC UNDERSTANDING

they spend significant time researching and capturing data, analyzing and contemplating the findings, documenting all that is known about the problem space, reframing the problem in different ways, and making decisions based on that information.

_ SYSTEMIC MODELING

to align viewpoints and bring together different conceptual domains, they use modeling and visualization techniques. They switch between considering the environment outside the core problem, and making different elements work together as an integrated system.

_ ENTERPRISE VISION

they generate ideas both by querying the environment and assuming that everything could be doable, identifying candidates for a powerful core idea that resonates with enterprise stakeholders, and seek to overcome constraints imposed. The results of this activity tend to look disturbing and provocative to the outside, but can be good starting points for radical innovation.

_ DESIGN IN THE ENTERPRISE

Design as a management paradigm, a problem-solving approach, and a professional discipline is a powerful instrument for business leaders. It can help enterprises find sweet spots of value generation in their wider ecosystem by imagining what could be, in an ill-defined and complex environment, and building on that vision to identify and convey a desirable change.

It helps reach out to people important to the enterprise, establish conversations and relationships and expose ways to delight them—as such, it validates and improves the logic of the business towards customers, employees, investors, and other critical stakeholders. It uncovers opportunities for improvement on a level of human experience, informing optimization and restructuring initiatives.

Design can be used as a way to link different viewpoints and domains, and align decisions towards a common vision that is shared and understood. It helps business leaders to break out of the current focus by introducing elements from outside into the organizational strategy. However, this means overcoming the idea of design as an afterthought, by making it a part of the way the enterprise thinks and acts. It requires going beyond the isolated, low-level design commonly practiced today. Design can realize its true value in the enterprise only if applied on a strategic level.

_ Embedding design in corporate culture

According to business school professors Fred Collopy and Richard Boland in their book *Managing as Designing*, a fundamental difference in the attitudes of business and design is how they deal with choice. Based on their research into the methods of architects from Gehry & Associates, they describe how those architects relentlessly kept creating new models, designs, concepts, ideas, and prototypes, until reaching a result that they considered to be truly great. They conclude that while the practice and thinking of management assumes the difficulty is in choosing between existing alternatives, design strives to create new and better options to choose from.

This assertion makes the case that to be truly embraced, design has to be positioned as part of the enterprise itself, the shared purpose that drives the activities, decisions, structures, and culture that surfaces in daily business. A culture that fosters a design attitude, that makes everyone involved partly responsible for this quest for the best possible way, to result in remarkable and delightful outcomes. Instead of putting it into another separate function, department, service, or other form of silo, business leaders have to consider design as a pervasive component of decision-making, part of every initiative and operation.

To the enterprise, design has the role of overcoming linear, incremental, and purely analytic approaches by facilitating sensemaking processes and generating new meanings for people. As such, it is used primarily as a source of innovation. An approach on how to leverage design as an enabler of business renewal is Design-Led Innovation.

As a way to achieve innovation, the practice of design in the enterprise cannot be exclusive to design professionals. To get it right, collaboration with design professionals is essential, but as an organizational competency it must be seen as a common mindset and approach, a collection of methods, and a shared experience. It can be applied by trained designers, by other professionals familiar with its essence, and as a paradigm in entire teams and organizations.

To be effective, it needs strong patrons and proponents—hybrid thinkers in connector roles—using it as a strategic instrument. These connectors can be trained designers themselves, or people from other backgrounds integrating design professionals into the enterprise environment, as a studio for business innovation.

DESIGN-LED INNOVATION

```
Market? What market!
We don't look at market needs.
We make proposals to people.
```

Ernesto Gismondi

A prevalent strategic role of design in the enterprise is one of purposeful, systemic innovation. Such an approach to design is described in Roberto Verganti's book *Design-Driven Innovation*. Roberto Verganti argues that any technology breakthrough is only the starting point of an innovation. In order to turn that into a business opportunity that changes the game of competition, enterprises have to redefine their meaning in people's lives, and offer this vision of a new meaning to them as a proposal.

As explored before, innovation can have its roots in many places, triggered by different members of the enterprise, and related to change in different domains. Design processes can be used to explore these places, develop understanding, and engage in a process of prototyping and exchange to generate innovation potential.

The idea of Design-Led Innovation describes the predominant role of design practice in the enterprise. Innovation in this sense goes beyond products and services offered to the market to include any kind of transformation that can be explored using design methods.

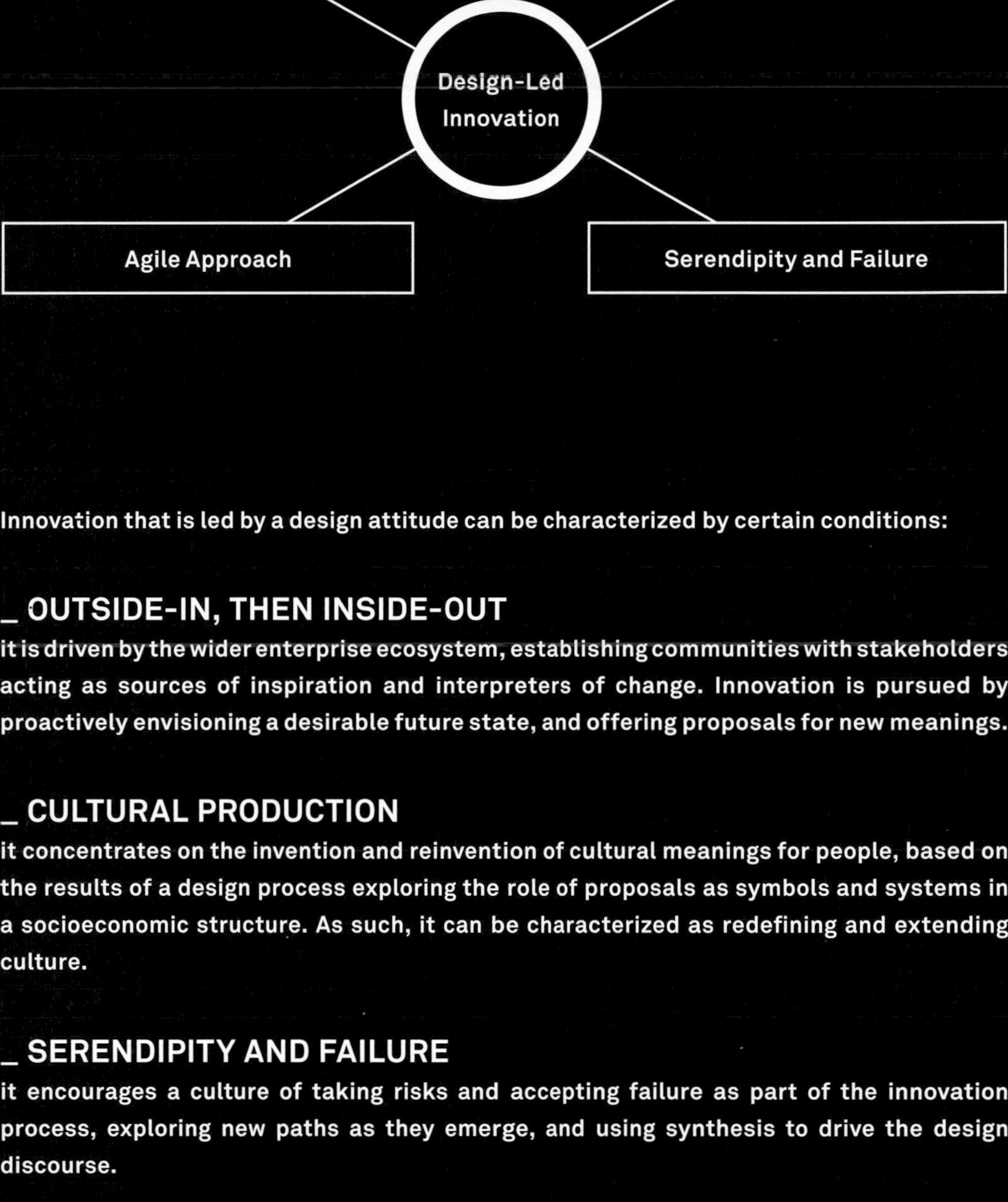

Innovation that is led by a design attitude can be characterized by certain conditions:

_ OUTSIDE-IN, THEN INSIDE-OUT

it is driven by the wider enterprise ecosystem, establishing communities with stakeholders acting as sources of inspiration and interpreters of change. Innovation is pursued by proactively envisioning a desirable future state, and offering proposals for new meanings.

_ CULTURAL PRODUCTION

it concentrates on the invention and reinvention of cultural meanings for people, based on the results of a design process exploring the role of proposals as symbols and systems in a socioeconomic structure. As such, it can be characterized as redefining and extending culture.

_ SERENDIPITY AND FAILURE

it encourages a culture of taking risks and accepting failure as part of the innovation process, exploring new paths as they emerge, and using synthesis to drive the design discourse.

_ AGILE APPROACH

it strives to get innovations out of heads and committees early, working with prototypes and models instead of abstract definitions, and implementing them iteratively and openly as a shared journey rather than as a planned process.

A lot of times, people don't know what they want until you show it to them.

Steve Jobs in an interview with *Business Week*, 1998

_ Design and strategy as a dialogue

Design as a strategic competency can help organizations to look at themselves from the outside, and transform the enterprise ecosystem from within. It enables business leaders to consider the enterprise in terms of the role it plays in people's lives, and to (re-)design it to create value for them. Faced with complex relationship challenges, this outside-in approach enables the enterprise to create new and better choices as a basis for better decisions.

To do so, the relationship of design and strategy has to be one of a continuous dialogue. Design has to inform strategy, by providing insights on potential innovations, and generating better options to choose from. Strategy has to be the basis for any vision made visible by design, from initial visual sketches that capture a future state to the redefinition of the enterprise as a modular system. The enterprise becomes a place to iteratively develop and pursue a portfolio of strategies informed and conveyed by design visions, fostering local experimentation and applying good practice to the wider organization. Design connects the enterprise with its cultural environment, and leads a discourse on a meaningful, viable, and feasible future.

This role leads to a new strategic positioning of design in the business, and also requires designers to expand their views beyond the individual products of their work, to all kinds of systems, transitions between system elements, and the big picture of the enterprise. It requires both design professionals and the business leaders working with them to take a step back and look at the enterprise in the context of its wider ecosystem, as a shape visible to all stakeholders.

Part 2 of *Intersection* introduces a common language for people involved in strategic design, capturing the enterprise as both the context of any design initiative and as the subject of its results.

AT A GLANCE

Using design as an approach to key business challenges enables enterprises to intentionally change the way they relate to people. As a strategic competency, it helps to bridge silos, capture relevant aspects, and synthesize a human-centric vision of the future.

Recommendations

_ Use design as a competency to develop a deep understanding people your enterprise addresses, and to develop and test responses to wicked key challenges

_ Instead of putting this competency in another silo, apply it as the linking force between the contributors to strategy and its implementation

_ Establish a continuous dialogue of interpretation and local experimentation between strategists, designers, and the enterprise ecosystem

_ Use an enterprise-wide design approach to drive change initiatives resulting in human-centric innovation as the focal point of your business strategy

CASE STUDY _ APPLE

APPLE'S ENTERPRISE

Creating integrated ecosystems of devices, content and platforms, for remarkable experiences in people's private and professional lives.

Businesses around the world look at Apple as the shining example of a design-led enterprise, and the company as the subject of a lot of conversations and articles on that topic. On the other hand, not too much is known about the way the company really works on the inside, and how they address the challenges of integrating design practice into a fast-paced business environment. This case study pulls together what is has been revealed about design at Apple, and looks to summarize key insights of the role, culture, and organization of design practice in that company.

One particular observation made by various commentators both inside and outside the company is the remarkable perception and culture of design, which can be considered a significant part of Apple's DNA. This goes well beyond an understanding of design as a function of Marketing, Product Management, or Communication. Championed by Steve Jobs until his death in 2011, Apple designed its own market instead of preparing for changing external conditions. By deliberately making huge leaps into the future and experimenting with potential shapes of their enterprise, they were able to make those states visible.

We believe that applying design strategically enabled Apple to draw a picture of their future that everyone involved could work towards, a shared idea about their enterprise. Therefore, they integrated design into their thinking and doing at a level only few other organizations do today.

In most people's vocabularies, design means veneer. But to me, nothing could be further from the meaning of design.

Design is the fundamental soul of a man-made creation that ends up expressing itself in successive outer layers.

Steve Jobs in an interview with *Fortune Magazine*, 2000

_ DESIGNING AN END-TO-END SYSTEM

Looking at the offerings Apple has put on the market since Steve Jobs returned to lead the company in 1997, its expansion into new market segments and industries is simply stunning. From being a computer manufacturer, Apple made its way into the areas of consumer electronics, communication, entertainment, and business solutions, becoming the world's most profitable business.

In this setting, the role of design has been a critical enabler for the whole enterprise. Starting with the users they made their products for, Apple naturally expanded their business into areas adjacent to their current business focus. Unlike expansion strategies of other companies of comparable size, such quests into new territories were not done in isolation from the current products, assets, and markets. Instead, every new offering Apple introduced was a logical extension of the portfolio, delivering another key part of the system. From computers to music players, from players to phones and tablets, from devices to content, from stores to cloud services—all these things are part of one platform, and contribute to one vision.

Design as practiced at Apple delivers both the insights and visions to conceive that system, starting at the user and step by step winning a larger part of their world. This involves experimenting with different materials, technology, and modes of customer interaction—the classic playing field of traditional design disciplines. But more than that, it also means designing new distribution channels, content structures, business models, system architectures, and other conceptual parts of the system. Instead of concentrating on one segment and neglecting the others, Apple's design approach enables systematic growth.

DESIGN AS A CULTURE _

Insights about Apple's corporate culture are largely based on anecdotal references. However, certain elements are emphasized repeatedly by various sources, and make the case that they approach product development more like a fine arts workshop than a classic industrial setting. Key to this is a sense of authorship, implying that there is one single, coherent vision about the final result of a design process and the way it should appear in people's lives. At Apple, this vision is part of the responsibilities of senior executives up to the CEO level, including the task to prevent any deviation or expansion that does not contribute to its achievement.

As a design-oriented company, many of the technologies and ideas Apple puts into implementing a vision come from somewhere else, and the company made a series of acquisitions to support that. In the end, however, all those parts fit seamlessly into one designed system, with a great appreciation for perfection in concept and execution. This requires bridging a lot of different viewpoints, be it different domains such as Marketing, Engineering, and Design, but also people from companies taken over, and different actors in Apple's ecosystem.

A core element of such an alignment effort is a strong attitude that drives the way an offering is envisioned and designed. Many observers stress the relentless quest for perfection, the constant reduction to the most essential. This also involves the hard work of refining every detail again and again in the service of the user, constantly questioning if something can be even better. Regardless of what part someone contributes to, this holds true for the work of teams in Experience Design, Software Development, and Mechanical Engineering.

Everything at Apple can be best understood through the lens of designing. Whether it's designing the look and feel of the user experience, or the industrial design, or the system design and even things like how the boards were laid out.

The user experience has to go through the whole end-to-end system, whether it's desktop publishing or iTunes. It is all part of the end-to-end system. It is also the manufacturing. The supply chain. The marketing. The stores.

John Sculley, former Apple CEO (Source: cultofmac.com)

We struggle with the right words to describe the design process at Apple, but it is very much about designing and prototyping and making. When you separate those, I think the final result suffers.

Jonathan Ive, Senior Vice President, Industrial Design at Apple (Source: thisislondon.co.uk)

Apple's product releases attract large crowds

Finally, Apple's culture seems to thrive on ideas. In his tribute to Steve Jobs, Jonathan Ive praised the way Apple's leader recognized that even the boldest and most powerful ideas begin their lives as fragile entities, so easily lost in a process of compromise, groupthink, and reinterpretation. Apple excels at making ideas fly, the very essence of design work.

DESIGN AS AN ORGANIZATION _

Although there is a lot to learn from good examples, we usually tell our clients at eda.c that there is no point in trying to be Apple (or SAP, Google, Facebook or Microsoft for that matter). In our experience, every successful enterprise is a unique entity. Organizations are certainly able to learn from role models, but they risk losing their core idea and unique identity when mindlessly incorporating elements from others.

One of the stories worth exploring for any enterprise is how Apple formally made design part of their organizational structures and business practice. Although the cultural basis and the systemic approach both are necessary ingredients of the way Apple embraces design, these elements seem to incorporate the underlying attitude into operational processes, reporting lines, and project work.

Apple's design group benefits from a direct reporting line to the CEO, giving design professionals a powerful position in the organization. What comes out of a design process is therefore part of the company's strategic direction, being the result of a dialogue between strategy and design to put that vision into a visible picture. This is radically different from most companies even in the same market segments as Apple, and a way to systematically circumvent design-by-committee, where too many people insist on including their own wish list.

To bridge the vision with the feasible, Apple puts a design concept into implementation even before a clear target state has been defined. In an interview for *Technology Review*, Apple's former Industrial Design chief Robert Brunner indicated that Apple sends teams of designers into factories sometimes for periods of weeks, pushing manufacturers to find new solutions and making technical decisions together. This expands the design task and the ownership to actors outside the boundaries of Apple's organization, and enables the joint experimenting and product making that is behind the company's remarkable products and services.

PART 2

The Enterprise Design Framework

The challenges enterprises face in a hyperconnected, highly dynamic, and technology-pervaded world require them to think holistically about their relationships to people. The design competency can help enterprises to consciously shape the way they are experienced over time, to provide a *face* visible and approachable to everyone interacting with them.

To tackle problems on this level, the practice of design needs to become more strategic and relevant to problem settings typical to the enterprise. This section portrays the Enterprise Design framework as an overarching design approach to address the diverse use contexts and business scenarios that can be used to facilitate enterprise-people interaction. It provides a common language that allows practitioners of different disciplines to converse about the problems and challenges of such a global design initiative.

Our framework is based on a blend of practice-oriented design approaches with roots in Systems Thinking, Design Thinking, and design practice, which we found to be particularly useful for interdisciplinary work in design projects on the enterprise level. It most likely contains many terms you already know, but also introduces some new ones. It attempts to bring these terms together as a way to design in large-scale, complex problem settings.

FRAMEWORK OVERVIEW

FOCUS	DECISIONS
visionary ↑	strategic
conceptual	tactical
applied ↓	operational

BIG PICTURE

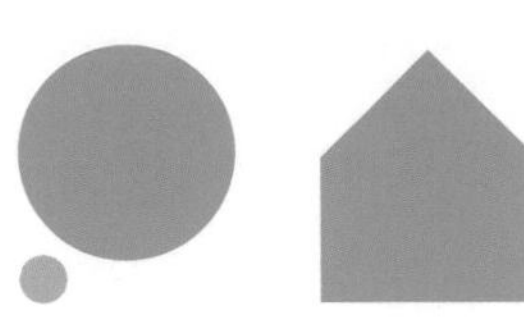

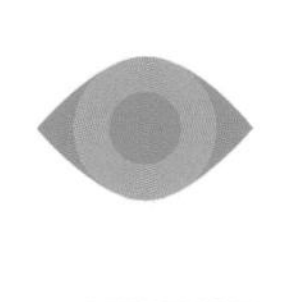

ANATOMY

FRAMES

DESIGN SPACE

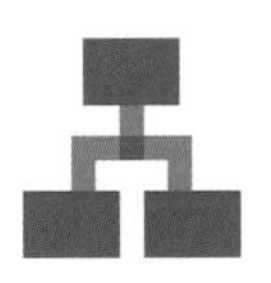

RENDERING

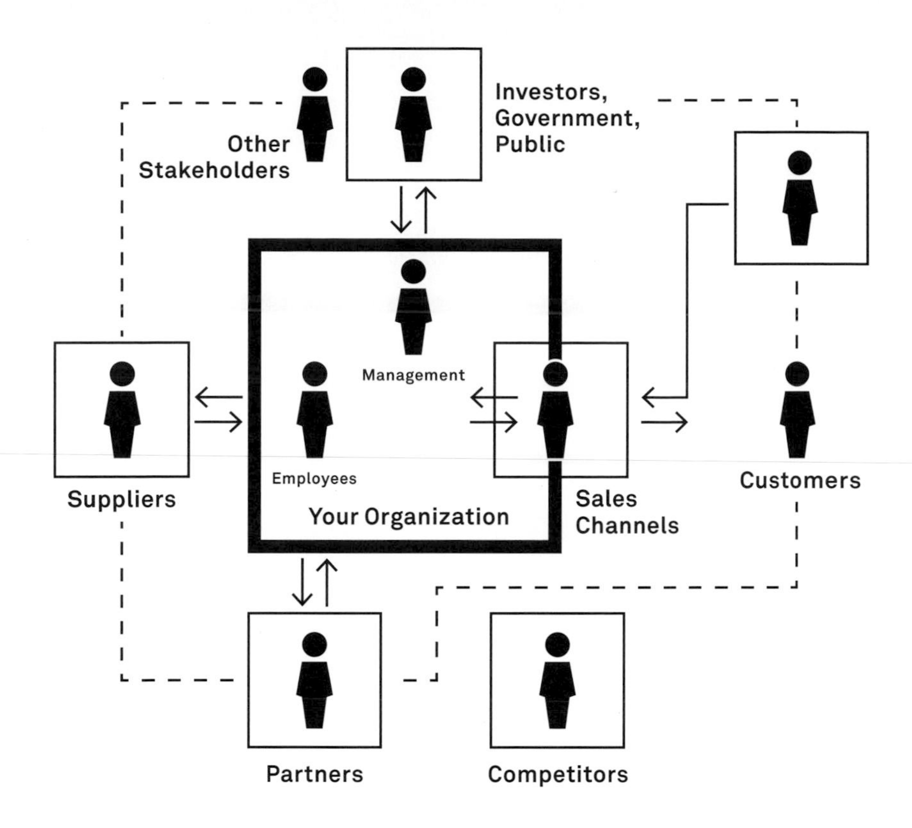

Scheme of a typical commercial enterprise

EN·TER·PRISE (PLURAL ENTERPRISES)

From Old French via Middle English and Middle French, feminine past participle of entreprendre ("to undertake"), from entre ("in between") + prendre ("to take")

_ *A company, business, organization, or other purposeful endeavor.*

_ *An undertaking or project, especially a daring and courageous one.*

_ *A willingness to undertake new or risky projects; energy and initiative.*

_ *An active participation in projects*

Source: wiktionary.org

4 _ BIG PICTURE

The Enterprise Design approach is based on the idea that in the individual perception of people, any visible manifestation of an enterprise is just a single element along a greater journey of interactions. This applies to everything people can see or use in the enterprise space, like websites, documents, communication channels, or retail stores. Any person in contact with an enterprise experiences a different subset of those elements, depending on the relationship to the enterprise and the activities being carried out.

The relationship between an enterprise and human beings vary in nature, and reflect various roles: people act as customers, partners, employees, or job candidates, sometimes on their own account, and sometimes representing a group or organization.

Instead of addressing each of these business cases, target groups, and communication channels one by one and producing more and more disconnected tools that add to the complexity, the Enterprise Design approach aims to respond to everything a person wants or needs in a coherent and integrated fashion.

By making those pieces work together in an overarching design, an organization can give itself a consistent shape across everything people experience when interacting with it, in a way that ensures no relevant aspect of that interaction happens without explicit intent. This provides a way to address a person's needs along the entire experience, and to purposefully connect people to each other. Anything an organization owns or influences in the enterprise should be integrated into that design.

© 2013 Elsevier Inc. All rights reserved.

_ EXAMPLE

Your relationship to a public transport service is determined by being a customer and a passenger. Every engagement with them—going somewhere from A to B—involves a series of steps: looking for a suitable connection, booking a ticket, travelling to the station, boarding the train, finding your seat, showing your ticket, arriving at your destination. Depending on your satisfaction with their services, your experience might also involve dealing with customer service to request compensation for a delay or other problems encountered, or participating in a reward program.

Examples for other relationships to that company include that of an investor holding shares, the clerk at the ticket counter or a train conductor—people fulfilling those roles are stakeholders of the enterprise, and contribute to what people within its realm experience.

This may seem like a huge, overwhelmingly complex task, and in some ways it is. To be able to craft a design for something as abstract and loosely bound as an enterprise, we need to take a step back to look at it again from some distance. By examining the way the enterprise is perceived by people being addressed, and decomposing the elements that constitute it in this perception, we can develop a better understanding of the problem as a whole.

This chapter describes three universal high-level aspects of any enterprise, as an approach to develop a holistic understanding about it as a complex system. It portrays today's established practices, addressing design concerns related to each of them.

_ Identity, architecture, and experience

Any enterprise of substantial size is by nature a complex entity. It carries out a huge number of transactions and operations every day, involving the assembly of work products, decision making and communication, in exchanges with customers, suppliers, partners, investors, and other stakeholders. It launches projects and programs, and develops elaborate structures to manage the various assets and relationships it needs to operate and evolve. These challenges apply to commercial businesses as well as to non-profit organizations, public bodies, and other forms of enterprises.

Consequently, addressing the concerns of an enterprise in its entirety requires looking beyond a single organization or market player at the larger ecosystem it is embedded in. Each actor of an enterprise has a different view on it, usually putting itself at the center and related parties at its periphery. All relevant parties like customers, external partners, and stakeholders have to be part of any comprehensive view on the enterprise, looking at that core organization together with every stakeholder it relates to.

Most approaches to thinking about an enterprise in a holistic way involve a certain degree of abstraction and the use of metaphors. This is needed to circumvent the inherent complexity in analyzing or modeling such dynamic and mingled structures. Such models picture the enterprise as a building or a city to be planned and constructed, as a personality with certain characteristics, as a living organism, or as a machine carrying out a given process.

To create a conceptual foundation for the Enterprise Design framework—to determine what actually is to be designed—we use the same type of abstraction to describe in Big Picture terms what constitutes an enterprise in the eyes of people. Using such simplified concepts as a paradigm to form a model of the enterprise helps to understand it as the subject of the design approach, and the context in which that design will be applied.

These aspects describe universal qualities that apply to any enterprise of all kinds and sizes, regardless of whether its management bodies are aware of them and if they have been consciously addressed. They provide a way to understand how an enterprise establishes human relationships, how it is perceived, experienced, and implemented by the people involved in its activities. Any design initiative carried out in an enterprise context aims to interpret and then proactively re-shape these qualities. On the following pages, each aspect is explored in more detail.

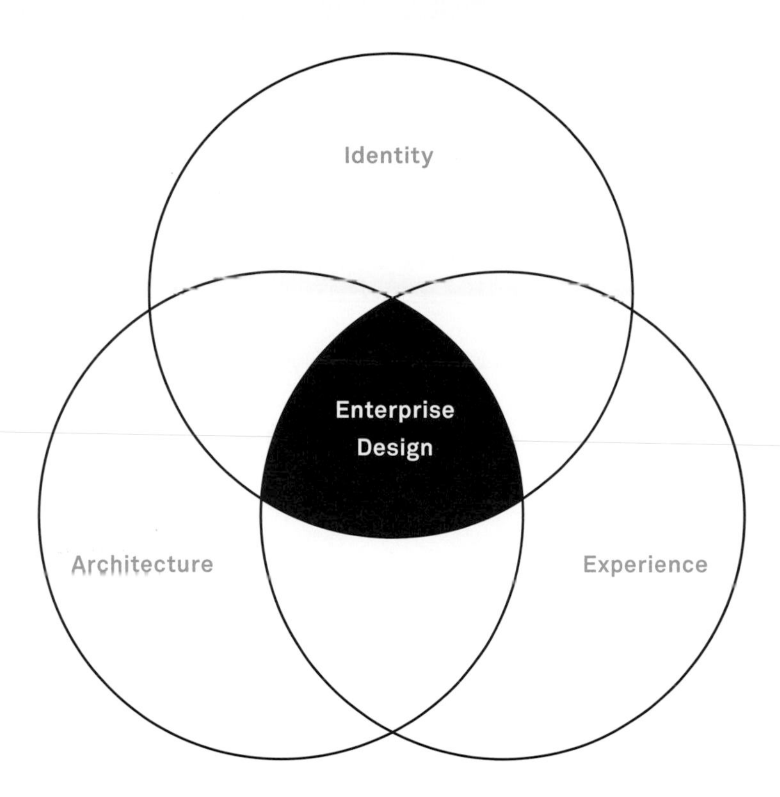

There are three universal qualities of all enterprises used as the Big Picture aspects of the Enterprise Design framework:

_ IDENTITY

what people think and feel about it — the enterprise as a mesh of personalities, impressions, and images in people's minds, expressed in symbols, language, and culture and reflecting a shared sense of purpose.

_ ARCHITECTURE

how it works and functions — the enterprise as a purposefully designed system of control structures, resources, assets, processes, and tools being put in place to make it work.

_ EXPERIENCE

what people get out of it — the enterprise as a space for interactions of people, environments, and artifacts, and how it addresses human motivation, perception, and behavior.

#1_IDENTITY

Whenever people engage in a conversation with others, they somehow present themselves as individuals with a distinct identity, having attitudes and values, conveying beliefs and assumptions. Likewise, when an organization interacts with the people and groups it is related to, it expresses an identity—in the way it looks and communicates, and in the way it behaves as an entity. Its effects range from the short relationship between a small shop and a one-time customer to the relationship of large corporations to entire societies. In the eyes and minds of people, that identity constitutes a certain image, which is the way identity is perceived by others.

In terms of identity, the enterprise is a space of actors in constant exchange. Organizations, individual people, teams or departments, services and functions —all of these actors have distinct identities and images of each other. In all interactions and conversations that happen inside the enterprise space, identities are communicated and used as differentiators and means of positioning. Just as people dress in a particular way, use language styles, or present themselves in social media to express their identities, organizations make use of visible attributes and messaging to constitute their identity. The identities expressed in an enterprise context reflect the relationships between the actors involved, their roles and responsibilities.

Expressing identity _

Over time, every actor in the enterprise develops a set of images in his or her mind, shaped by the impressions that identities made at each point of contact, or touchpoint. In reality, this is the only place where identities in the enterprise come to life: in people's minds. For an impersonal consumer product, this means the image the target audience projects onto its brand, and its expression of an identity that fits. In the case of a shared organizational identity, it is more about creating a shared understanding between its management, staff, other stakeholders and the public of what it is about and stands for. It is influenced by multiple factors such as the organization's past, things influential people say or do, and its accomplishments and failures.

_ EXAMPLE

The image of major oil companies has fairly suffered in the past decades due to some particularly ignorant behavior, scrutinized by the public—think of the Brent Spar oil platform disposal or the oil spill in the Gulf of Mexico. In both cases, the identity expressed by the organizations appeared to be largely ignorant of the negative effects of their work on the environment, resulting in a bad image and a seriously damaged relationship to a significant part of the population, leading the companies into serious economic trouble. While these companies might have been conscious of the they identity expressed towards their shareholders, they failed to behave in a way that would ensure a good relationship with the public.

The 2010 oil spill in the Gulf of Mexico

To understand the concept of an enterprise as a space of identities, we have to look at the way these identities manifest themselves in an enterprise's daily business and life. Based on a rather passive idea, identity takes on an active role as the culture emerging in and around an organization, lived in everyday practice. It manifests itself implicitly in the habits, assumptions, beliefs, and attitudes which are shared by its staff and which underpin the organization's activities. Culture becomes visible to people inside and outside its boundaries in emerging explicit expressions, as symbols, messages, conversations, and behaviors. Examples of visible elements are the way meetings are being facilitated, the design of office interiors, or writing styles applied to email messages.

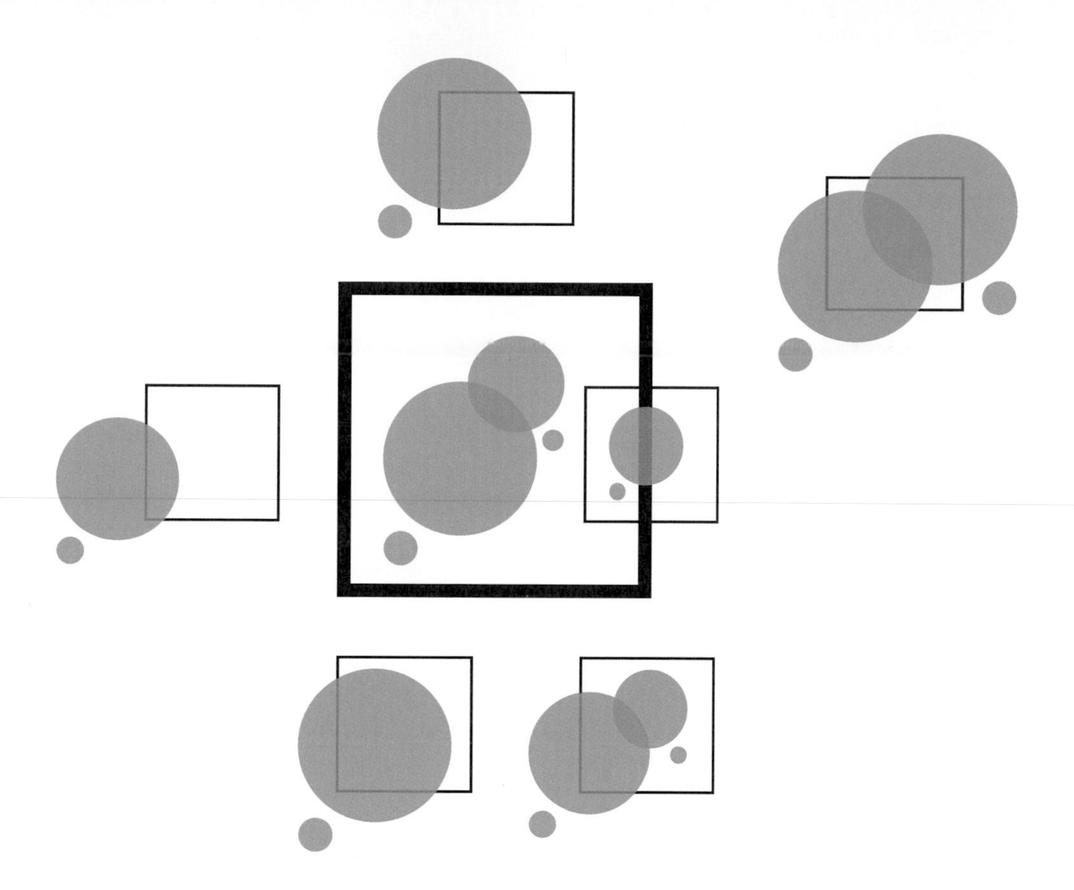

Neither organizations, nor groups nor individuals are necessarily conscious about their identity. To be effective, management has to align its strategy and vision of an organization's future with the identity shared by the people important to its success. This includes the culture that people within its boundaries are creating, as well as the images that external stakeholders have of it.

Shaping identity _

Identities are first and foremost abstract concepts inside of people's minds. As such, they do not only express themselves visibly, but they are constituted by perceivable elements. Messages, symbols and their meanings, behavior of role models and leaders—all those visible elements act as triggers for an emerging identity. They cause changes in individual and shared ideas, and broadly influence the resulting culture.

Any person having an idea of itself and the groups it belongs to has an identity, as have those groups themselves whose members share a common perception of their identity. It is the vision of who you are, who you want to be, and how you want to be seen by others. Such ideas exist for people, for companies and other types of groups, but also for services or products in the minds of their creators and consumers. In the enterprise context, different types of identities are relevant:

_ SHARED IDENTITIES

are those expressed by organizations and groups such as customers, suppliers and partners, but also by departments, countries or other types of groupings inside an organization. In small and young companies, these are largely influenced by the personal identity of influential leaders like a company founder. Shared identity is often subject to branding initiatives, aiming to actively shape the way that identity is expressed to the market and internal or external stakeholders.

_ PERSONAL IDENTITIES

of people are the conception and expression of individuality and group affiliations, influences by social, economic, ethnic, or religious backgrounds. In the enterprise, personal identities are intertwined with shared identities since people often act on behalf of organizations, and affiliation with a shared identity in turn becomes part of their personal identity — like being a Mac user or member of a club. The other way around, people express different identities or personas according to contexts, such as private or business communications.

_ IMPERSONAL IDENTITIES

are those attributed to an abstract concept or artifact, such as trademarks for products or services. These identities are often created to position an offering in the market or to communicate a certain quality of a product, attempting to give it a personality that is reflected in its brand identity. In some cases, impersonal identities can overlap with the shared identities of groups, such as a team working on a certain product.

BRANDING

A brand is simply an organization, or a product, or service with a personality. So why all the fuss?

Wally Olins, Chairman, Saffron Brand Consultants

Branding is the practice of actively shaping the way an identity is perceived by audiences important to the person or group it belongs to. While the concept of Branding can be applied to all kinds of shared, impersonal and also personal identities (*Personal Branding*), the predominant form of use in the enterprise is on the organizational level. This activity is usually referred to as *Corporate Branding*.

A successful brand project has to be deeply grounded in the strategic vision of its leadership while being aware of the culture and images in the organization. It strives to turn that vision into a comprehensive formulation that communicates it and makes it clear, audible, and visible as a coherent message. It reaches out to everyone important to the organization to shape their perception and image of it, and also seeks to *shape* the behavior and culture of all people involved in making it run—predominantly employees, but also external partners or even customers.

Branding initiatives have the goal to strengthen consistency, build reputation, and generate meaning for all stakeholders concerned. They typically result in a new or evolved visual design applied to visible artifacts produced by the organization, such as messages, products, and events. To be have an impact beyond the visible, they also address new guidelines for communication or personal conduct, such as the tone of voice used in messages.

To have profound impact on the daily business—for example to change the way customers are treated by staff or the quality of customer service—brand initiatives have to engage in extensive change management and influence everything and everyone involved in its activities.

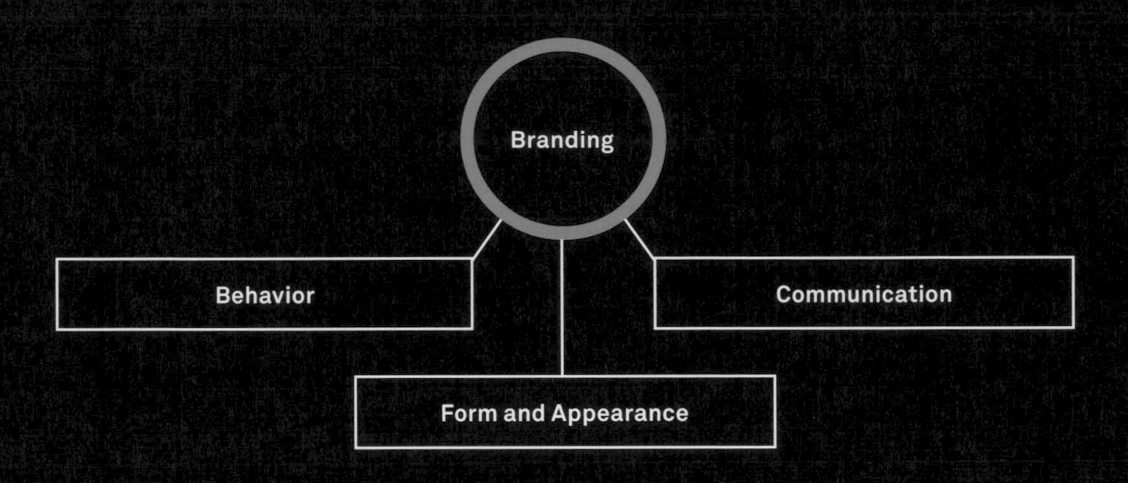

In particular, Branding initiatives usually attempt to give a shape to three different manifestations of identity:

_ FORM AND APPEARANCE

everything an organization produces that is visible to its various stakeholders, including products and services, physical and virtual environments, documents,and other kinds of artifacts.

_ COMMUNICATION

everything an organization says in public or in private to people inside or outside its boundaries, including press releases, advertising, and communication with individuals.

_ BEHAVIOR

everything an organization does when dealing with individuals or groups of any kind, including service execution, behavior of representatives, and operations visible to the public.

The central problem of brand building is to get a complex organization to execute a simple idea.

Marty Neumeier, author of *Zag* and *The Brand Gap*

Today, Branding is entering a period where it engages not just customers and employees, but all members of the enterprise of which it is part.

Mary Jo Hatch & Majken Schultz in *Taking Brand Initiative*

When leaders of an established organization perceive a gap between their vision and the reality of its identity as lived in daily business, they seek ways to recreate or clarify the idea shared by everyone of what the organization is about—whether consciously making this part of a corporate identity program or informally pursuing similar goals. Startup businesses face a similar challenge when they invent a new identity from scratch to establish a new enterprise as an ecosystem to do business with. This is the goal of branding initiatives, to make the vision visible for employees, customers, and other stakeholders in the enterprise.

Identities in the enterprise _

Identity is a ubiquitous quality of every activity in the enterprise. It is expressed as behaviors, messages, and symbols. It drives the creation and delivery of products and services, the content and style of communication activities, the design of physical and digital environments, and the behavior towards internal and external stakeholders. It touches every part of an enterprise, every audience, always and everywhere.

In its quest to define and shape identity, Branding as a professional discipline has overcome its fixation on the relationship to consumers and products, and evolved from being a pure Marketing function to a common effort that involves all parts of the organization. All activities, regardless of whether they are traditionally part of Marketing, HR, Communication, Finance, or IT, are building on the corporate brand as a channel to reach out to people, and as a way to establish a shared identity.

We believe rebrands should create symbols of change, not just a change of symbol.

Simon Manchipp (someoneinlondon.com)

Brands equals interface not surface.

Oliver Reichenstein (informationarchitects.jp)

To deal with the complexity and dynamics of the enterprise with its wide diversity of stakeholder relationships, Branding initiatives themselves need to become much more flexible and dynamic than before. As a static collection of visuals and guidelines, communication and messaging, or a series of internal events, they fall short of their promise of transforming the way the enterprise actually works and reaching and engaging all relevant stakeholders.

_ A platform for social exchange

Instead of a one-way communication activity, branding initiatives addressing the enterprise with all its activities have to establish identities in ongoing conversations. They have to become symbols exchanged in a constantly changing, highly dynamic environment of social relationships. Reaching beyond traditional control zones such as product packaging, the corporate intranet, or office interiors, they must be present in all conversations going on in the enterprise, among a variety of stakeholders, and across all communication channels.

From a people perspective, this means that brands are perceived more as an identity, as a person rather than as a style or property. Just as with a person you meet on the street, people have to be convinced and persuaded to engage in a relationship with a brand, and have a clear idea of what it can do for them. Therefore, brands must provide a context for interaction to encourage social exchange and participation, acting as welcoming hosts for the people they address.

A brand that acts as such a platform is embedding the question of *who we are* in daily operations and behavior, and encouraging information exchange among all stakeholders in shared spaces. It is visible in traditional media, physical environments, communication channels, and digital media such as Facebook or virtual workspaces shared with business partners. It seeks to influence rather than control, and flexibly supports Co-Branding with other identities, sometimes as host and sometimes as guest. It is fluid, adaptive to context, tangible, and approachable. In all these situations, it is consistent and clear in what it says and does, and reliably delivers on its promises.

In order to become a ubiquitous platform for everything an enterprise is doing, identities have to be implemented in concrete designs. The qualities to be expressed must be encoded in the messages, behaviors, and appearances experienced by people.

The identity aspect helps to understand how the intended result of a design process interprets, expresses and influences identity, and thereby shapes the way actors in the enterprise are perceived by others. By looking at the way an identity is seen by a stakeholder, a design can purposefully interpret and transform it into something visible and usable, to express its qualities in the distinctive signatures driving their experiences. Programs to shape identities are not easily implemented—after their definition and introduction, they have to be *lived* in everyday business, and constantly adapted to new conditions. This is where the Architecture aspect comes into play, the formal structures underpinning the enterprise and all its activities.

2_ARCHITECTURE

Any enterprise, regardless of type and size, is based on a complex set of fundamental structures facilitating its daily operations. The term *architecture*, when applied to an enterprise, refers to all the formal structures put in place to make it work. It represents the way an enterprise is constructed and functions as a consciously designed, man-made socio-technical system. Architectural structures underpin all activities going on between the actors involved in its endeavors.

Working with the architecture quality of the enterprise involves anything that enables organizing work and facilitating communication, automating business processes with the help of technical systems, establishing rules and policies, managing assets, resources, and events, as well as running the programs and projects to transform the way these elements work together.

Many of the people involved and impacted by these structures and dealing with the enterprise in some sort of stakeholder relationship, the architecture implementing the enterprise is largely beyond view. The results of its work, however, are quite visible: architecture determines how well an organization can assure service quality, how quickly decisions ripple through its reporting lines, or in what way it reacts to one out of hundreds of customer requests. Every time people interact with or within an enterprise, its architecture largely affects what they get back from it.

Exploring architecture _

The architecture aspect captures how the enterprise does what it has been created for. This includes a large number and variety of connected structures and substructures. Just as there is not only one identity in the enterprise, there is no single architecture, but a complex system of intertwined and overlapping structures, some of them describing the same elements but focusing on different aspects, and using varying terms.

Depending on the chosen perspective, an architectural description might encompass the structures of a single organization, or the entire supply and sales operation system including structures shared with customers and suppliers. It might describe how to produce and deliver products and services, communication channels, and management systems, or how the enterprise relates to investors and raises investments.

_ EXAMPLE

Think about the last time you used a logistics or postal service. Did they pick up and deliver your package on time without delays or errors? Did it arrive in the shape that you expected, or was it broken on the way? Were you able to track the delivery process on the web? All those aspects of their service depend on the formal structures that make the enterprise work, including vehicle and route management, data exchange, manual and automated workflows. To be successful, the architecture behind the service has to be designed for speed of execution, quality assurance, exception handling, and steady communication across the delivery process.

_ Working with architecture

Running an enterprise inevitably involves putting these structures in place, either consciously or just as they emerge in daily business. Processes, roles and responsibilities, information exchange, and some kind of office space are elements you will find in any organization. Their importance in individual cases depends on various factors such as the individual industry, the number of actors involved, or the size of the organization.

Enterprise relationships captured in architectural descriptions form a complex and intertwined structure of structures, undergoing constant change. Individual substructures cannot be viewed and addressed as static constructs in isolation, since they are difficult to separate and largely depend on each other to work—changing one means affecting others as well. Many formal structures in the enterprise, like social and organizational configurations, tend to change very quickly. Any attempt to capture a holistic enterprise-wide architecture can only capture a moment in this development or lay out the direction of this transformation, by modeling a current or a desired state.

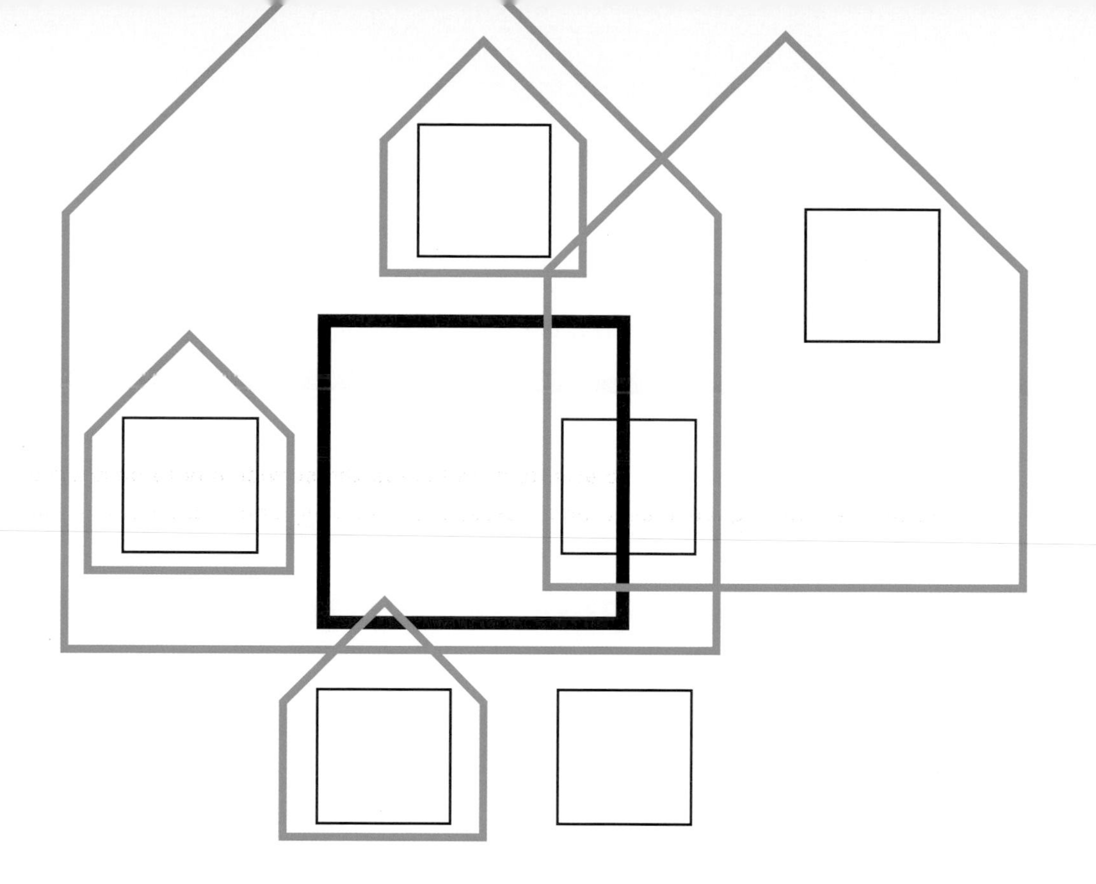

For those reasons, any work on architecture deals primarily with the complexity and ambiguity that comes with the large elaborate structures in constant transition that you typically find in an enterprise. While it is impossible to control such a system as a whole, initiatives aiming to shape architecture have to influence it by establishing controlled substructures, understanding the linkages between them, and managing change and development.

As the set of structures driving all execution, architecture is intrinsically subject to all transformation processes in an enterprise. To be aligned with strategic intent and vision, it needs to be understood as it emerges dynamically, and consciously transformed to fulfill the purpose of the enterprise. This is true even if the goal of this transformation is to maintain the status quo of the enterprise by adapting to changed conditions.

To understand these structures and inform management, decision making and governance, architecture models are used to describe or visualize all structures in the enterprise. They show stakeholders both a view on the current situation and alternative developments to be considered. The instrument of choice for this is Enterprise Architecture, a discipline that aims to make architecture visible to stakeholders involved in or impacted by a transformation.

_ EXAMPLE

IT Network architecture to enable data exchange between computers or other devices is quite important for most businesses to operate, but it is certainly more important for certain industries. For an Internet service provider it is at the heart of the business, while it might be tolerable for a farm to be cut off from the Internet for a few hours. For other industries, it might be physical facilities, financial resources, or other structures that turn out to be particularly relevant. The amount of effort and investment spent in designing, implementing, and optimizing a specific aspect of architecture therefore varies from case to case, and is dependent on the business model that applies.

CERN Computer Center

To develop an understanding of the structures relevant for a design problem, architecture has to be discovered and described, and focus on particular types of structures depending on what is relevant for that purpose:

_ MANAGEMENT STRUCTURES

and governance systems to steer operations, performance, and development of organizations and lead the enterprise to desired results, described as performance measurement and reporting structures and used in scorecards or evaluations.

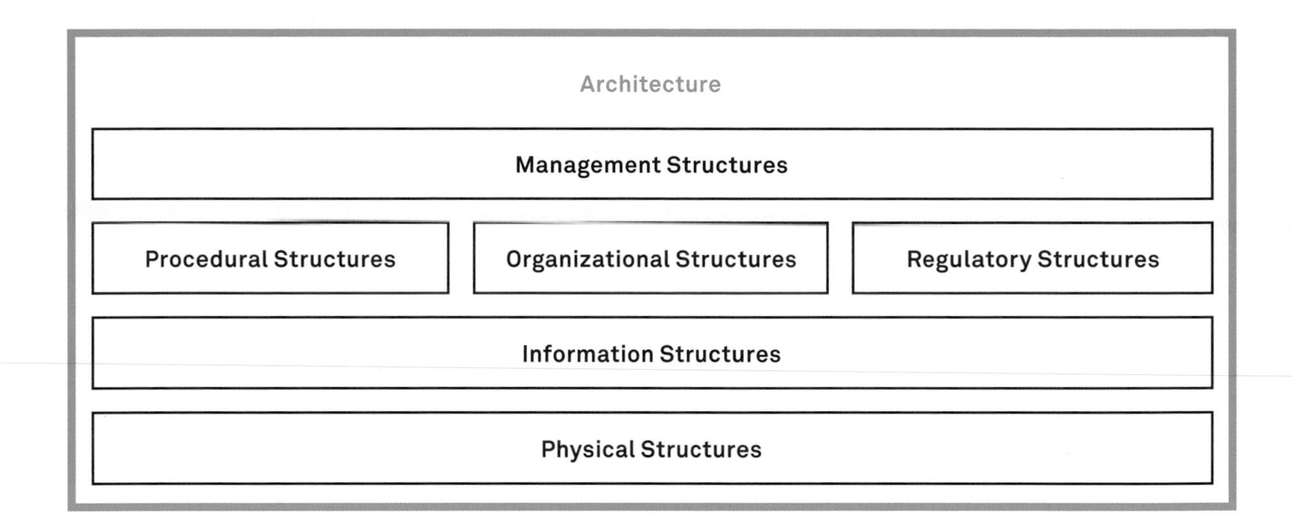

_ PROCEDURAL STRUCTURES

to execute operations, create products and services, collaborate with external partners, implement business functions, and take care of exceptions and emergencies, described as linked steps of business processes and used to manage activities.

_ REGULATORY STRUCTURES

establishing guidelines, policies, and rules described as decision statements and control structures as a basis to guide business decisions, and used to direct human and automated decision making and control financial transactions.

_ ORGANIZATIONAL STRUCTURES

dividing organizations into business units to manage work and choices being made, measure success, and handle processes with human involvement, described as formal relationships between people and groups, and used to define and manage roles and responsibilities.

_ INFORMATION STRUCTURES

implemented in systems and communication channels to handle and exchange data, implement asset inventories, messaging systems or knowledge repositories, described as technical components on different layers and used to manage information availability.

_ PHYSICAL STRUCTURES

used for operations such as buildings and facilities, machines, and energy supply and transportation systems, described as assets used to enable business operations.

ENTERPRISE ARCHITECTURE

That set of descriptive representations (models) that are relevant for describing an Enterprise such that it can be produced to management's requirements (quality) and maintained over the period of its useful life (change).

John A. Zachman (zachman.com)

Architecting as a discipline is probably among the most ancient activities of humankind. It has emerged to deal with the complexity of structural decisions involved in creating stable and durable buildings serving a wide range of purposes. Today, business leaders find themselves in a similar situation when planning and crafting their businesses, making decisions steering their development. Enterprise Architecture is an instrument to implement a strategy by exploring and reshaping structures in the enterprise. It translates business vision into execution by understanding how an enterprise works, modeling its potential future, and informing management and governance to drive decision making. The outcomes of Enterprise Architecture initiatives are maps of all kinds of systems in the enterprise and their relationships, along with goal-directed analysis, optimization, and restructuring to make the enterprise more efficient, robust, and integrated.

The use of multiple viewpoints enables those initiatives to address specific stakeholders individually and to deliver insights relating to their particular concerns. Such a view concentrates on a specific subset of details while leaving others out of the picture. Examples of such a view include models of the business processes, models on technology usage, or models illustrating information flows and critical paths. Viewpoints of Enterprise Architecture have the goal of capturing just the right amount of information to inform stakeholders without providing too much detail.

As an endeavor potentially touching every aspect of daily business, Enterprise Architecture initiatives must have the connections, authority, and alignment needed to bridge between organizational silos, and to trigger transformation processes just about everywhere in the enterprise. Business strategy and policies must form the foundation of the architectural principles used to inform the creation of those structures. Such initiatives face not only the complexity challenge of developing holistic and accurate models, but also of turning a redesigned model into projects, products, and processes that will have a real impact on the actual structures that drive the enterprise.

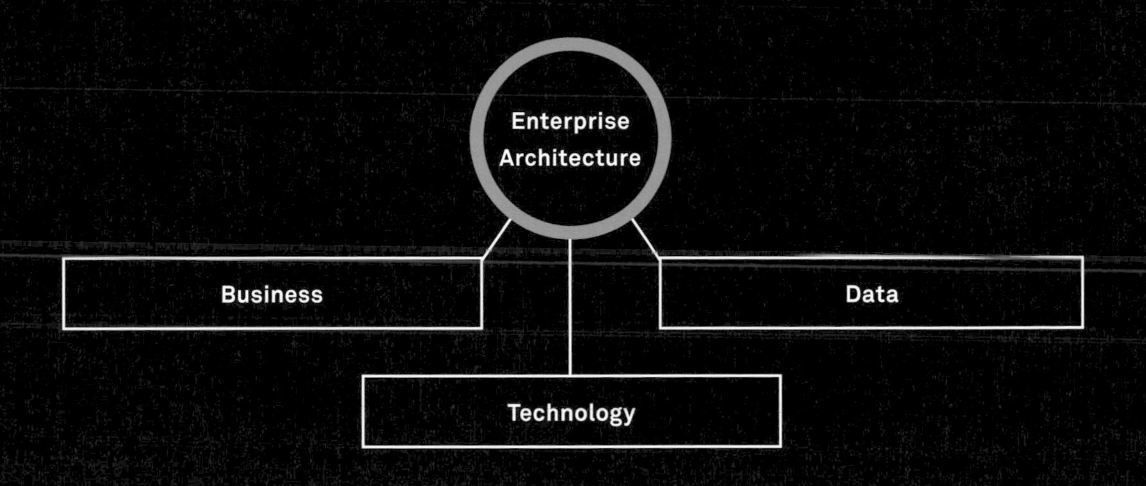

Today's frameworks typically concentrate on three major topics:

_ BUSINESS

illustrating the business model as system of financial streams, business capabilities, process and events, organizational structures and policies.

_ DATA

illustrating the use and flow of information in the enterprise, including consolidated data models on entities and their relationships, metadata and quality management, reporting and analytics.

_ TECHNOLOGY

illustrating the technical structures the enterprise uses to operate, such as applications, network infrastructure, computing resources, machines and appliances, or technical standards.

It's always about people. No matter how technical a problem may seem, ultimately it's always a people-problem.

Tom Graves, Enterprise Architect and Principal, Tetradian Consulting

_ The human side of architecture

Historically, the discipline of Enterprise Architecture concentrated on shaping complex IT and technology architectures in alignment with business requirements. Today's practitioners expand their scope to address all structures that constitute an enterprise and make it work. Organizational reporting lines, information systems delivering data, processes driven by automated systems and human decisions—all these structures are just different aspects of the same system. The resulting architecture is the foundation of every single step the enterprise takes, from sending a receipt to a customer to a merger with another organization. Just as buildings are structures made for people to live in and look at, enterprises are structures made by people for people as a space for interactions and transactions. In consequence, people themselves cannot be seen as assets to be incorporated in an architectural description, only the enterprise's relationship to them. Any person in touch with an enterprise is both a user of its architecture and a contributor to it.

Designing visible architectures _

The structures described in an Enterprise Architecture model ultimately enable the enterprise to reach out to people, facilitating all transactions and interactions with them. The systems constituting the architectural elements of an enterprise are driving these exchanges. They are visible to people as tools and services to solve tasks and make decisions, as information assets, communication channels, and workflows. They take concrete form in physical and virtual spaces, in personal conversations, phone calls, or web-based transactions, enabling people to interact with the enterprise.

Instead of being approached as a pure background process, Enterprise Architecture initiatives need to be aware of this role and design architectures that deliver real value to people in accessible systems. They should implement the strategic vision in a systematic fashion to address customers, employees, and other stakeholders by designing systems suited to their particular needs. They should result in concrete designs visible to and usable by people.

Applying the architecture aspect helps to explain how a design depends on the structure of the enterprise, and how it implements architectural decisions and principles to form a part of the same structure. Architecture as universal quality of the enterprise is an inherent part of every design that largely influences how it works and what people get out of it in individual experiences.

3_EXPERIENCE

The identity and architecture aspects described earlier are both attempts to understand and describe the enterprise as an entity, the way it works and the way it expresses its personality. In the real world, the enterprise as a whole remains an abstract concept, largely invisible to people. The only things people in touch with or related to an enterprise actually get to see are a subset of its concrete manifestations.

The term *experience* refers to the individual interactions with these manifestations. Every time a person is exposed to the products and services, buildings and spaces, messages and dialogues, or websites and information systems of an enterprise, those offerings become an element within that person's experience. The experience aspect of the enterprise defines it as a space wherein people are experiencing its concrete appearances, represented by other people, technology, or artifacts, everywhere and all the time. In that space, every stakeholder, every customer, employee, investor, or other person with a relationship to the enterprise is experiencing it in some way through these appearances. This viewpoint is not about how the enterprise works on the inside or what values and qualities it is associated with, because in their daily lives, people rightfully don't care much about these things. The experience aspect allows us to look at the enterprise as the direct impact it has, on real people, and in the real world.

Understanding experience _

Everything going on in an enterprise has an impact on and is impacted by people's experiences. When a person engages with a service, organization, brand, environment, or product—virtually everything visible that the enterprise produces—his or her experience is determined by the way those offerings are designed, and how well these designs address human qualities and context.

Experiences are highly subjective impressions, so that their nature, duration, and perceived quality vary from case to case, as does their individual relevance and impact in the enterprise. A single experience might take a few minutes, days, or even much longer—it depends on how long the person involved experiences its duration, and when a new experience begins. Regardless how well established and intense an enterprise's relationship to a person might be, its appearances will only play a limited role in that person's experience.

_ EXAMPLE

Everything is experience—consider experience in the context of the previous examples:

An Enterprise Architecture function planning to redesign a company's network architecture strives to ensure continuous availability of information, just as redesigning the logistics structure of a postal service has the goal of ensuring timely deliveries. Both are ultimately put in place to impact the experience of customers and employees.

Equally, a Branding initiative that positions an oil company as a responsible actor with regard to environmental issues is only visible in actual behavior and messages, again experienced by employees, customers and other stakeholders such as the public.

Also, when using a public transport service, people usually do not care very much about exactly how trains operate or are maintained, or how the transport company thinks about itself. They just want to get from A to B, comfortably, on time, for a fair price and without trouble. Using a train for this voyage is just a means to achieve the real goal, like visiting some friends.

People are having experiences with the visible appearances of the enterprise, everywhere and all the time. Every activity being carried out, every dialogue or decision by anyone related to the enterprise, qualifies as such an experience. In turn, every interaction, transaction, or other endeavor depends on the associated experiences of the people involved. Consequently, every offering made available to people addresses some kind of experiential concern.

_ Shaping experience

In business reality, most projects to transform the enterprise start with an experiential concern in mind—ultimately, some person or group is the target audience, the people who see and use the results. But instead of deeply exploring and analyzing this target experience to come to a design that fits, many projects lose sight of it early in the process. More often than not, an initially clear vision of the intended experience gets replaced by more and more product features together with technical considerations, derailing the original goal. Decisions are made without considering their impact on the experience, resulting in offerings that fail to adequately address the human qualities and context in which they are intended to be used.

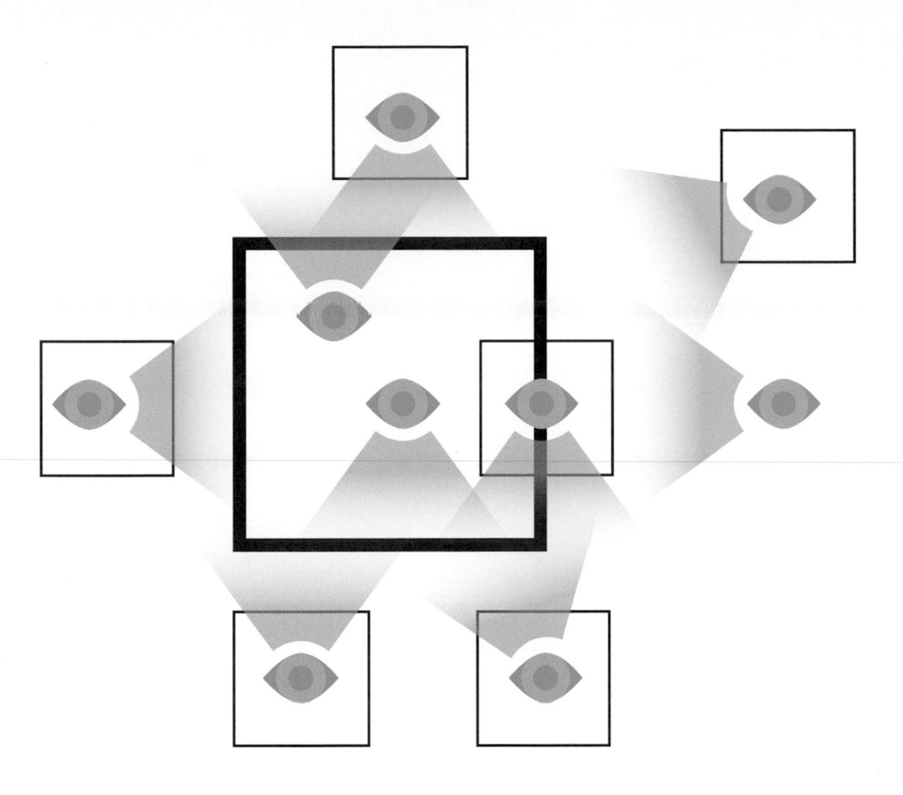

To overcome this issue, decisions about and within transformation projects in the enterprise must be made with their experiential impact in mind. The design of a product or service itself is not what matters in the end, but its meaning and usefulness for the people it addresses. All technical systems and their interfaces, all products, artifacts, and visible elements, are just there to support that goal.

There is a range of emerging practices aiming to proactively understand and shape those elements that influence people's experiences for the better. This activity is commonly referred to as Experience Design.

EXPERIENCE DESIGN

While everything, technically, is an experience of some sort, there is something important and special to many experiences that make them worth discussing. In particular, the elements that contribute to superior experiences are knowable and reproducible, which make them designable.

Nathan Shedroff

As explored in the previous part, design always seeks to have an impact on people's thinking, feeling, or activities. As such, all design projects are transformations that seek to influence experience. The emerging practice of Experience Design is a process with a specific regard to this impact on experience, and differs from traditional design processes in that it puts a lot of effort in understanding the factors that make a great experience, and does not necessarily focus on a specific medium or technique.

In the enterprise context, Experience Design is a strategic instrument to establish and transform different kinds of relationships in the enterprise. It allows us to make design decisions based on a deep understanding of the influence a product, service, process, or other kind of solution to a problem has on the experience of its audience.

Design practices applied in this field are based on a deep understanding of the people you are designing for, their lives, characteristics, needs, and aspirations. Therefore, designers employ a human-centered attitude and use methods to engage in an active exchange with people. They employ extensive research techniques and attempt to develop empathy for the audience of a design, and to capture the emotional, aspirational, and contextual parameters that will make an experience a great and memorable one. They synthesize these findings into tangible prototypes of the final design that serve to validate and iterate it before anything gets actually implemented.

User Experience

Customer Experience

Experience Design

Employee Experience

Brand Experience

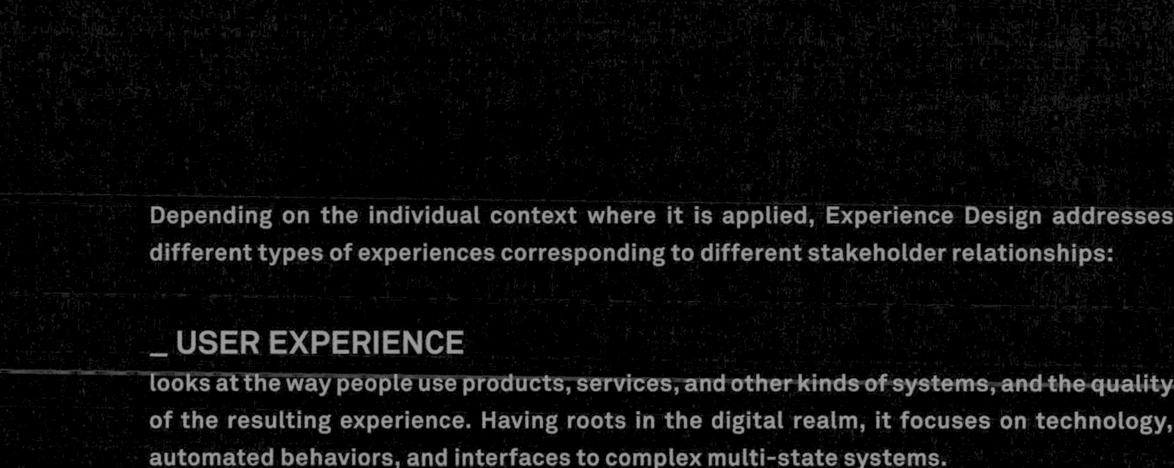

Depending on the individual context where it is applied, Experience Design addresses different types of experiences corresponding to different stakeholder relationships:

_ USER EXPERIENCE

looks at the way people use products, services, and other kinds of systems, and the quality of the resulting experience. Having roots in the digital realm, it focuses on technology, automated behaviors, and interfaces to complex multi-state systems.

_ CUSTOMER EXPERIENCE

addresses commercial relationships in the enterprise, from being aware of a need to its fulfillment using a product or service. Having emerged from Marketing and Customer Service, the discipline focuses on customer satisfaction and retention.

_ EMPLOYEE EXPERIENCE

seeks to optimize the relationship between an organization and its employees by addressing their experiential needs and concerns along their career. This internal view of Experience Design is often seen as prerequisite for a good Customer Experience since employee behavior has a big influence on its success.

_ BRAND EXPERIENCE

has emerged from the more traditional practice of Branding, and attempts to look at a Brand from the perspective of the people addressed and transforming their experience when interacting with it.

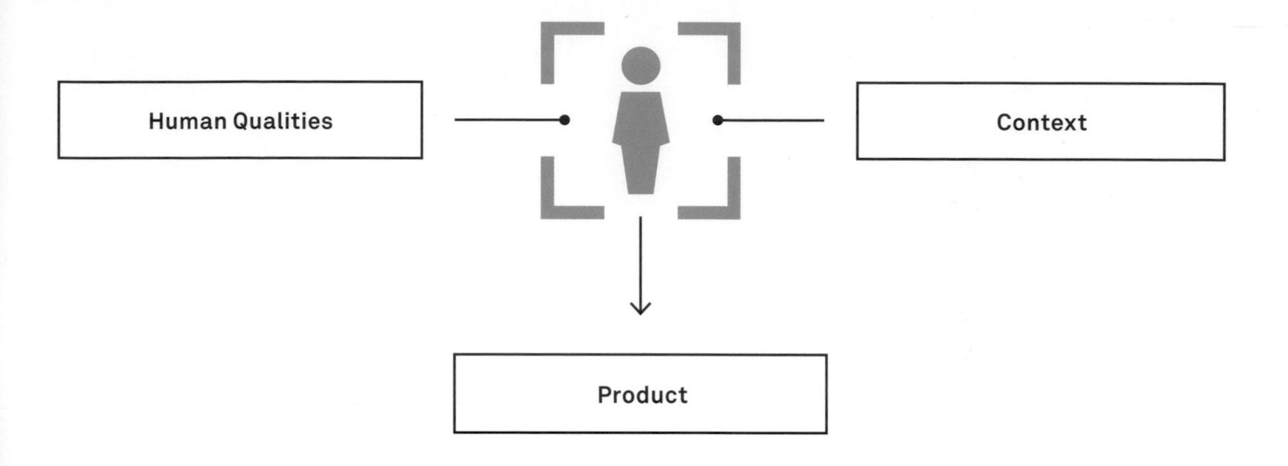

Whether or not an experience with a product—anything an enterprise makes available is valuable and desirable to people depends on how well it addresses its audience and the context in which it is used. The better the enterprise deals with these factors, the more useful and delightful it becomes for the people it addresses:

_ HUMAN QUALITIES

how a design addresses a person's perceptions, individual preconceptions and motivations, cognitive and physical abilities as well as attitudes, emotions, and affections. In the mind of people, it's not the offering itself that counts, but the meaning, utility, and subjective satisfaction it adds to their lives.

_ CONTEXT

how an offering integrates in individual contexts of use, such as social structures and culture, physical contexts and individual situations. This includes elements such as language, symbols and codes, and behavioral norms, and whether people are in a private or a professional context, on the move, at home, or in the office.

Experiencing the enterprise _

Seen from the perspective of a person, an enterprise can be described as a sequence of experiences. Everything going on in the enterprise ultimately impacts the experiences of those it is related to. For people, there are no business processes and technologies, just activities and tools they use to accomplish tasks. Content, information, and messages are merely means to inform choices and generate ideas or knowledge. Collectively shared identity, symbols and meanings inspire the emergence of culture and the creation of visible artifacts. These are produced and lived by people every day, becoming part of their experiences.

Every business objective or technical concern driving a project or process also can be expressed as an experiential goal of a specific stakeholder group or an individual. Successfully addressing its implications means influencing the experience of those people. It requires understanding their needs and aspirations to design a solution that helps them to achieve their goals.

Consequently, experience is a universal quality of the enterprise that impacts its relationship to anyone involved in its endeavors. Every project, every solution to a problem can be seen as part of a design to influence the experience of people, even if that experience is not consciously considered.

Experience Design applied in the enterprise context aims to proactively influence the relevant experiences of important stakeholders, equally applicable to a manager, a customer, or another person related to the enterprise. Using it successfully means making every factor of a design that influences the human experience a conscious, explicit decision. It is dependent on insights about the people you are designing for, and empathy is a prerequisite for developing this understanding.

Because of its broad scope, the results of an Experience Design approach are manifold. They range from the basic considerations of how a product or service integrates into a business strategy and meets human needs, to all the visible and usable elements that people interact with, like the conduct of a phone agent or the structure and appearance of a website or piece of software.

_ EXAMPLE

A manager responsible for a purchasing department of a large furniture company is a typical internal stakeholder. His or her job is to ensure availability and quality of materials, good prices and efficient supply chain operations. This translates to experiential goals such as the need to be informed about the market situation and prices, to be able to control and prioritize deliveries, and to establish a trustful working relationship to people inside and outside his organization. Part of an Employee Experience initiative would be to find out about this experience, and designing everything to support these goals. A possible outcome could be a new reporting system, that delivers all relevant information for specific business roles in a dashboard.

A dashboard for a purchasing manager role, a specific part of a larger Employee Experience

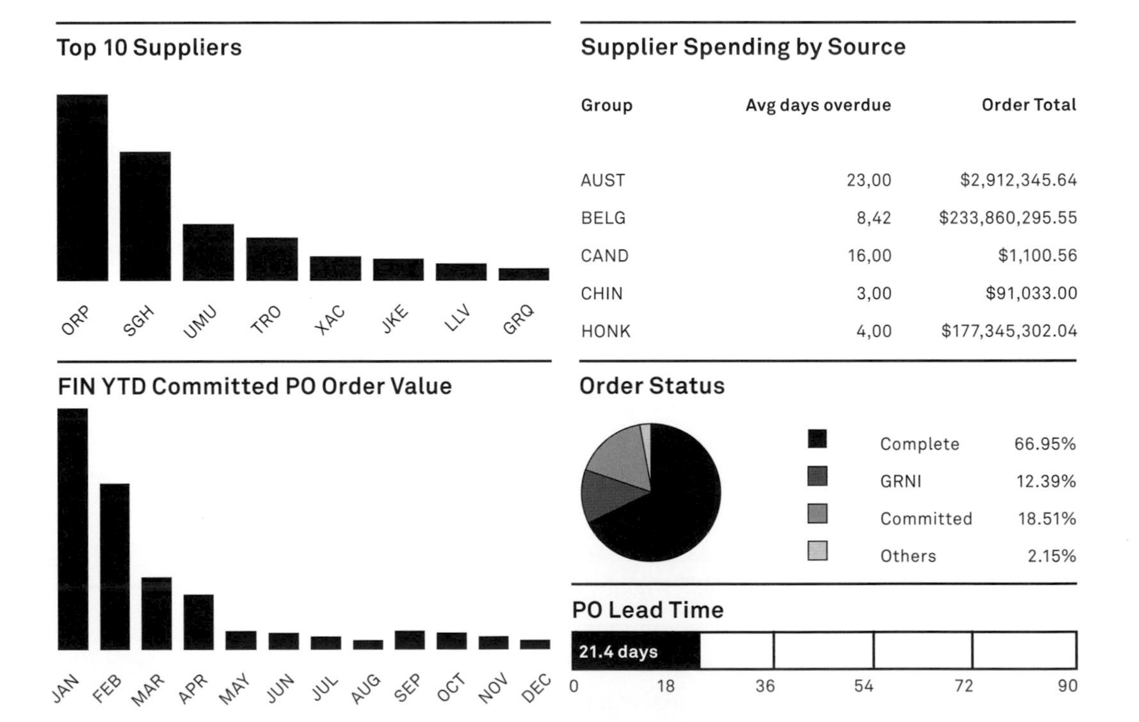

Experience strategy _

Because of this ubiquitous presence and fundamental relevance of experience in the enterprise, it is an important success factor for any organization, but largely ignored today. Deliberately addressing this factor as part of an Experience Design initiative can help organizations to gain competitive advantage by optimizing their relationship to people and the interactions with them.

On an organizational level, this can be achieved by formulating a globally applicable experience strategy. Such a strategy expresses what stakeholder experiences in the enterprise should be like, and guides all decisions about elements influencing these experiences.

The individual objectives and qualities addressed vary widely depending on the particular stakeholder relationship. An experience strategy targeted at your customers starts from an external perspective, the result of your work, and can be used to transform your organization to make their experience with your products and services a great one. This in turn might depend on an internal experience strategy that seeks to integrate people into the business, to give them what they need to do their jobs, to encourage their personal growth, and to inspire the working culture as it is lived every day.

_ EXAMPLE

An interaction between a customer refueling a car at a gas station and the company behind that service usually lasts just a few minutes. It is embedded in a broader context of getting from A to B, and is just one element within that experience. The company has only that short time span to influence the customer's experience.

A clerk at the gas station on the other hand shares the whole work day with that company, so that the influence on his experience is much higher. How well the particularities of his or her professional life are addressed is in turn influencing the experience of the customer.

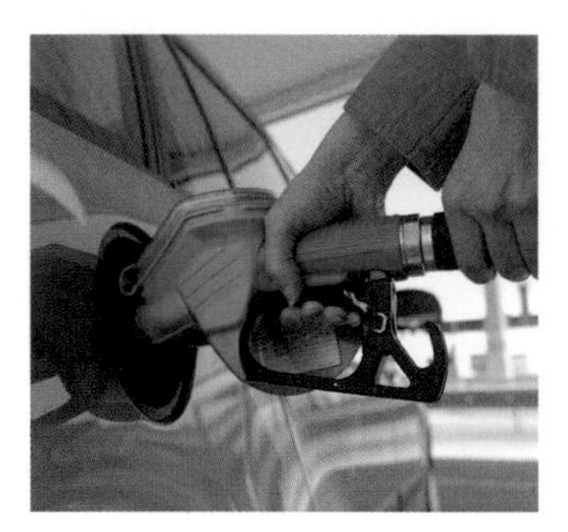

Every person or group involved in activities in the enterprise is a stakeholder and potential target group, such as customers, suppliers and partners, or investors. The strategy may be applied to designs targeting very narrowly defined groups, limited to specific roles such as journalists or job candidates, or temporary relationships like visitors or readers. It may also attempt to very broadly influence the experience of everyone in touch with an organization. As it is being applied today, however, on the level of individual products, media, or services, Experience Design often falls short of its potential to have a real, business-critical impact on key stakeholder experiences.

Used on the enterprise as a whole, considering the experience quality allows us to understand it truly from the perspective of people, to explore their experiences and to design for them on a strategic level. Accordingly, it has to be used in conjunction with identity and architecture, to look beyond the individual experience at everything that constitutes the enterprise behind it.

DESIGNING WITH BIG PICTURE ASPECTS _

After exploring the nature of enterprises in some detail based on universal qualities, it is important to point out again that all of them are describing the same things, just with a different focus. Identity, architecture, and experience are always there, in any enterprise. In other words, regardless of what element you are focusing on, it is still the same enterprise.

The Enterprise Design approach places design at the intersection of identity, architecture, and experience. Looking at the enterprise from these perspectives allows us to explore it as the context wherein a design initiative is meant to be applied, and as the system that it seeks to reshape and transform.

Considering the enterprise in such a way allows a design to act as the *face* of the enterprise to people. It can implement identity as a visible and approachable personality, turn architectural systems into usable and useful interfaces, and integrate these elements into people's experiences.

A strategic design initiative is an attempt to align these qualities on a common course. It allows us to approach enterprise transformation processes from the outside in, starting with the intended experience of the people you are designing for, and subsequently influencing identity and architecture to achieve outstanding experience qualities.

Successfully designing for these qualities requires revealing them to people in many concrete manifestations, and orchestrating these solutions to play a meaningful role in people's experiences. In the next chapter, we will continue with the individual elements that compose such manifestations at a more detailed level.

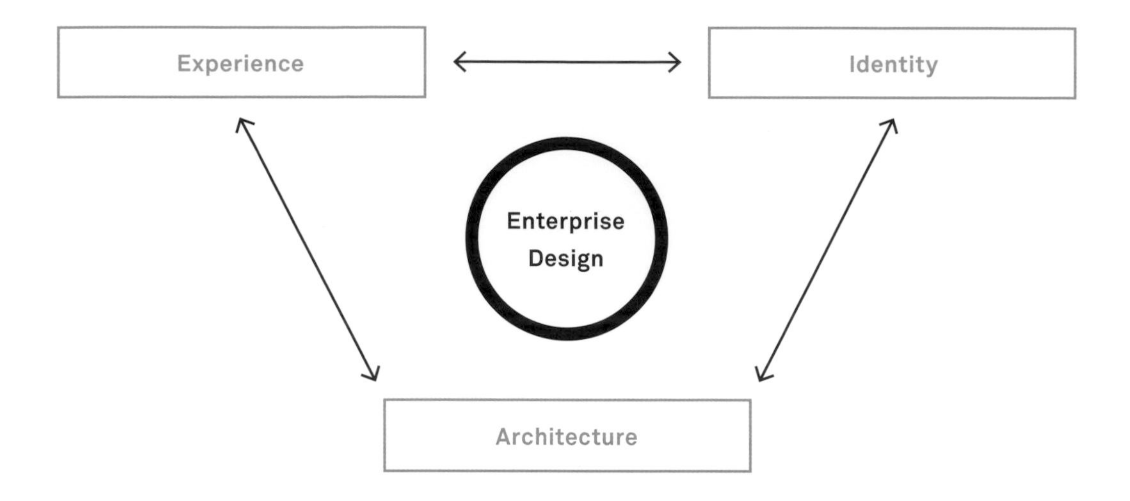

In practice, any transformation of one of the three qualities means also (often unknowingly) changing the others as well:

_ IDENTITY

has to be perceived and lived as concrete manifestations in people's experiences to come to life as shared ideas, which depends on behaviors, messages, and values expressed in business operations facilitated by architectural structures. In turn, experiences then shape these ideas and thus influence identity.

_ ARCHITECTURE

describes a set of artificial structures, put in place around a group of human beings. In order to work, architecture has to reach out to people's experiences, produce useful, valuable systems and interfaces, and address and implement identity in daily operations as concrete behavior, appearance, and communication.

_ EXPERIENCE

can be seen as a short-lived, volatile space in which the results of architecture and identity become visible manifestations. Achieving a positive experience depends on the orchestration of these visible elements, in a way that makes sense and creates value for people, becoming a part of their lives.

In consequence, these qualities have to be addressed as one.

AT A GLANCE

Identity, Architecture, and Experience are aspects that capture universal qualities of any enterprise, touching everything it does, and being both the context of any strategic design initiative and subject to its outcomes. They determine how the enterprise is seen by people, how it works and achieves its goals, and the role it plays in people's lives.

Recommendations

_ Use branding to transform your space of identities, designing brands to be dynamic platforms for social exchange rather than static symbols or messages

_ Develop a thorough understanding about the architecture that makes your enterprise work, and design to leverage and evolve the underlying structure

_ Base all design work on an experience-driven view, designing to enhance and reshape people's experience and transforming the enterprise to match this

_ Use the Big Picture aspects to explore the enterprise as an overarching, dynamic ecosystem, and to put designs and their results into context

CASE STUDY _ IKEA

IKEA'S ENTERPRISE

Becoming the world leader in functional furniture and accessories, combining design excellence with accessible pricing.

IKEA is not completely perfect. It irritates me to death to hear it said that IKEA is the best company in the world. We are going the right way to becoming it, for sure, but we are not there yet.

Ingvar Kamprad, Founder

To many, IKEA is one of the world's most fascinating companies. Founded in 1943 in Sweden, IKEA is today the world's largest and most successful retailer in the furniture sector. In Europe, it is hard to find people who do not own at least some IKEA products, and for most students it seems to be the only place to go for equipping an apartment. When looked at from some distance, the company seems to be a stunning, well thought through system of elements that work together seamlessly. Everything fits together, and the elements of the enterprise work the same way across the borders of countries, languages, market segments, and cultures. Even across its stores, catalogue, and website, IKEA appears, communicates, and acts the same way. To achieve such a remarkable level of coherence, IKEA puts an equally remarkable effort into the systematic design of its enterprise, and excels at aligning different aspects and components on a common course.

IKEA perceives the design competency as the key enabler to its business, and applies it to everything that can possibly be designed, bringing together the purpose, values, and strategic goals of the group. The underlying IKEA concept is above all well thought through in its very details. It is a complex system of systems that took years to create, refine, and grow, which involved a myriad of conceptual design decisions and concrete transformation projects to get to its current state. In these efforts, the universal qualities of identity, architecture, and experience are addressed as the universal qualities behind the enterprise, reflected in everything the company is, says, and does. All operational and development activities contribute to embedding and sustaining these qualities in IKEA's enterprise.

_ IDENTITY

The group's enterprise identity is based on a strong core idea, which is communicated to every important stakeholder—clients, staff, suppliers and other actors in the enterprise environment. IKEA is following the vision of contributing to people's lives by being your partner in making yourself a home. This idea is conveyed in a way that is simple and clear, and is represented in the variety of identities that the group creates and communicates.

A well-organized system of brand identities holds all of this together. The binding element is the IKEA brand, aiming to represent the sum of ideas, offerings, and values the company represents. Unlike other corporate brands in the consumer world such as P&G or Daimler, the IKEA corporate brand takes a strong position in every instance of enterprise-people touchpoints and is a focal point for IKEA's consistent perception by people. The name is associated with the underlying concept, holds the different sub-brands and identities together, and is a vehicle to convey the company's key messages to its audiences.

The ubiquity of the IKEA brand elements is visible in the application of its characteristics. Drawing upon the group's origin from Småland, IKEA appears openly Swedish, reflected in the choice of colors and other visual elements. But reaching far beyond the visual, this heritage is also represented in the choice of product identities based on terms from the Nordic countries. Each category draws on a different set of words, such as places, islands, or given names, sometimes with a semantic connection—for example, bathroom articles are named after Scandinavian lakes, rivers, and bays. This system gives the portfolio a coherent identity system, reflecting the brand's characteristics and being easy to remember.

Moreover, this aspect of the brand is communicated by using speakers with a Swedish accent in advertising, and putting Swedish books into all shelves in the stores. Other values of IKEA's brand identity are visible in every aspect of the business. All channels are seen as a means to convey and reinforce IKEA's identity to all stakeholders. The uncomplicated, direct, and friendly appearance is reflected in all messaging, from official communications to error messages in their online kitchen planner. The culture promotes open exchange and empowers co-workers, making everyone a brand ambassador in communication and behavior.

IKEA combines self-service elements with a very personal communication style.

ARCHITECTURE _

The IKEA enterprise works like a highly structured and thought-through system. The architecture considerations touch many different types of structures, but their interconnections reveal an overarching, enterprise-wide system of related system components and interactions. Key areas of IKEA's architecture include the product portfolio, logistics and supply chain, finance and cost control, information systems, and operational business processes. Following the same core idea as the brand identity to be conveyed, this architecture is the prerequisite to execute on IKEA's promise of excellent products at low cost.

All products are based on a set of highly standardized parts and construction principles, allowing the group to benefit from huge economies of scale. Moreover, the assembly of the product itself is done by the customer, turning the principle of shared responsibility between the customer and IKEA into a collaborative model—we deliver all the parts, you build it yourself. Products delivered in parts can be optimized for efficient packing and transportation to the store network. This illustrates the idea of leveraging capabilities of external actors in the enterprise ecosystem, in this case the customer providing an expensive capability that is considered essential for a furniture business. To enable this shared model, the construction itself is simplified and described in visual manuals, again following strict standards to be easy to use. Besides the clear instructions, using visual explanation makes a single manual internationally applicable.

Customers encounter the same product multiple times across their journey.

With the same rigor, IKEA addresses architecture aspects related to physical, operational, financial, or IT structures. Money flows across a network of companies, separating the IKEA concept from the store operations and background processes. Core processes of production, logistics or sales are supported with a role-based and localized intranet to give every Co-worker the information and tools he or she needs. Customers are part of this system, contributing essential processes such as check-out, assembly, and home transport. Virtually everything is organized in a coherent way—this is visible in the universal information architecture underpinning the store layouts, web navigation menu, printed catalogue and IT systems everywhere in the world.

EXPERIENCE _

Given the level of rigor and attention to detail IKEA puts into defining and applying its Architecture and Identity, it is easy to assume that these two elements are at the heart of their enterprise. In reality, however, they are merely the effects of designing everything coherently, and of the purposeful execution of a strong vision. And that vision is fundamentally about Experience.

IKEA defines its mission in terms of the Customer Experience it wants to create —contributing to people's lives with its products, and partnering with customers to save money together. All elements of the enterprise contribute to this very well-defined experiential vision, applying design disciplines and related areas of creative transformation to make it reality. IKEA's products are designed to include all functions of home furnishing, thereby providing a large diversity of styles, variants, and price ranges to have something for everyone. The high regard for product design at IKEA is visible in the systematic way of making items fit together in product families and combinations. The do-it-yourself philosophy is an essential part of this experience, turning the shortcomings of assembly and transport into a benefit, and making items cheap and easy to transport. Digital Self-Services make it easy for clients to transact with automated tools.

IKEA's Experience Design efforts reach well beyond their products. The stores are designed around the activities of people who want to make themselves a comfortable place at home, facilitating a journey across a set of well-designed elements. This includes presenting products not as objects, but as cozy spaces with a character—connecting to people's ideas about their own home and giving them inspiration, but also bringing together products in designed arrangements. With the well-known restaurants and cheap hot dogs, IKEA's food division cares for people's breakfast or lunch. The company also provides the Småland day care during the stay, extending well beyond the core offering of furniture based on identified experiential concerns.

Design at IKEA centers on the customer, but extends to brand, employee and user experience approached in unison. In conclusion, it is not just Identity and Architecture producing these solutions—it is the other way around, making their strategy tangible as a global experience.

5 _ ANATOMY

The aspects explored in the previous chapter form the basis to capture and describe what an enterprise actually is in the eyes of the people it is related to. They determine the overall subjective judgment from a human point of view—what it is about, how well it works, and to what extent it contributes value to people's lives. Although they apply to the enterprise as a whole, having identity, architecture, and experience qualities, there is no single person who has a full view of their entire range. The enterprise as an entity is always quite fuzzy, without clearly defined boundaries and with a lot of inherent complexity and hidden parts obstructing the view.

While business success depends on the delivery of information and services to all of these critical stakeholders in the enterprise, their individual needs, aspirations and levels of insight vary widely. Anyone in touch with the enterprise experiences just a certain part of it, having a limited view on its architecture and a particular image of its identity. Consequently, overarching enterprise qualities manifest themselves differently in individual cases, and become meaningful for people only in the context of a specific relationship to the enterprise. In such instances, the visible and tangible elements of a relationship are perceived and put into use by people as part of their greater experience. They equally play a role in the larger architecture of enterprise operations, and they constitute identity and image. The strategic value of design comes from influencing and transforming these qualities from a human point of view. Key to this is the idea of the enterprise as a shared sense of purpose, with the goal to design every element that occurs in a relationship to contribute to that larger vision.

© 2013 Elsevier Inc. All rights reserved.

_ EXAMPLE

Consider the different stakeholders of an airline, each having a particular view and level of insight on the enterprise. A business passenger only experiences the flight service itself, having some superficial insight into the operations, but not into the background processes. The travel agent who booked the flight only experiences the web-based booking service, the experience with the airline being determined by a completely different set of factors. Inside the company, an operations manager may actually never experience these parts of the operations, but needs detailed data about the activities involved in their execution, and their profitability. In the enterprise as the wider environment, a staff member working for the airport again has a completely different view, facilitating a part of the operations as a service that the airport makes available to the airline.

Because of the specificity of every relationship, this involves looking at the details, and aiming to consciously shape the small elements that make a human interaction a positive and memorable one. It means a shift in perspective, from a macro-level of enterprise-wide qualities to the smaller scale, the enterprise as a microcosm. The elements portrayed in this chapter can be seen as constituent parts of an enterprise-people relationship, to be taken into consideration in any strategic design project.

_ The elements of enterprise relationships

In their book *Pervasive Information Architecture*, Andrea Resmini and Luca Rosati argue that today's space of interconnected information systems, media, devices, and tools makes us constantly jump among the elements of a complex system of items and environments, resulting in dynamic and unforeseen behaviors. They describe such systems of connected elements as *ubiquitous ecologies*, as open systems of linked items in a space with physical and digital layers. Instead of being designed in isolation, elements which act as part of such a system have to support human activities that span all of them.

Classic design disciplines tend to deal with only a distinct and clearly separated subset of those items, visible in the dominance of specialized knowledge areas like web, interior, or editorial design. While such specialization is necessary to account for the particularities of a specific design medium, in a hyperconnected world any outcome will just one part of a complex space of interconnected elements, with people using different subsets and configurations. It requires designing beyond separate categories, towards an integrative and hybrid approach to design. This is particularly true in the enterprise context, where so many people get in touch with each other, with various artifacts and procedures, interacting in so many different ways.

When approaching a wicked design challenge in an enterprise environment, the number of types of elements that may be relevant quickly becomes overwhelming. Depending on the chosen models, it might involve customer segments, business processes, activities and tasks, media and devices, digital information systems, and a multitude of other elements.

Similar to the universal qualities explored previously, these elements appear everywhere in the enterprise, at all levels and in any activity. Design initiatives can be seen as endeavors to positively affect the system of these elements: enhancing the relationship to certain actors by reaching out to them via relevant touchpoints, defining and delivering new and better services and content. Considering these elements explicitly in the design process is a way to broaden the view beyond the typical low-level focus on a particular artifact or medium, user, or target group, and to understand the broader context of the enterprise-people relationships behind the design.

As every item, be it product, piece of information, or service, is quickly becoming an ecosystem or part of an ecosystem, as such it has to be approached, designed, and consumed.

Andrea Resmini and Luca Rosati

In our consulting work, we found the following generic definitions to be particularly useful to approach a design challenge in the enterprise context with regard to its constituent parts:

_ ACTORS

the various stakeholders related to the enterprise, addressed or impacted by its activities, or involved in their execution.

_ TOUCHPOINTS

any moment in time and space a person gets in touch with the enterprise in some individual context.

_ SERVICES

value propositions offered by the enterprise to customers and other stakeholders, and made available as activities and their results.

_ CONTENT

pieces of information or data being produced, exchanged and received in all kinds of communication processes in the enterprise.

On the following pages, each aspect is explored in more detail.

#4_ACTORS

Any design initiative, be it just a small website or a large program in the enterprise, ultimately targets people. The term *actor* refers to subsets of these people playing a certain role, reflecting a certain type of relationship—as users, customers, officials, or other types of roles that an organization may choose to explicitly address. Whenever a person acts and does something, he or she incorporates one or more of these roles, representing an actor of a certain type.

Because design is fundamentally about successfully addressing people, many design philosophies base the design process on their study and involvement, apparent in paradigms like user-centered or participatory design. In the enterprise context, designers find themselves facing a wide range of potential actors, so that the question of who to address can become quite difficult to answer. Depending on the level, context, and objective of a design initiative, the number and definition of actors to be taken into account vary widely.

Selecting and defining actors therefore is a conscious decision, and can be regarded as a fundamental act of designing. Based on this decision, the intended outcomes can be based on a strong understanding of the actors addressed and their role in the enterprise. This section introduces the concept of enterprise stakeholders as a basis to define, select, and address actors, and make this the foundation for all further design decisions.

From stakeholders to actors _

Organizations are dependent on the societies in which they operate. These societies are the market for their offerings, their source for capital, labor, and knowledge, and the larger ecosystem they contribute to. Many management approaches focus on customers and shareholders as their most important stakeholders, yet other significant stakeholders include prospects, internal staff, suppliers, partners, or the public. Only recently businesses are shifting their attention to the full range of stakeholders, realizing that the concerns of those interest groups in the wider enterprise translate directly to their own concerns.

The concept of stakeholders was described by Edward Freeman in 1984, arguing that corporations are established and operated according to a set of social contracts, with the purpose to meet the expectations and demands of their stakeholder community. Every person having an interest—or a stake—in the enterprise, its goals, activities, or results, qualifies as such a stakeholder. According to Freeman, strong business performance depends on honest, open, and balanced relationships to all stakeholders, instead of concentrating too much on one particular interest group.

Stakeholders become actors addressed by an enterprise when they are consciously considered as people targeted by its outcomes. Generic concepts like *user* or *customer* often fail to address the particularities of a certain actor. People and their roles are so different that they have to be regarded as separate actors, to achieve a design that recognizes and enhances the relationship to them. The people viewpoint portrayed in the next chapter explores a human-centric design approach in more detail. Here the focus is on identifying the right people to address prior to exploring their context and characteristics.

_ Actors in the enterprise

Most classic design initiatives happen on a low level, where the scope and target group have already been decided and are part of the design brief, so that they target just one major stakeholder group. In contrast, taking an enterprise-wide point of view often results in a large number of diverse, interrelated actors to be considered.

All of these actors have different values, goals, and concerns to be taken into account. They reflect different roles in the enterprise, different levels of control and influence, and of insight and confidentiality. A business depends on the successful integration of many actors and their conflicting priorities and interests. Consequently, an enterprise design initiative needs to be aware of the stakeholders it considers relevant and addresses as actors. The space of actors to be addressed can become quite complex.

The definition of actors depends largely on the individual context. A single person might incorporate several actors, that are closely interrelated and depend on each other. There might be secondary actors, whose actions are mediated through other actors. Some stakeholders make use of automation to execute activities, so that there are machine actors, which act on behalf of the human actor behind them.

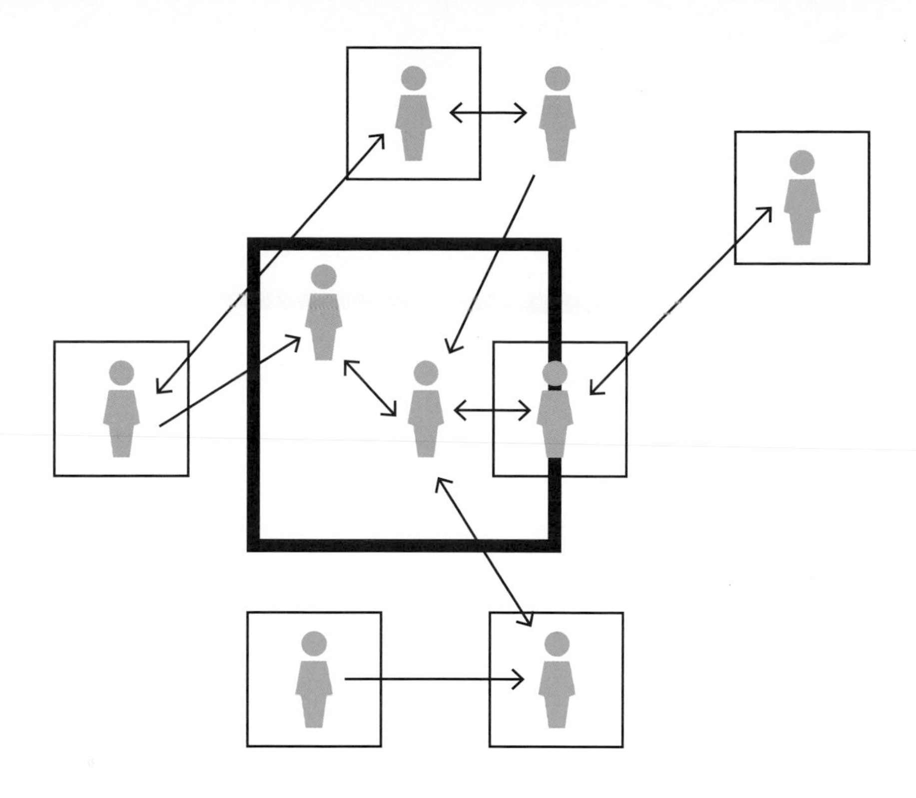

To develop a broad picture of the stakeholders in the enterprise and identify relevant actors to address by a design initiative, think of them in terms of the Big Picture aspects:

_ IDENTITY

think of people, groups, organizations, or brands as images in people's minds and participants in a dialogue, and consider all of them to be potential actors to involve. Design your own brands to be actors on their own, and match their role in that dialogue, including appearance, behaviors, and messages to convey.

_ ARCHITECTURE

think of people involved in executing business activities—communications, management, production, and delivery—and identify actors based on the different roles people play in those activities. Design enterprise structures to integrate these actors and accommodate their particular context.

_ EXPERIENCE

think of people interacting with the enterprise and the unique journeys they are taking. Each particular type of relationship or set of experience attributes is a potential starting point to identify and define a distinct actor. Design for actors that share a certain type of relationship, or experience a similar journey.

ROLE MANAGEMENT

The West conducted a nuanced discussion of identity for centuries, until the industrial state decided that identity was a number you were assigned by a government computer.

Bob Blakeley

Enterprises today face the challenge of managing a wide range of actors, each with a set of roles attributed to his or her identity. People act in the context of job roles, customer segments, project teams, or other types of actor definitions. In order to address them proactively, they have to be identified both in automated architectures and manual interaction, and represented in artifacts and systems. This enables an organization to tailor its services, communications, and offerings to the people it is related to, taking into account the individual relationship.

The need to connect to various actors using digital systems, and to control the communication on a very detailed level has led to the emergence of a knowledge area known as Role or Identity Management, which focuses on identifying and managing how individuals access any sort of system. It is based on the practice of managing identities in digital directories, and using these definitions as a means to be responsive to the individual actors and people incorporating them. In practice, abstract actor definitions then translate into role concepts, groups and individual accounts, being used for identification, personalization, and customized service provisioning.

Applying Identity Management enables the enterprise to explicitly address actors, and to tailor its offerings and communications to the needs of individuals and groups. To be successful, such an initiative must not just tackle the inherent issues of privacy and data protection. The challenge of choosing the right entities, identification levels, and processes is fundamentally related to the enterprise-people relationship to be reflected in the managed identity space. The goal is to capture identities so that the enterprise can to connect people to brands, organizations, services, and each other in a meaningful way.

Managing roles and their identity typically involves multiple aspects, which together can be used to identify and address people in the enterprise:

_ DISTINGUISHING ENTITIES

a role is associated with an entity of some sort, which can be a person, a group, an organization, or even an abstract concept like a brand or a fantasy figure. The level of identification ranges from total anonymity to customizing on an individual level.

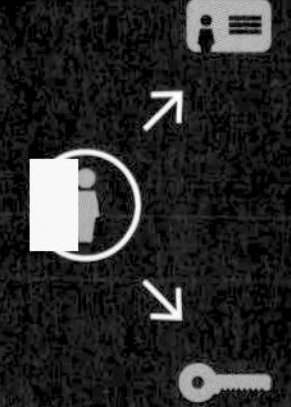

_ ATTRIBUTING ROLES

information about identity can be attributed to physical objects such as a card, to data such as a password, or based on apparent attributes, like a person's gender or the country a website is accessed from. Identities may be self-attributed, such as a user registration, or assigned by others. The attribution can be purely symbolic (like a logo), or include an identification mechanism.

_ MANAGING ACCESS

identities are used to grant and restrict access to virtual and physical spaces, setting permissions and privileges. It allows validating attempts to access a resource against an identity and its validity range.

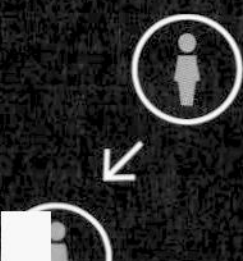

_ COMMUNICATING IDENTITY

identities are made explicit and visible in profiles and messages, and actively communicated. This is the basis for gaining legitimacy and managing reputation, and positioning identity as means to be perceived by others and create trust and awareness.

_ EXAMPLE

On the most global level, an organization must involve people acting as customers, suppliers, managers, shareholders or journalists. In a particular context, for example investor relations, these global definitions translate into more concrete actors, such as organizational and private investors, the company's own financial managers, external analysts, or funds managers. Applied to a concrete outcome such as a website dedicated to the financial community, the choice and definition of actors might be used to conceive different sections suited to particular needs, or role-based access to offer personalized information.

Enterprise stakeholders can be classified into three major categories, all of them to be considered potential actors in the enterprise design:

_ TARGET GROUPS

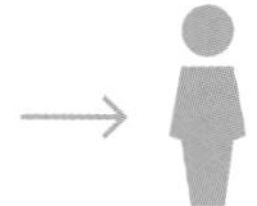

people acting on a market the business addresses, such as prospects, customers, job candidates, or investors. The enterprise reaches out to them to place offerings and value propositions.

_ AGENTS

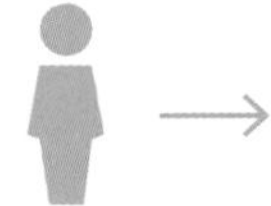

people involved in core business activities, such as managers, staff, contractors, service, or outsourcing partners. The enterprise depends on them to operate and evolve the business.

_ ENVIRONMENT

actors in the wider periphery of interest groups, such as press, public, authorities, alliances, or neighbors. The enterprise strives to establish and maintain good relationships to them.

A challenge and prerequisite to design for actors is therefore to develop a broad view on stakeholders, and identify characteristics to differentiate them properly, from a major role on the enterprise level down to the individual role an actor fulfills in an activity. This in turn can be used by the enterprise to be responsive to the set of actors incorporated within an individual person. Many organizations struggle to find the right level of granularity to identify, define, and address their actors. The definitions must be fine enough to explicitly address an actor relationship, but also generic enough to achieve this within a reasonable amount of effort.

Working with the concept of actors is a powerful tool, and lets us be conscious of all the actors that an enterprise has to take into account. It allows us to explore their roles, relationships, and individual journeys when dealing with the enterprise. More than that, when applied to managed identities, designing with actors enables organizations to treat people individually, by delivering personalized and customized access channels.

Role Management is a field full of technical challenges—making identities work across several systems, ensuring data quality and security, and dealing with the inherent complexity of individual access levels. To deliver value to people, it must be based on a design that bridges the technical and the social sides. Questions related to confidentiality and information needs, data ownership and privacy, communicated and perceived identities, and reputation and trust are the drivers behind such a design.

The emergence of managed identities in information systems as profiles, permissions, and roles is merely a digital representation of a social reality that needs to be explored. Designing for a variety of actors therefore requires a multi-disciplinary and multi-faceted approach. An enterprise design initiative must be based on a deep exploration of stakeholder relationships, to come up with a set of actors that reflects their structure and relevance. Such a model of actors enables the enterprise to reach out to people with a tailored design, one that is suited to their particular relationship context, and built on managed identities.

The actual communication and interaction with actors usually happens through in a wide range of channels, both direct and mediated. A key challenge is to understand these connections, and to leverage them to reach out to actors.

#5_TOUCHPOINTS

Every time someone is in contact with an enterprise, this encounter is mediated by a *touchpoint* of some sort. Here, a touchpoint designates an interface between a person and an enterprise. The term refers both to the event itself and the physical context where it happens. Touchpoints facilitate activities such as using a service, receiving a message, or seeing an advertisement somewhere—therefore, they apply to brands, services, products, and other types of visible offerings made available by an enterprise to people.

The origins of the concept can be traced back to different practices, such as Service and Brand Management, multi-channel Marketing, and Customer Relationship Management. The emergence of new kinds of embedded digital media and the areas of ubiquitous and pervasive computing further widened the scope of the concept to include human-machine interactions across several devices. Today, the touchpoint concept is being adopted in many design-related knowledge areas, and is being used for all kinds of interactions. Depending on the individual focus, touchpoints might also be called situations, or interfaces. It is this universal applicability and wide scope that makes touchpoints such a useful concept in the enterprise context. It allows us to explore, improve, or completely redefine how a person experiences the enterprise in all its facets. This chapter explores touchpoints further, as the set of real-world links between people and the enterprise.

Mapping stakeholder journeys _

The relationship of people and the enterprise can be described as a journey across touchpoints. This concept has been used for quite a while in the context of Customer Experience, but essentially applies to all types of relationships. The journey itself in turn largely depends on the individual relationship: some stakeholders, such as employees, regular customers, or shareholders experience a long and intensive journey, while others just briefly touch the enterprise. All of them can be quite important experiences—a brief encounter might even have more influence than a protracted episode, depending on the context. Looking at encounters enables us to optimize the individual touchpoint, while dialogues and episodes are useful for the more intensive interactions. The lifespan view is useful for exploring how the relationship evolves over time, and how to engage people to expand it.

The focus of the touchpoint itself is the individual exchange, rather than the larger relationship or dialogue. It looks at a single instance for such an exchange to take place, focusing on the particularities of that individual context. But depending on what scale of journey you are focusing on, the touchpoint might play a different role. Consequently, the enterprise must see their touchpoints as individual links to people, while being aware of their role in the context of sequences.

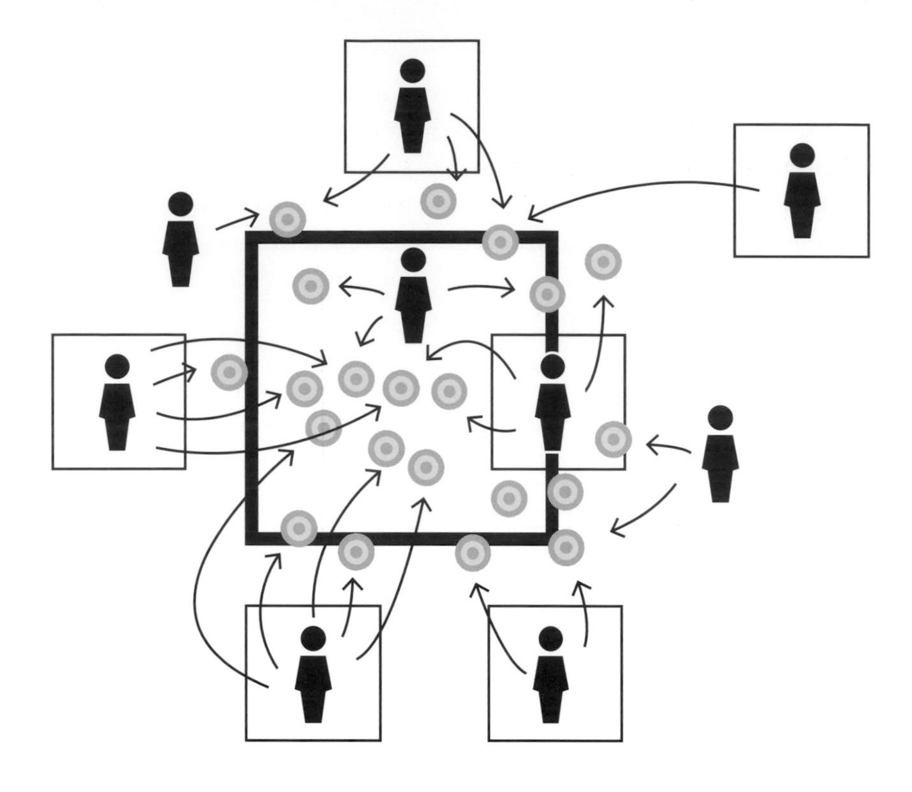

Depending on the context, journeys across the enterprise involve different quantities of touchpoints, have varying durations, and can be seen as different levels of the underlying relationship:

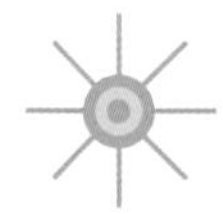

_ ENCOUNTERS

or single touchpoints, are very brief and isolated moments where a person is in contact with the enterprise.

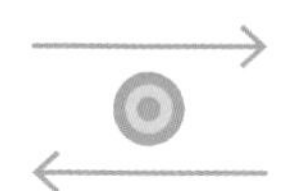

_ DIALOGUES

or short term journeys, are interactions with a person or a system of some kind, but usually without major interruptions and lasting from minutes to hours.

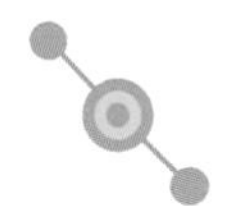

_ EPISODES

or long-term journeys, are part of a larger activity, and involve multiple dialogues or encounters with interruptions in between. They can last from days to weeks.

_ THE RELATIONSHIP LIFESPAN

or overall journey, covers all touchpoints between a person and an enterprise as a whole. It reflects the entire relationship as a sequence of exchanges.

_ EXAMPLE

Coming back to the example of booking a flight for a business trip, the first encounter could be an advertisement seen somewhere, reading about an airline in a newspaper, or having a flight popping up in an online search. The activities involved such as booking a flight and the activity of travelling itself could be seen as dialogues. This series of dialogues then represents an episode, from awareness to service delivery. In the long run, the relationship lifespan captures aspects like overall customer satisfaction addressed with loyalty programs, which in turn create new touchpoints and dialogues. To switch perspectives to other stakeholders, you might think of the journeys of the traveller's personal assistant booking the trip, or the one of a commercial pilot, all actors essential to the airline's business success.

Designing with touchpoints _

Designing with touchpoints always involves a strong sensory dimension, because they will be perceived by people. Depending on the nature of the touchpoint, that might involve designing with visual, tangible, or auditory elements. More deeply, the touchpoint exchange also involves intellectual and emotional qualities, if it enables understanding, usage and utility and helps to achieve a good experience. This requires different design aspects to work together to form a coherent whole. The result of designing for a touchpoint can be a tool, environment, or information medium, or any combination.

Used for design work in the enterprise, the concept of touchpoints is a very valuable tool for analyzing the contexts wherein an outcome will be applied, and demands a strategic choice of the touchpoints to consider and explicitly address. Because of their nature as interconnected elements, considering touchpoints makes the designer conscious of the different aspects of the larger system that the design controls, influences, and integrates in. Design that considers the journeys of different actors across touchpoints inevitably also involves the transitions and relations between touchpoints, bridges that enable both linear conversion and non-linear jumps.

In their book *This is Service Design Thinking*, Marc Stickdorn and Jakob Schneider compare the idea of such journeys to other forms of time-based media, such as music, movies and stage plays. Multiple touchpoints are experienced as a sequence of elements, much like an animated movie is made out of static pictures. To explore the particular qualities of such a journey, it can be useful to look for the typical characteristics of music or film—rhythm and melody, message and drama, in order to capture and influence the experience of a certain actor.

TOUCHPOINT ORCHESTRATION

The term *orchestration* with respect to touchpoints was used in 1984 by Lynn Shostack in her concept of *Service Blueprinting*, and implies that it deals both with the analysis and evaluation of existing touchpoints, and with the transformation, creation, and alignment of new or changed touchpoints. It borrows from the world of music, emphasizing that different touchpoints work together like musicians, to bring a service to life.

In his book *Brand-Driven Innovation*, Erik Roscam Abbing defines the practice of touchpoint orchestration in more detail. It breaks the journey of actors into different steps and contexts. Depending on the nature of the touchpoint, it allows us to repartition the tasks of designing the details across different design professions, while being conscious of the interrelations between those individual contributions. It permits designers to innovate and bring the brand promise to life at individual touchpoints. Although a dedicated approach to touchpoint orchestration is a fairly young professional practice, it will have to become part of the toolkit of anyone dealing with design on a strategic level.

As a system of interconnected elements, touchpoints appear as a network, an infrastructure for an individual journey. In order to achieve such an orchestration, a first step is to determine which touchpoints to consider a relevant part of the journeys of a particular actor important to the enterprise. A way to do this is to explore which ones are representative for the whole enterprise in the eyes of people as key steps in their journeys, and to connect and enhance these touchpoints with particular attention.

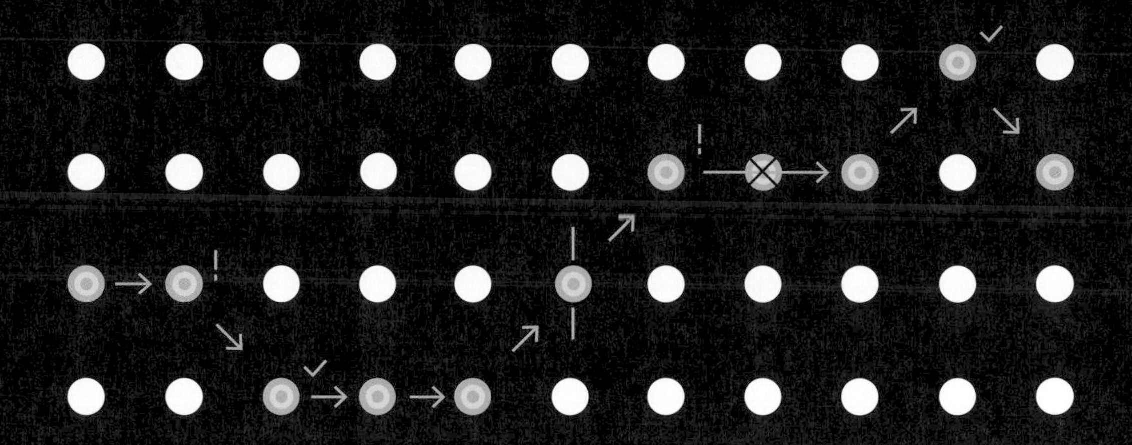

In the literature on the subject of Touchpoint Orchestration, practitioners are particularly interested in identifying touchpoints that stand out for people in some way:

_ CRITICAL INCIDENTS

touchpoints that make a strong impression and result in a memorable experience, both in a positive way (pleasure points) or a negative way (pain points).

_ MOMENTS OF TRUTH

touchpoints that provide an opportunity for the enterprise to achieve an outstanding performance in an exchange with an actor, leading to a better relationship.

_ CONNECTION POINTS

touchpoints that are used to transition from one state into another, taking the journey on a new level, such as a customer conversion or a switch to another medium.

_ REDUNDANT POINTS

touchpoints that could be avoided, unnecessary steps that could have been saved, or encounters that place needless burdens on the actor.

_ The enterprise as a contextual universe

The idea of touchpoints is fundamentally about the context of experiences. Any design, at any level, integrates into a given context, and creates new contexts with its perceivable outcomes. This makes them comparable to the physical places the enterprise makes available to facilitate its exchanges with people.

Many touchpoints actually happen in physical environments, such as stores and offices, while others exist in a virtual space and can be reached using communication technology—from calling a phone line to a web-based project space. In the enterprise, touchpoints therefore correspond to the physical and virtual places people use to get in touch, and the context they provide to the exchange. It is where prospects become aware of offerings, buy or use them, where employees work, meet, exchange, and collaborate—virtually and physically, along shorter and longer journeys.

Designing for enterprise touchpoints involves exploring the community of actors, to bring them together in shared contexts. It also implies a careful prioritization to determine the relevant subset, and striving for convergence and divergence where it makes sense. This enables the organization to introduce new touchpoints and remove weak ones, to make them work together as a strong chain. In that sense, the concept of services is useful, since it focuses on the greater purpose beyond orchestration.

The quality of actor journeys across the touchpoints with your enterprise usually varies widely. This has to do with the complexity involved in addressing so many different ways to exchange, and varying degrees of understanding about the touchpoints' characteristics:

_ INFLUENCE

on some touchpoints, the organizations can exercise full control, such as their corporate stores. In other cases, like a social network, the control is much weaker. In other instances, like press articles, there is little influence at all.

_ MATERIALITY

a touchpoint may be primarily of virtual or physical nature, of a mixture of both. It involves engaging different senses, like the smell of a store or a product, or a specific way to interact with a interactive system.

_ CONNECTIVITY

touchpoints may appear to be rather separate, or deeply embedded into a network of other touchpoints. They might connect various people and enable social interaction or be used by a single person in isolation.

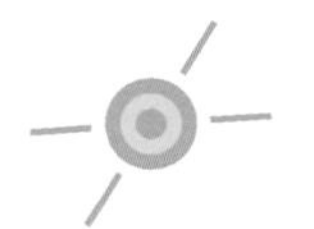

In the enterprise context, touchpoints can be described as exchanges incorporating the universal enterprise qualities towards actors. To explore them further, look at them in terms of these qualities:

_ IDENTITY

the way you express your identity at touchpoints with actors as perceivable symbols, apparent in artifacts, messages, and behavior. This translates to the sensory, behavioral, and communicative implementation of brands in individual instances.

_ ARCHITECTURE

the way you consider touchpoints in the architecture of your enterprise, plan to use them to deliver information and value, and make them part of your operations. Architectural elements are used to implement touchpoints as communication channels.

_ EXPERIENCE

the part the touchpoints play in the individual experiences of people related to the enterprise, and the way you attempt to influence that experience by redefining their journeys. This is the basis to decide what touchpoints to consider and address in a design.

6_SERVICES

An idea which has gained a lot of traction in recent decades is the notion of services as the basis for business activities. It is reflected in the idea of a service economy—a constantly expanding business sector that dominates Western economies—and also a development where the service part of more traditional business models becomes a more important and relevant aspect.

There are numerous examples for the domination of the service paradigm in today's markets. Beyond purely service-based industries like finance, logistics, or the public sector, other economic sectors are adopting service orientation as a way to change and extend their offerings. Hardware and software providers turn their products into platforms and markets for virtual goods—by establishing service ecosystems around traditionally manufactured products, by integrating service provision to complement a product, or by switching to a rent model. In any case, the offering shifts from a transfer of ownership to an integrated product-service continuum that focuses on an ongoing managed relationship instead of a one-time sell.

Another case of service orientation applies to the field of information technology. The idea of *Service-Oriented Architectures (SOA)* describes the concept of breaking big information systems into a network of small, connected services. Following that technical definition, a service executes a single function that may be needed to run an automated part of a larger business process. It makes this function available to other services, so that an entire automated process can be executed by a chain of service interactions. Compared to traditional applications, services in an SOA can be recomposed in new ways more easily, reused for new purposes, and replaced with better or different services when the need arises. This is the basis for all the automated services offered by technology companies, both large and small, now commonly referred to as being in *the cloud*.

The different views on services mentioned above demonstrate a large variety of different concepts and definitions. Regardless of the respective idea and focus, all of them reflect the same basic idea of encapsulating and specifying a certain activity, be it automated, manual, or both, to generate a benefit for some actor other than those providing the service.

By that definition, services can be described in terms of the benefits they realize for an external beneficiary, the service consumer (result-oriented view). Those benefits and the way they are produced can be subjected to planning, analysis, engineering, and design approaches, undertaken by the service provider to accomplish the best possible service.

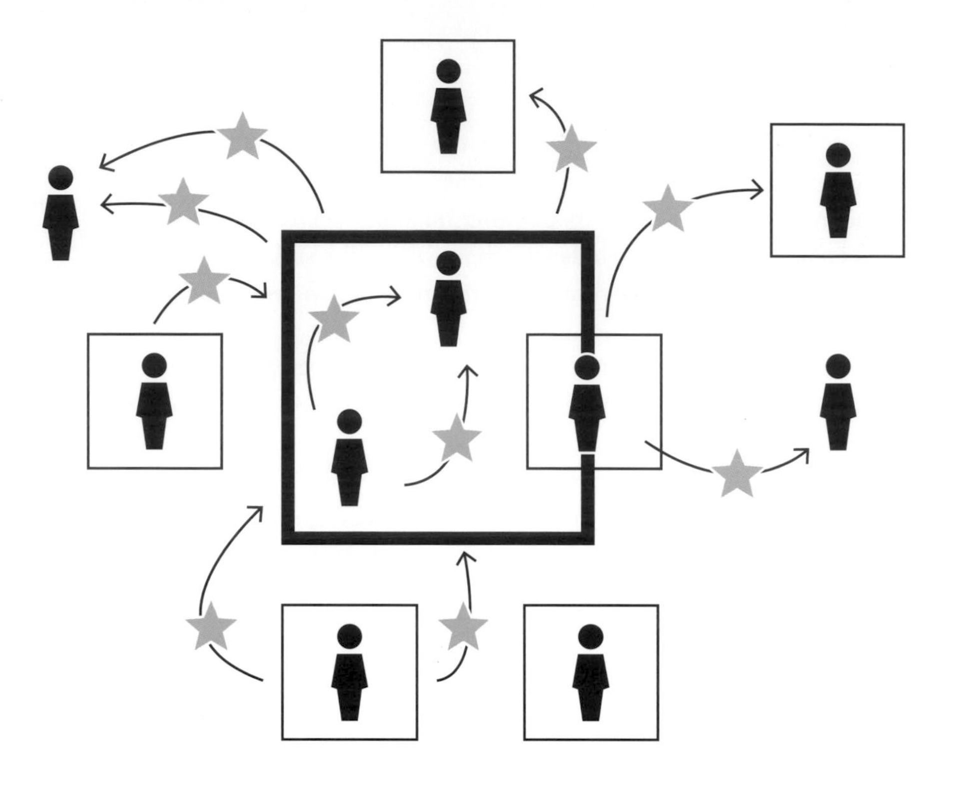

This interpretation of services as *benefit generators* makes them a valuable concept in any strategic design initiative, aiming to deliver a benefit to actors in the enterprise across different touchpoints. They are dynamic processes, being produced at least to some extent the same time they are consumed. To develop a more detailed understanding about services as design elements, you have to look at their inherent characteristics.

The service qualities described here allow an understanding of the large variety of forms services can take. In terms of the qualities, any combination is possible—services directed at people or machines, services with any degree of automation and visibility, isolated or highly interdependent services. In reality, there are few examples of extremes, and many services show mixed characteristics.

The attributes of an individual service have great impact on its success in delivering the benefit it was created for, and can make it work or break in certain situations. While automation and the use of technology are critical success factors for many services, it is important to keep in mind that any machine involved works on behalf of a person, so that the final actors appearing in a service are human.

_ EXAMPLE

To illustrate these further, think of these qualities as they become apparent in gastronomic services. Depending on the individual business model, you might find yourself dealing with people or machines, have little choice, or be engaged an extensive dialogue. A restaurant can be based on fast automated delivery of standard food like McDonalds, or careful preparation of hand-made food relying on the chef's distinctive talent.

The enterprise as a service delivery system _

The notion of services has proven quite valuable as a paradigm to design complex socio-economic and socio-technical systems, mainly because the inherent people, business, and technology aspects are all very present. This makes the concept of services also applicable beyond classic service offerings targeted at customers, to the enterprise as a whole. Essentially, all business offerings and activities considered part of the enterprise can be captured in terms of services. The concept is useful as a way to think about the enterprise and what it does to achieve stakeholder benefits.

Services can be understood by a set of distinct qualities, although they vary widely depending on the individual case. Applying Lynn Shostack's model, these qualities allow us to make conscious design decisions, and to develop a blueprint of the service production process:

_ INTERACTION MODE

the type of interaction between service consumers and service providers during the course of the service, represented by activities that involve customer tasks. Variants range from personal services delivered by people to people, to automated services where both actors are represented by machines, and various mixed modes. A very common mode is self-service, where the service consumer interacts with an appliance that acts on behalf of the service provider.

_ CONSUMER VISIBILITY

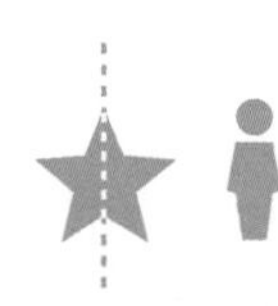

the degree of visibility to the service consumer of activities involved in service production. Extremes are back-office services that consumers are completely unaware of, and those services with a great involvement of a front-office, dependent on a lot of exchange between provider and consumer. Services can be delivered as a black box whose inner functions are hidden from the consumer, or completely transparently, or anything in between.

> It's no surprise that companies outside the economic service sector explore services as a paradigm to deliver experiences, bring identity to life, and connect products and environment in a meaningful way.

Erik Roscam Abbing in *Brand-Driven Innovation*

_ PRODUCTION AUTOMATION

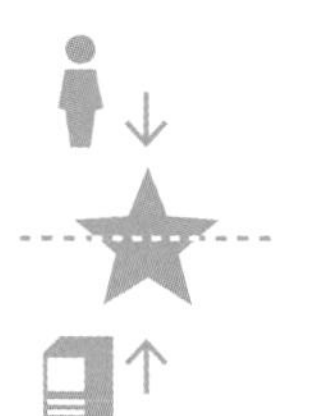

the degree to which the production of a service is automated using information technology or other means. There are services completely delivered by machines and computers, while others depend on a large degree of manual work to come to life. Many services designed for mass consumption today rely on a mixed strategy: as much automation as possible for the standardized delivery, while relying on manual interventions to deal with exceptions.

_ INTERDEPENDENCE

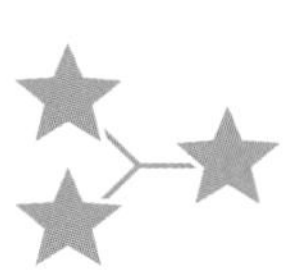

services are never provided or consumed in isolation, but as part of a larger environment of activities, interactions, and other services. This makes them interdependent with elements outside their own scope and reach, and beyond their control. While services can be designed to operate as independently as possible, at the very least they depend on their consumers to be successful. In other cases, services rely heavily on their integration with an outside context.

In his book *The Service-Oriented Enterprise*, Tom Graves argues that you could be rethinking any activity in the enterprise as a service. Such a definition would include services provided to customers, internal services such as HR and the canteen, but also yearly audits or broad business activities such as Marketing, or management in general. Services add the time dimension to the relationship between enterprises and the actors they address, making their offerings and activities part of their larger journeys across touchpoints. They are the conceptual frame that turns these touchpoints into the service-related business events that appear everywhere in the enterprise, in all exchanges with actors. They bring the enterprise to life—including the architecture used to deliver the service, the identity conveyed with it and the effects on people's experience.

_ Designing with services

In the context of an enterprise design initiative, services can be used to express what an enterprise offers—or should offer—to its various stakeholders, which take the role of the consumers of a particular service. It allows us to define the final benefit, the value an enterprise realizes for people. It also helps to avoid losing sight of this goal because another intermediary goal gets in the way, such as the manufacturing of a product, or the construction of a machine.

A design approach applied to services has to be cross-disciplinary and adaptive to the individual context in which the service is produced and consumed. Depending on that service context, it might require a wide range of different business, design, and engineering skills to define the service offering, involving physical environments, interactive media, or communications—the scope covered in the area of Service Design.

A service-oriented view on the enterprise can be applied to virtually any offering, capturing it as the set of purposeful activities it carries out to provide benefits to its actors. Within this context, service design is a powerful thinking tool for complex relationship challenges, because it applies a holistic view on the service to be designed, starting with people's experiences. Service production and consumption are the exchanges between actors in the enterprise, represented by the content being exchanged across touchpoints.

People don't want to buy a quarter-inch drill. They want a quarter-inch hole!

Theodore Levitt

Services are omnipresent in the enterprise, since they are the activities that allow delivering the stakeholder benefits that the organization has been created for. As with actors and touchpoints, consider the role services play in the manifestation of universal enterprise qualities:

_ IDENTITY

in terms of identity, services are instances of organizational behavior planned in advance, and executed to deliver a benefit to stakeholders. Services make an organization act — think of it like a person serving other people in the enterprise, as the proactive implementation of your brand, both shaping and shaped by behavior.

_ ARCHITECTURE

services are connected to the architecture of an enterprise in two ways: they are dependent on the structures as a service delivery system, and in turn become a part of that structure themselves. Service architecture refers to the way a service is implemented and operates as part of the business activities.

_ EXPERIENCE

all services contribute to the goal of achieving a benefit for some human actor in the enterprise, resulting in a particular service experience for the consumer. But the experience quality also applies to actors involved in the service provision, both success factors usually being quite closely related and dependent on each other.

SERVICE DESIGN

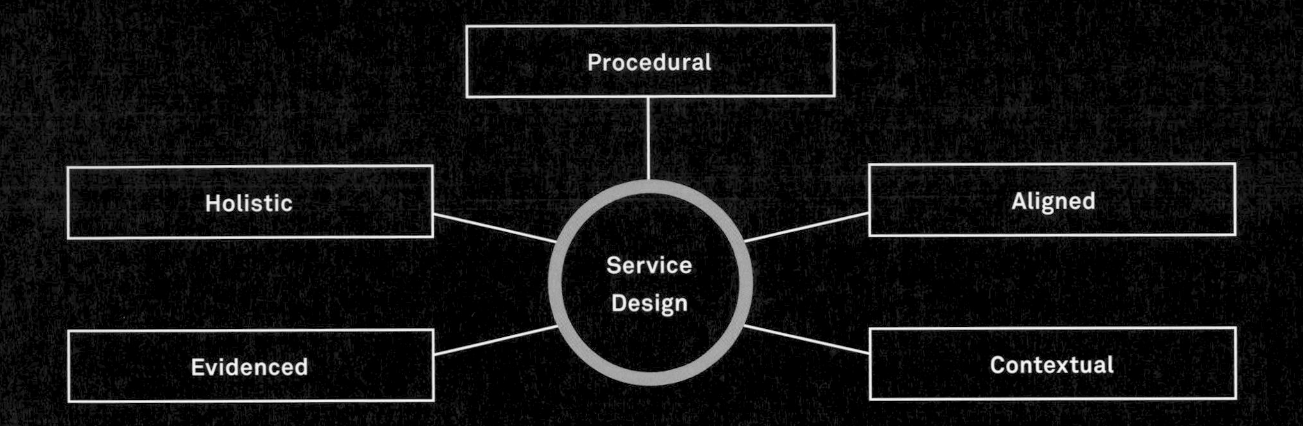

The emerging approach of Service Design aims to respond to the specifics of designing services, regardless of their individual implementation in artifacts, messages, and environments. It concentrates on the specific characteristics of service provision and consumption, and the aspects of services as procedural designs involving both tangible and intangible aspects. Service Design embraces the difficulty of applying a design approach to something as complex and hard to outline as services.

Service Design is a cross-discipline approach uniting many design skills involved in conceptualizing, implementing, and delivering services. The service design itself happens on a meta-level, using methods and approaches from related disciplines and adding them to the blend. Designing services involves a high degree of variability and ambiguity, because the end is unclear at the beginning of such a project. Moreover, unlike manufactured products a Service Design cannot be completely prescribed—they come into existence at the same time they are provided and used. It is this comfort with ambiguity, combined with high-level thinking, that makes the Service Design approach so well suited for strategic design in the enterprise.

Practitioners are applying an inclusive and open view:

_ HOLISTIC

designing services involves a mindset beyond predetermined media of established design disciplines. It requires thinking of the design as a sequence or process, as a system with states and layers, with parts delivered by automated machines as well as human elements, and considering this to be part of a larger environment. It also takes the wider cultural and social aspects into account.

_ ALIGNED

because of the important human element in the creation and delivery of services, Service Design needs to consider every person involved in service provision and consumption, both as stakeholder and as part of the system to be designed.

_ EVIDENCED

any artifact that appears as part of how a service is viewed is considered with regard to the wider role it plays in the service choreography, instead of as an isolated entity. Service Design makes use of prototyping to capture these tangible evidences of a service.

_ CONTEXTUAL

because services are experienced by people in a certain spatial environment (known as *Servicescape*), Service Design is based on the physical context and environment in which it happens.

_ PROCEDURAL

it uses time-based techniques such as role-play or staging to capture service consumption and provision as a sequence over time, and specifies processes of interrelated activities performed in a certain sequence. Services are defined as scripts or scenarios, together with environments and artifacts.

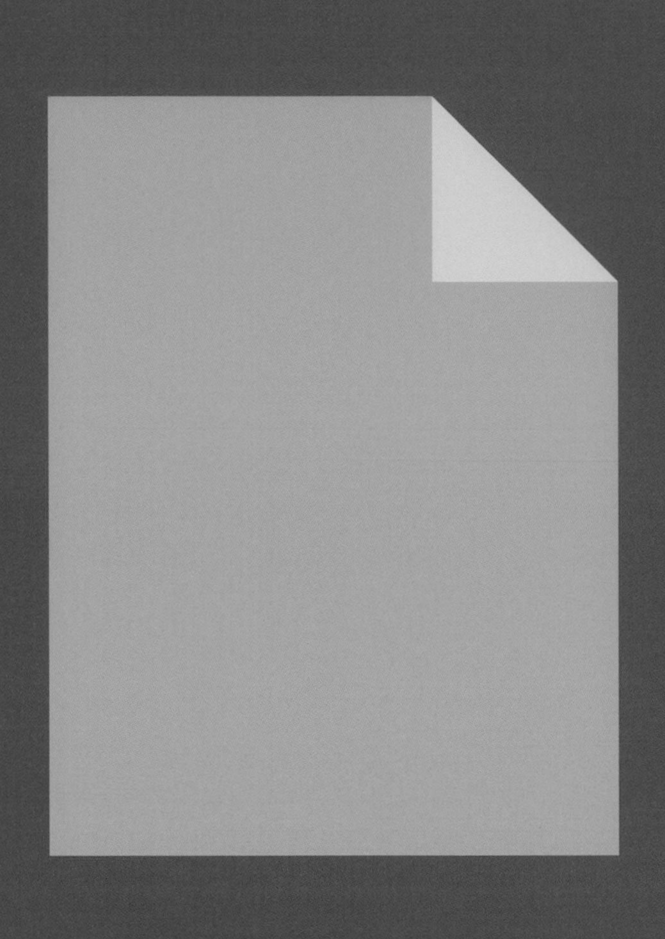

7_CONTENT

Content is not a feature.

Kristina Halvorson

The term *content* is widely used today in various professional areas, such as web communications, social media, and knowledge management. In all these contexts, it refers to any piece of information or data that is contained in some sort of medium, exchange, or system, and made available to people. It covers all messages, documents, and other types of packaged information being produced and exchanged in some sort of communication process.

Because of the sheer mass of data and content appearing and the difficulty of working with it, the ways to manage content for websites have made significant progress in the past two decades. Based on the idea of separating the content from its delivery and presentation, there is a whole industry that specializes in storing, managing, and delivering content in *Content Management Systems (CMS)*. Building on the same idea, managing content in the enterprise is an even more profound challenge. Large organizations today use sophisticated *Enterprise Content Management (ECM)* systems to control and use all their content, including systems for document and records management and data warehouses, to control and use all their content. These systems are used in professional areas like Knowledge and Information Management, and form the basis for any managed use of content in the enterprise context.

However, these approaches are based on the assumption that there is a mass of content that needs to be managed and governed. To understand content in the context of strategic design in the enterprise, we have to take a step back and look at the overall nature of content in business, and the role and purpose it fulfills for people in touch with the enterprise. This detailed understanding allows us to create and sustain content that is meaningful and useful to these people, and is a valuable business asset to the enterprise.

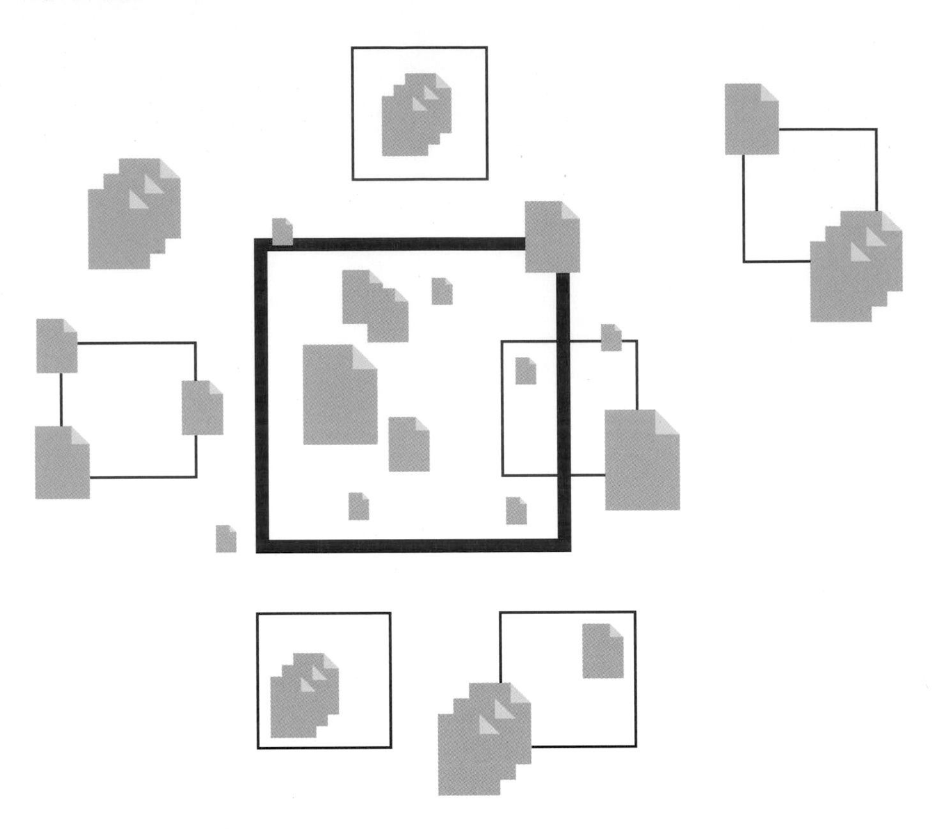

_ Content as a business asset

Based on such a comprehensive view, all bits of information that appear in a business context can be considered content, and potentially as a part of the efforts to manage, develop, and utilize it. It includes documents and files, messages and conversations, and all sorts of data stored in databases or other kinds of repositories. It can be calculated numbers and structured data, but also any kind of textual, visual, audio, or video information.

In an organization, content elements are assets that are vital for its success, both for internal activities and the relationship to the outside world. On the inside, it is the intellectual capital that supports all kinds of business activities with context and insights, touching every department and role. On the outside, it is everything the organization communicates to its external stakeholders.

The example illustrates the wide range and tremendous amount of content produced in a typical business environment. The different types, formats, scopes, and media reflect the various purposes content is produced for, to play some role in a communication process in the larger business context. How well an individual piece of content fulfills that purpose depends on its ability to create meaning for those being addressed.

_ EXAMPLE

Think of information generated around the products of a car maker. The production process comes with internal policies, guidelines, procedure descriptions, and information about the products themselves, stored in some database used for internal purposes. The same information is partly also used on the corporate website or in catalogues, together with written text, photos, videos, tables, and figures.

Moreover, products ship with technical manuals, and are subject to advertising campaigns and promotion, again producing content. They are referred to in press releases and communications, but also in transactional messages around the sales process such as phone conversations or emails. There is content produced by the carmaker itself, but also a whole sphere of media coverage, conversations in social media, and descriptions of cars on trade platforms. In the car itself, there are signs and signals, symbols and labels, content produced to help drivers use it.

To have an impact on the daily business, content needs to be defined first and foremost in terms of the meaning it creates for its audience. Meaning is inferred not from the content itself in isolation, but also from the context wherein it is perceived by people. That context includes what the content is about, who the originator and recipients are, as well as the chosen medium and structures to communicate it.

Beyond the pure data, content is a representation for real-world objects and activities. Think of financial values represented in an invoice or an account statement, representing the transfer or possession of actual money—having as much "real" impact as doing the same transaction with physical banknotes. For people dealing with content, these referents therefore become a part of the things contained in the medium, and are subject to exchange processes together with the information or data portion of the content.

Therefore, content can be seen as the substance of any exchange going on, in every business activity. Anything that generates meaning for the actors addressed by an enterprise can be considered part of its content, regardless of the medium or format, amount of data, or level of editorial care. Turning data into content meaningful for stakeholders therefore requires looking at the roles and functions of content in the enterprise, the larger context where it is put into use.

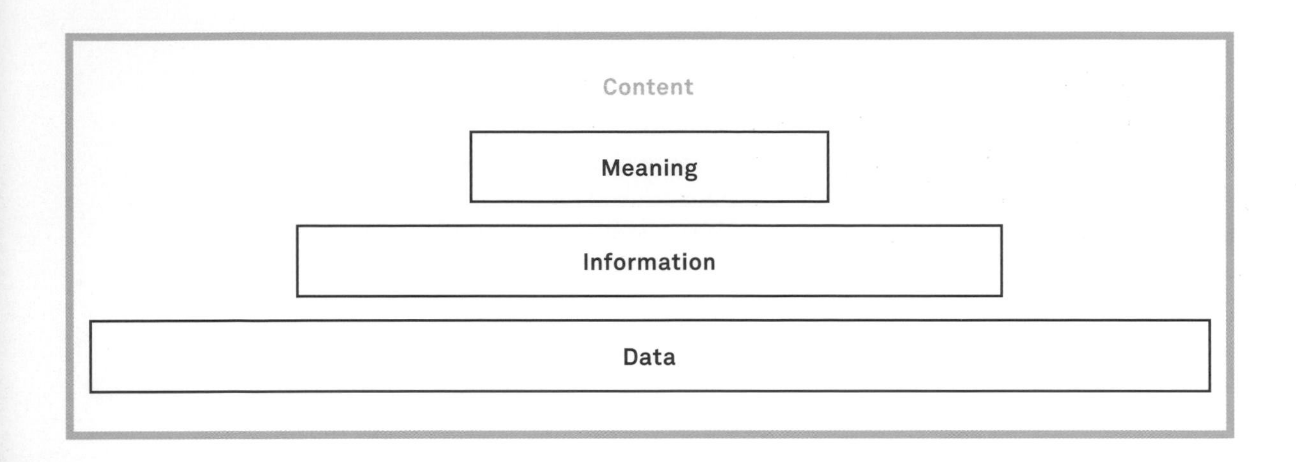

The purpose of content in business contexts can be described using three levels of usage, loosely inspired by the *Data / Information / Knowledge / Wisdom hierarchy (DIKW)* known in information science:

_ MEANING

is the content interpreted and implicitly applied in a specific context, used to generate knowledge through validation, inspiration or reasoning. It is used to form opinions, make decisions, and guide actions.

_ INFORMATION

in this definition is the level where content is made explicit to deliver insight to people, by informing people about something. On this level, content needs to be expressed and presented in a way that enables understanding by the target audience.

_ DATA

is a group of attributes and their values made explicit on the lowest abstraction layer. On this level, content is being captured, processed, consolidated, and stored. It is about getting the right data, and improving its quality or consistency to prepare for a specific use.

CONTENT STRATEGY

Whether content fulfills its purpose depends on a wide range of strategic decisions, from topics and messages to the tools and processes used to acquire, create, and manage it. Content Strategy helps to make these decisions, in order to manage the creation, use, and lifecycle of all kinds of content.

It is a young field that follows the idea that what enterprises should be after is not more content, but the right content. Content Strategy is about understanding, strategizing, planning, and controlling how, when and why what content is being generated. Content Strategy professionals turn the set of core topics and key messages into a coherent approach, analyze and audit current content, and use these insights to formulate goals, and to define structures, formats, attributes, and styles.

From a content audience point of view, Content Strategy addresses the quality aspects of message and meaning, format and type, as well as access and media in a holistic fashion. To achieve such high quality content, practitioners develop a plan for the use of tools and workflows for content creation, acquisition, publishing, management, and governance. Although it touches technical questions on content interoperability, storing, aggregation, and transformation, these aspects are merely means to an end: achieving better content with less effort.

Moreover, putting a Content Strategy in place helps organizations to communicate in a world where the line between authors and readers, or producers and consumers, becomes blurrier and blurrier. Actors in the enterprise are now contributing participants in content ecosystems, and actively produce new or repurpose existing content by composition or commentary. To leverage and benefit from social content as a strategic resource, it has to be addressed with a strategy.

Content Strategy can be universally applied in many areas in the enterprise, including communications, Knowledge and Information Management, social media and collaboration. With the ever-growing amount and influence of information, its greatest potential lies in the application to all content in the enterprise.

A content strategy typically consists of several aspects:

_ CREATION

generating, structuring, and formatting content; defining its attributes, styles, and ownership depending on the individual context.

_ COMMUNICATION

defining key topics and messages, communication channels and comprehensiveness; ensuring findability, usefulness, and consistency across all instances.

_ GOVERNANCE

auditing and cataloging content; publishing workflows, editorial calendar and continuous improvement; managing the content lifecycle, transformation, and aggregation.

Enterprise content _

Content is the basis for all understanding, collaboration, and decision making in the enterprise. It is the substance in all exchanges between the different actors, and across all touchpoints. It is part of any service or offering, and is both by-product and resource of any activity. As a concept, it enables the capture of any kind of encapsulated information, regardless of its individual purpose, format, or origin. Anything the enterprise makes available to people related to it depends on the quality of the content used and generated.

Today, content elements are only seldom regarded as assets for a potential enterprise-wide use. It is much more common to produce or acquire content in an isolated fashion in various places and groups across the enterprise. Even the content of a single department or around a single service is often generated in isolation, for a particular purpose or medium. This is particularly visible in the mass of potentially relevant content lying asleep in archived email threads.

The effects of this lack of coordination and alignment are not only the inefficiencies of redundant content produced and maintained in parallel, but also inconsistency for the actors addressed with a particular content. Key to achieving high quality content is to address it as part of a strategic enterprise-wide design initiative.

Designing with content _

More than just the technical issues of storing content in a flexible way, a design addressing content on a global level has to make a wide range of strategic design decisions. The objective of design with regards to content must be to enhance the meaning of a content element. Content must be considered as the substance of services provided for and by enterprise actors across all touchpoints, striving for high quality content in every instance.

Applying content strategy globally in today's world of constant content production and exchange offers a tremendous opportunity to organizations. Thinking about content strategically enables us to explore and leverage synergies and cross-content effects to get key messages across to all audiences in a coherent way. Moreover, it permits us to reduce cost of creating, managing, and distributing content through reuse and alignment.

To better understand the role of content in the enterprise, think of it in terms of the universal enterprise qualities:

_ IDENTITY

the messages you convey in communication processes toward actors in the enterprise. It describes what a brand *says*, how to engage audiences, how people are involved in content creation and dialogues. Content production as well as its tone and style reflect the organizational culture, and constitute identity.

_ ARCHITECTURE

the operational side of the enterprise as a working system depends on communication processes among its constituent parts. Content in that sense is about the formal structures put in place to create, manage, analyze, and publish information, and how the enterprise depends on the exchange of content elements.

_ EXPERIENCE

the way a person receives, perceives, and understands content to derive meaning from it in the context of his or her particular experiential context. It refers to the subjective nature of content, its individual quality and success depending on the interpretation of a person and the meaning it creates in that context.

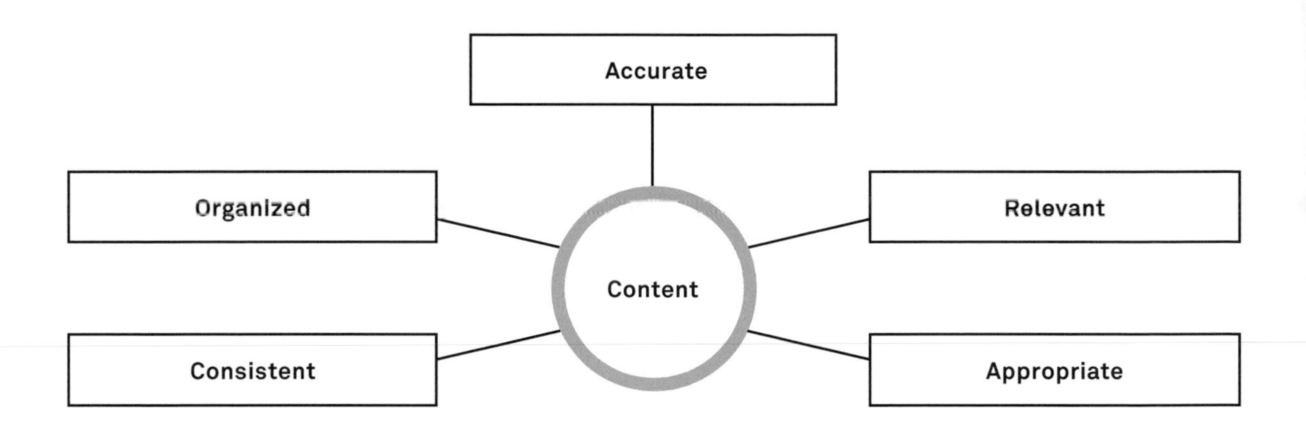

High quality content in the enterprise can be characterized by a set of characteristics that make it meaningful for people exposed to it:

_ ACCURATE

content elements should be current, not contain false or misleading information, and convey these messages clearly and unambiguously.

_ RELEVANT

content elements should be valuable and useful for their audience and be usable by them to achieve that purpose.

_ APPROPRIATE

content elements should be adaptive to the individual use context and media, and convey messages in a comprehensive, well-presented, and digestible way.

_ CONSISTENT

content elements should fit into the larger system of information exchange, integrative and coherent with complementing or related content elements.

_ ORGANIZED

content elements should be easy to find and self-describing, as well as efficient to produce, manage, organize, and reuse.

_ DESIGNING WITH RELATIONSHIP ELEMENTS

The aspects explored in this chapter are the building blocks of an enterprise-wide design initiative. They are modular, interconnected parts of any interaction with a system which constitute the underlying relation. Organizations provide, reuse, and recombine relationship elements to shape and transform their enterprise. They are the building blocks of the enterprise as a system of relationships.

The result is a constant exchange: the enterprise makes services available as content exchanges via touchpoints, to play a role in the journeys of actors. In that dialogue, elements are always co-produced with the people you address, in the moment they come to life. Engaging in a dialogue with the enterprise requires embodying a certain role, and equally there is no service without interaction between provider and consumer, no touchpoint without an environmental and physical context, and no content without meaning for someone. Therefore, a design can only partially control and predict the configuration and implementation of relationship elements in an individual exchange.

In the enterprise, these elements appear as fractal, recurring structures, regardless of what level you look at. They play a role in human interactions with small systems such as websites, but also in large scale systems such as entire markets or the criminal justice system. As such, they form a hierarchy of larger and smaller elements that have to be orchestrated for a design to succeed on the enterprise level. They are intangible in terms of the means of bringing them to life, but also appear as tangible, visible elements which are perceived and used by people.

Unlike the Big Picture aspects described in the previous chapter, relationship elements can be actively incorporated into a design initiative, and addressed with a transformation. They are among the topics most design initiatives address in one way or another, although sometimes in an implicit or accidental way. When designs get more strategic and large-scale, it is quite valuable to make them explicit. Outcomes can be expressed in terms of actors addressed across a set of touchpoints, services provided, and content produced. They are design materials you can collect, catalogue, monitor, analyze, orchestrate, and reshape.

The elements portrayed in this section are so useful because they are universally applicable in different situations and levels. How to interpret, collect and translate them depends on the individual context and frame applied. To put them into perspective, we need to look at a design challenge from different angles, which is what the next chapter is about.

AT A GLANCE

Relationships are determined by the interplay of various elements over time, coming to life as part of a larger dialogue. These elements recur across all scales and domains. Incorporating these elements explicitly as part of strategic design initiatives permits us to actively transform that dialogue, thereby transforming the relationships it embodies.

Recommendations

_ Consider and explore the actors critical to your business, and their journeys across various touchpoints, to set priorities for developing the enterprise

_ Understand and redesign the activities of the enterprise as a set of services across those touchpoints, engaging with actors in exchanges of content

_ Use the fractal and co-creative nature of enterprise elements as a strategic tool to transform the enterprise, in small steps but on all levels

_ Use these elements as the raw materials in your designs, establishing a system that makes sense to people by constituting and shaping identity, architecture, and experience

CASE STUDY _ VDA

VDA'S ENTERPRISE

Supporting member agencies and their talent pool by facilitating the exchange between different actors of the film industry

The Verband der Agenturen (VdA) is a non-profit organization working on a national level, and regroups talent agencies all over Germany working for actors, directors, and authors. The organization understands itself as a trade association, representing its member agencies to the government, other associations and international counterparts. The sector is characterized by the collaboration of many actors and specialized professions in the context of film and theater productions.

Like any other industry, the entertainment sector is undergoing a transformation in light of digitization. Companies store profiles, pictures and videos, and other information on actors and other professionals in digital databases, communicate largely via email, and support process with information systems. Member agencies make their websites a catalogue of the clients they represent, along with content on their projects and news.

In 2007, the VdA approached the web agency °visualcosmos with a powerful idea. In their strategic planning meetings, the association's board members had noticed that most of the content its members managed, published, and exchanged was comparatively static content, since profiles, curriculum vitae, and media changed only relatively slowly. Yet the effort to keep this content up-to-date was quite high, due to all different places and formats where it was needed. More than that, the other actors in the industry such as casting networks used the same data, resulting in a lot of replication and manual updating. The VdA wanted to expand its role beyond mere representation, and create a platform for data exchange.

The outcome of this project, the VdA Pool platform, set the standards for an automated exchange of talent profiles. It transformed the German entertainment industry, and shifted the effort of agents away from data maintenance to more valuable activities. Its design was the result of a deep dive into the Anatomy of the enterprise.

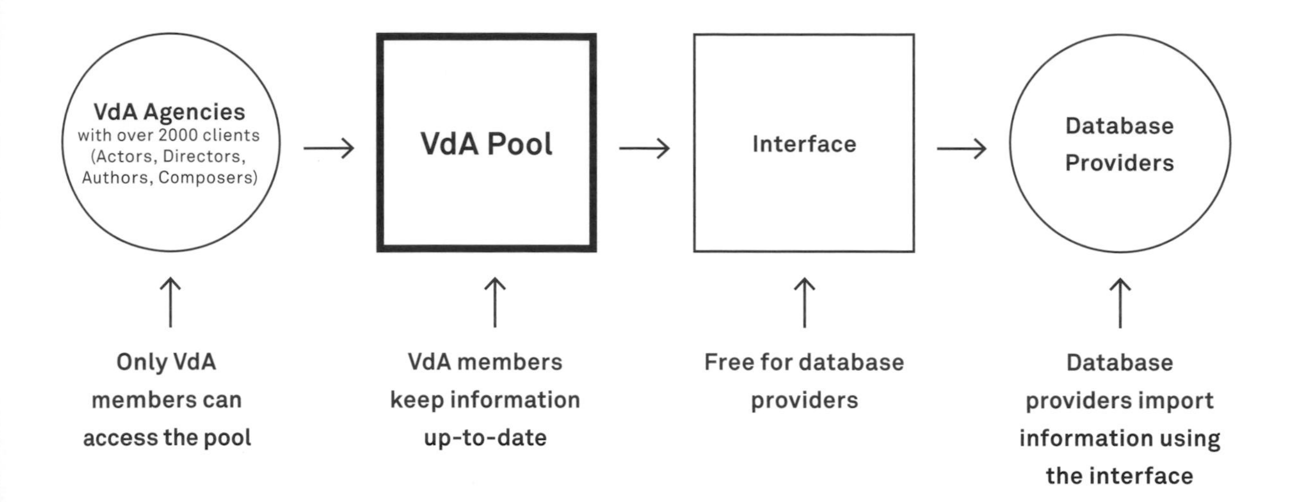

_ ACTORS

Together with the client, we started investigating the stakeholder landscape, to develop an understanding of the actors the VdA needed to address with a platform for digital data exchange. A key user group would be the member agencies, who would provide the input of data into the system in collaboration with their clients, update information, and make it available to the entertainment market.

The market actors to be addressed then included a wide range of actors with different needs. The core target group was comprised of casting agencies and production offices trying to find talents for their projects. Looking at the wider ecosystem, there were also providers of third party web portals, theaters and independent film companies, and TV channels.

The list of stakeholders was then used to create a detailed mapping of roles and relationships. In a later phase of the project, this mapping provided a basis to develop use scenarios and business cases, to inform the technical implementation of user roles, and access permissions and profiles.

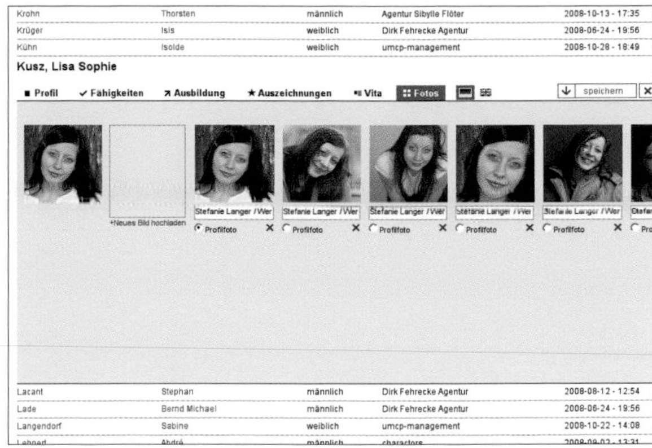

TOUCHPOINTS _

As the VdA Pool was conceived to drive all communication processes related to client profiles, there was a wide range of touchpoints to be considered. All actors should be enabled to participate in these processes according to their roles and particular needs. The choice of touchpoints to be included was largely influenced by business considerations. For example, it was a conscious decision of the members that the VdA as a non-profit organization would not make the data available publicly on the web, thereby engaging in a competitive business. Also, it was in the interest of member agencies to drive traffic to their website and sharpen their own brand identity rather than being just a data provider to a shared industry portal.

This led to the decision to make the data available to market actors only in an XML format via a digital machine interface. To download data into their own database, users would have to develop their own import mechanism. The web was used to present that offering to the market, and to allow member agencies to enter and maintain their data into the VdA Pool. The VdA also provided support by phone, email, and web documentation for all actors directly accessing the pool.

_ SERVICES

As the VdA decided to evolve from a pure representative function to the facilitator of a data exchange process, it also had to adapt its service offerings both to member organizations and other actors in the industry. Once the target state was defined as a vision prototype of the future service, we collaboratively developed a new service model and shifted some of the existing member services.

The core services of the VdA Pool could be defined as a web interface to manually enter and update client profiles, and a machine interface to export datasets to load them into another database. In the course of the project, however, it became apparent that this scope would be extended by a machine-based service to import data into the VdA Pool, to be used by member agencies already using their own internal databases. Later, also the export mechanism was used by the association's members to load their own data into other systems, such as databases to drive their public websites.

These new services enabled the VdA to deliver a benefit to both its members and the market they served. The organization decided to outsource service provision related to technical assets and data maintenance, and concentrate their own efforts on driving adoption of the new services.

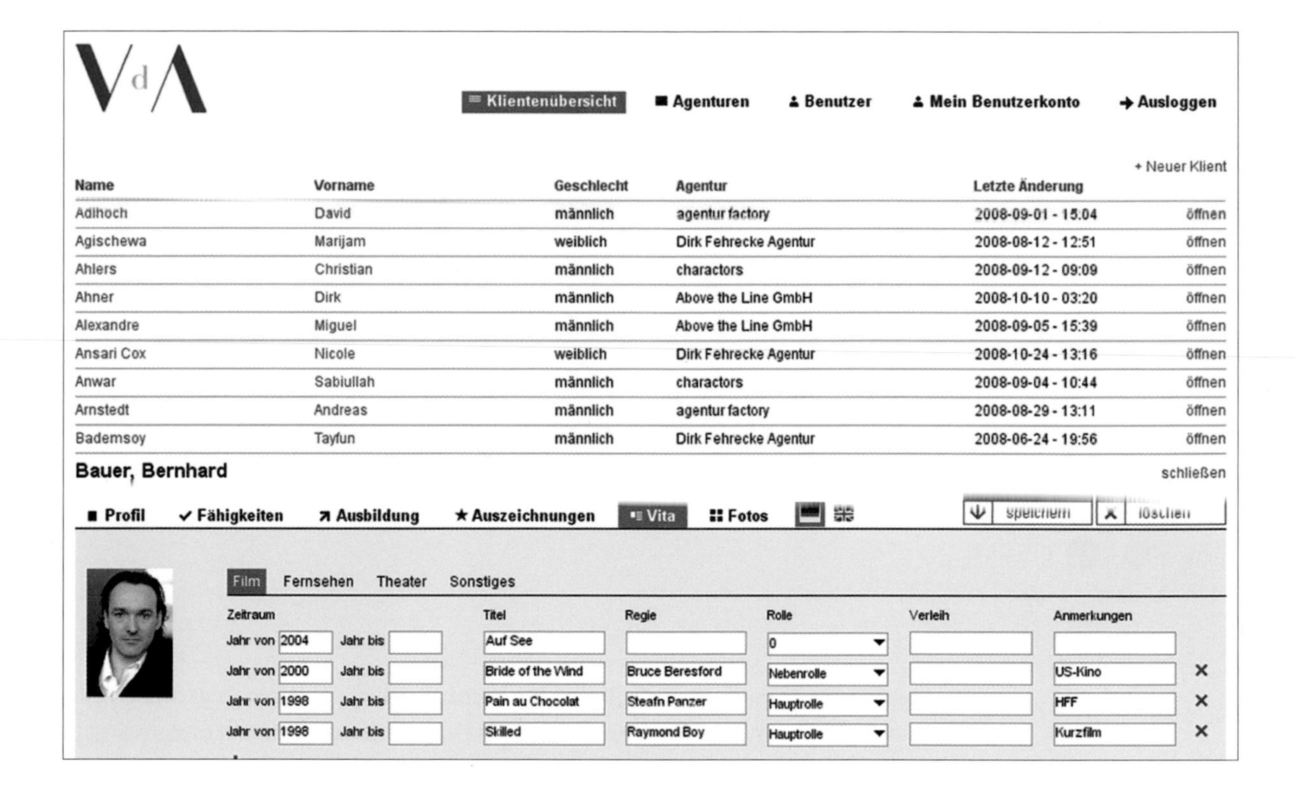

The VdA Pool initiative focused on the client profiles as the content central to the new offering, being exchanged and shared with the market. Consequently, addressing the issues of content structure, presentation and governance were a key part of the design process.

The design team conducted a series of workshops with the different stakeholders to elaborate a content model that worked for all actors, aligning a large variety of formats into one standardized model, validated in the subsequent phases of the design process and also undergoing several iterations after launch. By now, this model has advanced to be the industry standard to document talent profiles, for both presentation and electronic exchange.

Together with the VdA, we elaborated a Content Strategy approach to ensure the quality, timeliness, and consistency of the pool's content, adopting a mix of member self-service and professional governance.

6 _ FRAMES

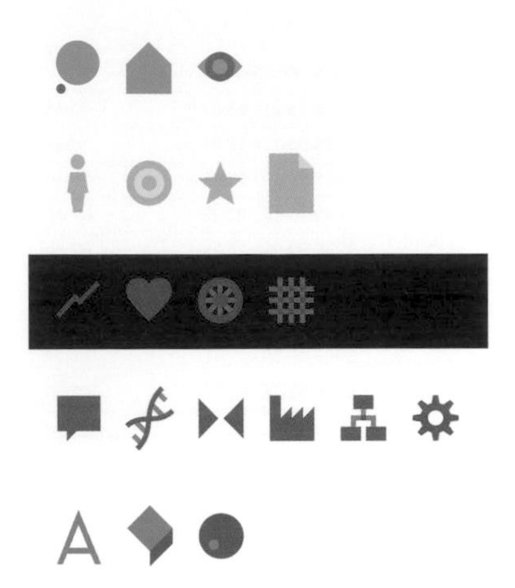

The two previous chapters explored Big Picture aspects pertaining to everything being said and done in the enterprise, and the Anatomy elements implementing those qualities in relationships to people, appearing on all levels as a fractal structure. Their ubiquity and dynamic interplay illustrate well the large degree of complexity design initiatives face on the enterprise level.

Any design challenge deals with the ambiguity of an underdetermined problem setting. Applied to an enterprise context, design has also to deal with a particularly complex environment, a *wicked mess*. The number of elements and interactions to consider, the diversity of stakeholders and their concerns, the many structures and systems a potential solution depends on, and the soft factors of culture, identity, and change, all make it difficult not only to develop a solution, but also to define the problem to be solved. That complexity makes it impossible to understand an enterprise in its entirety just by breaking it into smaller parts, even for a relatively small company—there are just too many variables which constantly change. Just about everything could be considered an influencing factor relevant to a design process.

We found that design practitioners sometimes struggle with such high-level, complex and ambiguous problems beyond the classic tangible realm of their fields. They tend to leave problem definition and formulation to other disciplines such as management or engineering, and prefer to come into a project at a stage when *the client knows what he wants*.

© 2013 Elsevier Inc. All rights reserved.

Using design strategically requires stepping out of that comfort zone, and applying the solution-oriented mindset of design practice to the larger context of the enterprise that the client seeks to transform. It involves exploring the problem space, questioning and enhancing given problem definitions and constraints, and influencing strategic decisions instead of taking them for granted. The intent of the Enterprise Design approach is to capture the enterprise as both design context and subject in a holistic fashion. To envision a future state beyond an isolated problem setting, designers need to be aware of the enterprise context it is embedded in. The 4 aspects described in this chapter help to develop conceptual models of the enterprise, to capture the details that are necessary and useful during the design process by looking at the enterprise from a particular viewpoint.

Designing means modeling _

Because of its inherent complexity, exploring the enterprise as a field for design requires looking at it from different angles. In design theory, this activity is known as *framing*. The rationale of taking a certain perspective is to generate a frame, a particular viewpoint on a given context and environment, in order to make sense of it and develop a model of elements to consider—either implicitly as a mental model, or externalized in a visual representation or other artifact.

Creating models from different perspectives (also known as *framed models*) is part of any design activity. In the most general sense, models are representations of something—such as a given situation or context, an idea for a new or evolved state or system, or visible objects like a new building or website. They can be represented in drawings, diagrams, physical models, or written scenarios. Usually, models are either a simplified picture of reality or an illustration of a potential future. They are used to make thoughts and ideas explicit and create maps of related elements, as communication vehicles, and to gain insights into the expectations and needs of stakeholders.

Externalized in documents or other artifacts, framed models help us exchanging with people relevant to the design project, and to develop an understanding of their particular views. Those perspectives enable them to judge what is important, and what change to consider as a success. Driven by facts observed, personal experience, and our own thinking, frames in design combine objective and subjective reasoning. As abstractions, they concentrate on certain elements while deliberately leaving out or radically simplifying others, providing ways to look at a system from different perspectives and through different lenses.

Beyond just exploring the existing, models are used in the process of design synthesis and ideation to connect different aspects of a problem, and to explore and elaborate possible future states to resolve it with regards to these aspects. They permit us to make change visible as descriptions or prototypes that concentrate on one particular viewpoint while keeping other things abstract, and to try out and share different directions and options. They help us make difficult decisions in a space of concerns, become aware of tradeoffs between those decisions, and uncover opportunities for innovation. Furthermore, they are used to define the result of a design process, and to make detailed plans for implementation or production.

_ Design viewpoints on the enterprise

To understand the enterprise in a way that allows picturing the impact of a purposeful transformation, designers need to deeply explore it and envision its possible future states. Because of its inherent complexity, this means working with a particularly ill-defined problem setting. Any approach to gain insight and develop understanding therefore involves creating several simplified models of the enterprise as a system, with the concerns to be addressed guiding the choice of frames.

The aspects covered on the following pages help us to think about a given problem setting, in determining things to consider in design research and subsequent phases of the process. They enable us to envision potential outcomes with regards to stakeholder concerns, and the target state of the enterprise as a system. They are universally applicable to any kind of strategic design project, and provide a solid basis for custom, project-specific models.

The four framing aspects used in the Enterprise Design framework suggest a set of fundamental perspectives which have their origins in design and systems thinking. They guide conceptual modeling and help when deciding on a direction according to strategic choices:

_ BUSINESS FRAME

captures how market actors create value, operate, and interact, and allows expressing in business terms the objectives the client or business owner intends to accomplish with a design initiative, the customer value gained, and the change desired.

_ PEOPLE FRAME

captures who in the enterprise is being addressed with a design initiative, and permits grounding all design decisions in the individual goals, characteristics, needs, expectations, and context of individuals and groups in a human-centric way.

_ FUNCTION FRAME

captures the purpose the enterprise fulfills and the behaviors it exhibits towards its stakeholders. The functional viewpoint helps the understanding, prioritization, and selection of a set of requirements the outcomes of a design initiative is expected to meet.

_ STRUCTURE FRAME

captures the objects and entities relevant to the enterprise, how they interrelate and form a structure. It enables an exploration of the problem domain in conceptual models and the transformation of those models in the course of a design process.

8_BUSINESS

Really, what we're doing as designers is, ultimately, and inevitably, designing the business of the companies that we're working for. Whether you like it or not, the more innovative you try to be, the more you are going to affect the business and the business model.

Tim Brown at the *Rotman Business Design Conference 2005*

Part 1 of *Intersection* looked at the way the design competency can help to create value and sustain relationships with the people in touch with the enterprise, and to support an organization in developing and growing its business. Any design initiative is a form of investment in a transformation, to do business in a different and better way than before. Therefore, such an initiative always has a business objective, regardless whether that objective is stated explicitly, or exists only as an implicit understanding in the heads of stakeholders.

Looking at the enterprise from a business perspective as a playing field for design enables the design team to figure out and to further the rationale behind a client request, and to understand the context and objectives of the project. It helps to clarify why you are doing that project at all, and determines the conditions that make it a success or a failure.

Traditional design practice tends to consider these dimensions of the work only in a superficial way. Design goals are seldom expressed in terms that can be easily traced back to the business objectives of a client, making clear in business terms what a project should achieve. To make an impact in the enterprise context however, designers must challenge the status quo and extend their constant inquiry for the best possible outcome to the business results it generates. It becomes vital to develop a profound understanding of the business context, to reach out to all stakeholders and appreciate their views and concerns, and to design in a way that proactively contributes to strategy and objectives.

Both areas of focus have the goal of creating competitive advantage, by producing a compelling offer, outperforming the competition in dimensions relevant to customers, and reducing effort and cost. Taking this perspective allows us to apply business thinking to the design process of improving and reinventing, and addressing business concerns beyond a narrowly defined design brief.

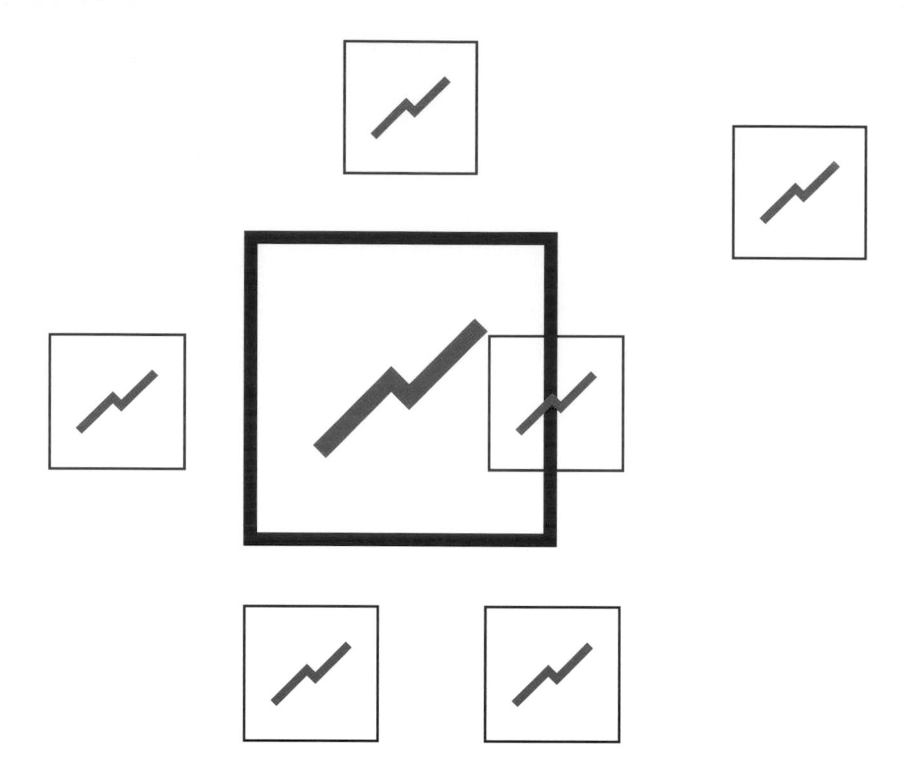

_ Understanding the business context

In our experience, the most common reason for failure in a design project is not related to the quality of the ideas or the execution, or the talent of the design team. Many projects fail because the business objectives and the strategy driving them were unclear or misaligned with the design process. The Enterprise Design framework makes the business frame a core part of design research and modeling activities.

A thorough understanding of the business context behind everything a client or owner wants or considers necessary is an invaluable resource for designers. It allows us to align proposals according to the priorities and perceptions of stakeholders, tracing back requirements and wishes to business concerns, and coming up with solutions which pertain to this environment. More than just integrating into a given context, it allows for modeling and communicating a potential future in business terms, and envisioning the transformation triggered by the outcome of a design initiative. Modeling a business context is a process of inquiry, of capturing the existing, deriving constraints and chasing opportunities for improvement. It is an expression of the business contribution of a design.

_ EXAMPLE

As an example for the business dimension in design, think of a company asking a design studio to create a website for them. In their briefing, the company might state requirements such as having a modern web presence or being available to customers around the clock. Applying the business frame allows translating these intermediary goals to strategic objectives that reach beyond an isolated website: supporting customer relationships with an improved reputation and service, leading to more revenue and better retention.

As soon as these business goals are known and agreed on, the design team can consider options holistically across the entire Customer Experience without restricting the project to just one channel. An outcome designed to contribute to the business objective might address service operations and after sales, brand considerations, or other elements outside the initial scope of designing a website.

A business view on the enterprise _

As explored in Chapter 3, design is used mostly within the context of the products and services provided to the customers of an organization, or the communications around those offerings. As a strategic competency for responding to complex problems, it can be used for a much broader range of areas beyond product development. With all the actors and things going on in the enterprise, there are numerous opportunities to drive change by design, following the strategic direction of a Big Picture view on the enterprise.

Exploring the enterprise from a business viewpoint provides tactical direction where design initiatives as investments in transformations can have the most significant influence to help a business strategy come to life. It changes appearances and behaviors, initiates different ways of doing business, and uncovers new opportunities to generate value for people. This also involves introducing new elements, removing others, and rearranging their interplay in business activities.

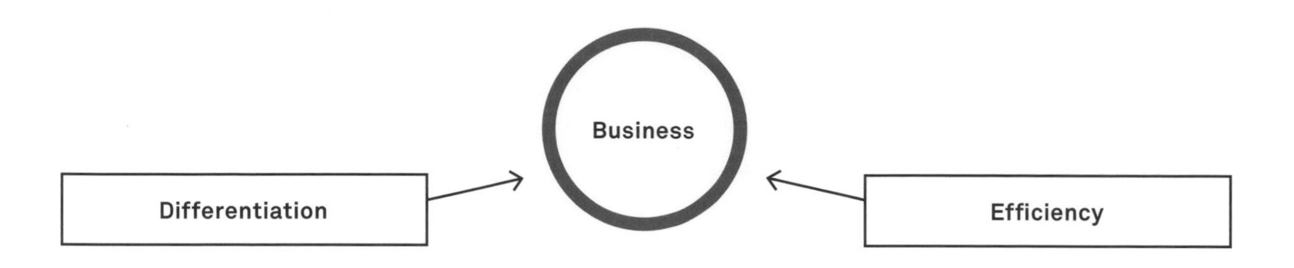

The business frame focuses on how a business works and how it engages with the markets it addresses, driven by two considerations:

_ DIFFERENTIATION

the way a business creates value for its customers, and fosters and sustains relationships to them, and makes a unique and distinguishable proposition to the market.

_ EFFICIENCY

the way a business maximizes profitability, manages resources and infrastructure, and achieves its objectives with the least possible amount of monetary engagement and human effort, and achieving a sustainable development of the enterprise as a whole.

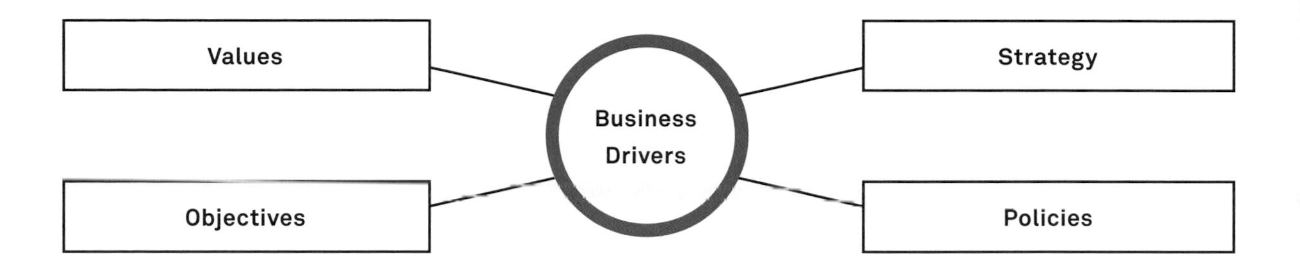

In exploring the drivers behind design initiatives, we found the following elements to be particularly relevant to the business context:

_ VALUES

pertaining to the fundamental idea of what the business is about, what it stands for, and how stakeholders see it now and in the future. The goal of design synthesis is to capture and express this idea, to help developing a vision of the future by making it visible.

_ STRATEGY

related to the way a company seeks to achieve competitive advantage over other market players by differentiation of its offering and maximizing efficiency. An awareness of how it contributes to these efforts makes a design initiative strategically relevant.

_ OBJECTIVES

or goals related to the business success as well as metrics suitable to measure whether those goals have been reached. They form the business case of a design project, describing the goals and, if possible, quantifying the return on investment.

_ POLICIES

are constraints and rules either self-created by the business following strategy and values, or imposed on it by regulation or law. They state the way things have to be done by the business, so that they affect design decisions and should be clearly defined.

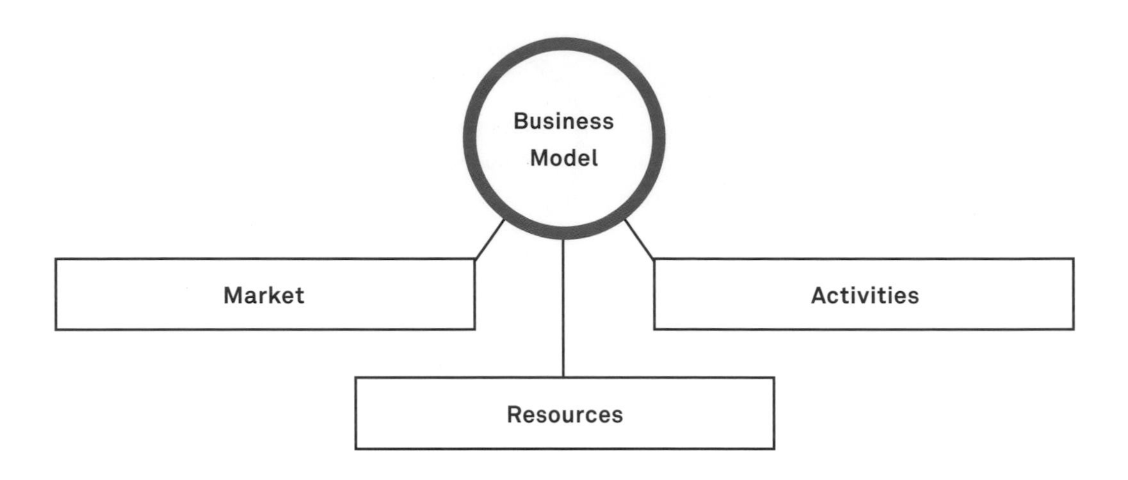

When modeling from a business perspective, the following elements are often of interest to research activities and solution definition:

_ MARKET

the larger environment to be addressed, with competitors and substitutes, customers to be served and offerings for them, prices and cost structures for customers, as well as market trends and issues. Design outcomes should achieve a better positioning for the organization on a market it is engaged on, or enable it to address a new market.

_ ACTIVITIES

activities carriend out thoughout the business, that span organizational units, people and tools for automated processing, exchanges via channels used to interact with customers or other organizations, transactions, and flow of revenue. Design outcomes should enable faster, simpler, and more robust and reliable processes.

_ RESOURCES

assets important to the business and the cost they produce, technical or physical infrastructure elements used as platforms to automate and manage work, external partners involved in business activities, communication channels, and tools. Design outcomes should achieve more with fewer resources and less cost.

As a starting point for the business frame applied, look at the enterprise as a whole in therms of its Big Picture aspects. This allows identifying opportunities for innovation and improvement, and developing the key qualities that help an organization pursue its strategic direction in daily business:

_ IDENTITY

the identity conveyed by the organization, brands, and informal and personal identities, and the way these identities are reflected in image and reputation are essentially how the enterprise creates trust toward its stakeholders. Brands are organizational resources and assets, and should be designed to help do business and develop it.

_ ARCHITECTURE

the combination of formal structures put in place and managed to execute and operationalize the business strategy. The business focus looks at the role of these structures for its activities, how they are maintained and developed, and how to improve business performance by achieving better results with less effort.

_ EXPERIENCE

the way the business appears in the lives of its customers and other stakeholders as a service provider, employer, or in other roles, and the noticeable value it creates and provides to them. Exploring the experiences of people in touch with the enterprise allows identifying business opportunities to provide more valuable benefits.

BUSINESS DESIGN

Key Partners	Key Activities	Value Propositions	Customer Relationships	Customer Segments
	Key Resources		Channels	
Cost Structure		Revenue Streams		

The Business Model Canvas (businessmodelgeneration.com)

A very recent addition to the area of design disciplines is Business Design, a term used more and more as a job role by companies in the field of strategic design such as the design and innovation consultancy IDEO, as well as in management graduate programs. The essence of this approach is designing and redesigning businesses expressed in models applicable to both new companies and established organizations. It is about applying design to problems related to generating strategic options, defining a good core idea of a business, and modeling how the business generates demand and provides value, and finally about operationalizing this model.

An approach to innovate with business models is outlined in Osterwalder's and Pigneur's book *Business Model Generation*, applying design to inform strategy, define and reshape business models, and plan their operationalization. It portrays the *Business Model Canvas* as a tool to generate a map of a business, and to blend ideation, prototyping and storytelling with analytic and strategic tools such as the framework described in the book *Blue Ocean Strategy* by W. Chan Kim and Renée Mauborgne. Practitioners of Business Design combine a design-led approach to achieve business innovation with proficiency in core business domains, including Marketing, Operations, and Finance. Key to this combination is applying a spirit of entrepreneurship that bridges an emotionally strong business idea with a viable business model.

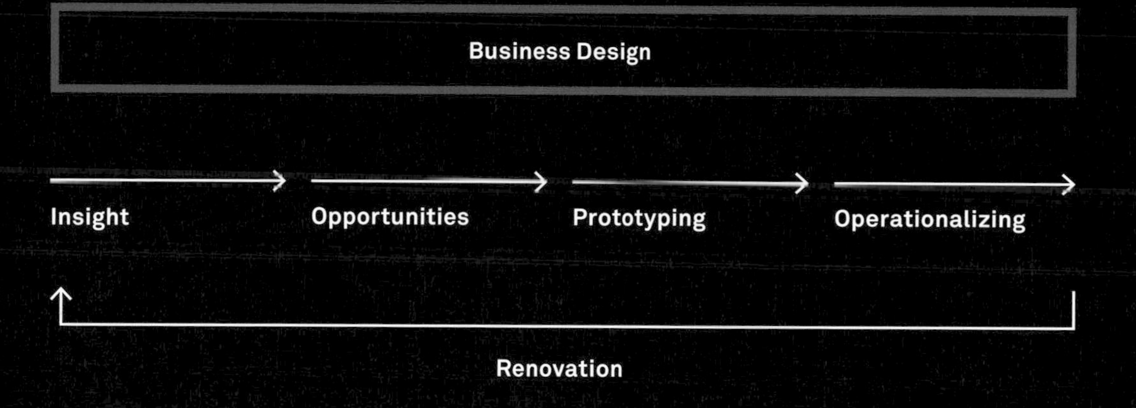

Business Design encompasses different steps of analyzing, designing, prototyping, and introducing new and evolved business models, with a focus on achieving viability:

_ DEVELOPING INSIGHT

scanning the environment using quantitative analysis and visualization, to understand the drivers behind a business model as well as external and internal success factors.

_ DISCOVERING OPPORTUNITIES

identifying and applying recurring patterns in business success, in order to discover opportunities for growth, and explore options for a new competitive positioning.

_ PROTOTYPING AND SIMULATING

making models visible as evaluation and communication devices such as stories, sketches, financial scenarios, or small scale prototypes to be tested with customers.

_ OPERATIONALIZING AND ITERATING

defining and refining management systems, organizational structures, capabilities, and resources behind a business model, to determine the critical factors that make it work.

Anatomy elements that form relationships in the enterprise are the necessary ingredients to do business:

_ ACTORS

stakeholders and the business roles they play in the market and the enterprise, most importantly customers, but also investors or (potential) employees. Depending on the level of design, they can be further segmented into different customer segments, investor types, or organizational divisions.

_ TOUCHPOINTS

channels used to reach actors, offer, deliver, or acquire products and services, and exchange money, information, and goods. They provide the interface to do business with people, both as consumers or representatives of customer organizations, and can be of physical or virtual nature, direct or indirect.

_ SERVICES

business activities that generate a value that can be expressed in monetary terms, and create or sustain relationships to customers or other actors. This includes the offerings produced and made available for sale and consumption, as well as the ongoing processes creating or supporting stakeholder benefits.

_ CONTENT

messages and communications with regard to business decision making, within the organization as well as with other actors in the larger enterprise. This includes customer communication along the commercial process, as well as information used for management decisions for management and operations.

Designing business _

Approaching a design challenge from a business perspective requires designers to think like an entrepreneur, or—in the case of an established business—like an internal or external investor in a business transformation. Key to this is the concept of business modeling, the practice of understanding and redesigning the way an enterprise creates value and generates revenue. Based on the business view of the enterprise and individual business context described before, it involves modeling the business and its potential evolution.

Although applicable for any design challenge, the business frame is often considered only superficially in today's design practice. In order to make an impact on a strategic and enterprise-wide level, it is crucial to develop a deep understanding of the way business is done, both currently and in a changed future setting. A strategic dialogue with business leaders that have commissioned the design work is only possible with a design practice that aims to inform strategy. Designers have to do so in business terms, making explicit their plans to transform enterprise qualities, and how design outcomes can contribute to business success.

A prerequisite for such a dialogue is to have the design team work closely with business stakeholders when preparing and running design initiatives, especially those working on a plan or vision for the future of the enterprise—entrepreneurs and executives, strategists and Marketing experts, owners of projects and programs, investors, customers, and suppliers—essentially anyone tasked with innovation and change, both inside and outside the organization.

Applying the business frame thoroughly is essential for design teams to position themselves, as innovation drivers in the enterprise instead of as merely executors of what others have defined. But this also results in new challenges. It requires discovering business opportunities in design research, and generating choices as investment options. These options have to address how the outcomes of the design process will contribute to business success by generating customer value or improving operations, requiring both analysis of the current business and operating model and a quantification of success conditions. Likewise, they have to be grounded in a well-defined business strategy, translating values, strategy, objectives, and policies into design principles to be applied.

#9_PEOPLE

Think of two customers. Both were born in 1948, male, raised in Great Britain, married, successful and wealthy. Furthermore, both of them have at least two children, like dogs and love the Alps.

One of them could be Prince Charles and the other Ozzy Osbourne.

Marc Stickdorn and Jakob Schneider in *This is Service Design Thinking*

Applying a human-centric perspective today is probably the best known way to approach a complex design challenge. This follows from the idea that in the end, it is always people you are designing for, people probably very different from yourself and your team. The goal of any design initiative is to achieve an engaging experience, by providing outcomes which are considered valuable by their audience.

Just as the other aspects described in this chapter, there is no definitive collection of tools or methods, but a range of different approaches that can be used to better address people with a design initiative. A core principle is the goal to understand and relate to people, beyond traditional market research approaches as collecting demographic data about them, or asking them for their opinions. Genuinely designing for human reality requires the designer to find out about people and their lives—attitudes, concerns, aspirations, daily routines, and contexts—in order to to generate the insight knowledge and empathy that drive design decisions. Applying the people frame therefore means approaching both the design problems faced and the potential results envisioned in a humanistic fashion. The insights gained about people allow us to find opportunities to make things better for them and to define and predict the role the outcomes of a design initiative might play in their lives.

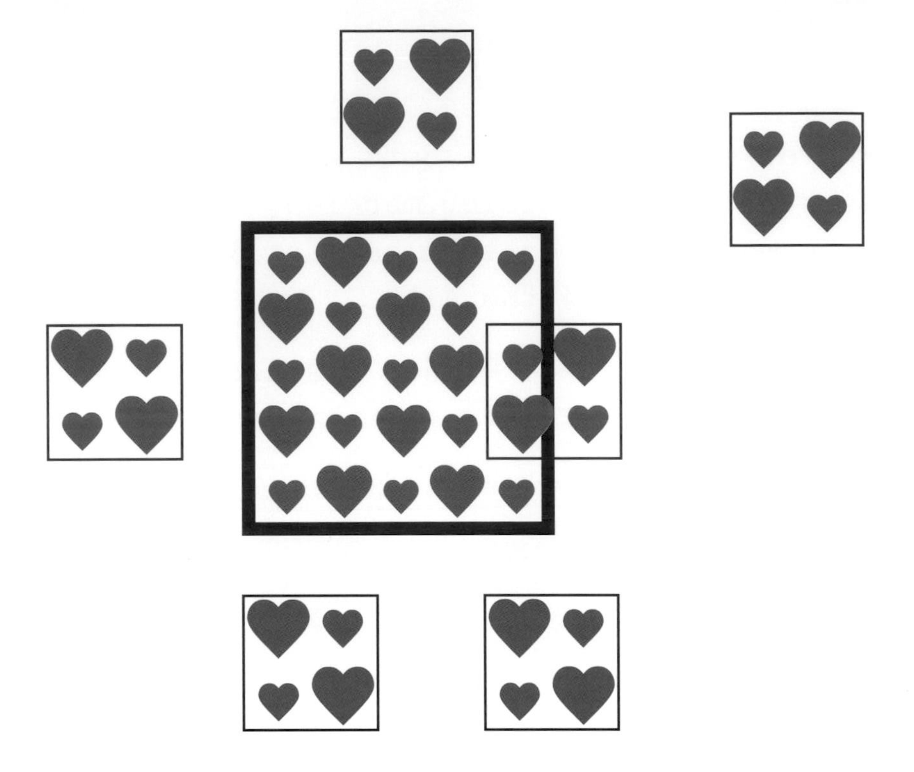

_ Developing knowledge and empathy

In our daily lives, we encounter people we know well and others we do not, people we do or do not understand when we talk to them. We consider some people friends and are indifferent to others. This aspect of social reality is closely related to design work and the role of the people frame. People are special—addressing them requires designers to appreciate and embrace what makes us individual, to come up with concepts and ideas that fit.

The people frame is based on a straightforward assumption: if we take the people that a design is made for into account and relate to them on a human level, it is much more likely that we will meet their needs and expectations, with the objective of making them happier than before. This in turn is an implicit goal of any design project, since making these people happy means engaged employees, satisfied customers and investors, other stakeholders to the business. While this principle sounds quite simple, holding to it in real life turns out to be somewhat complex. It means making people and their lives an integral part of our creative thinking process, people who in many cases we do not know much about.

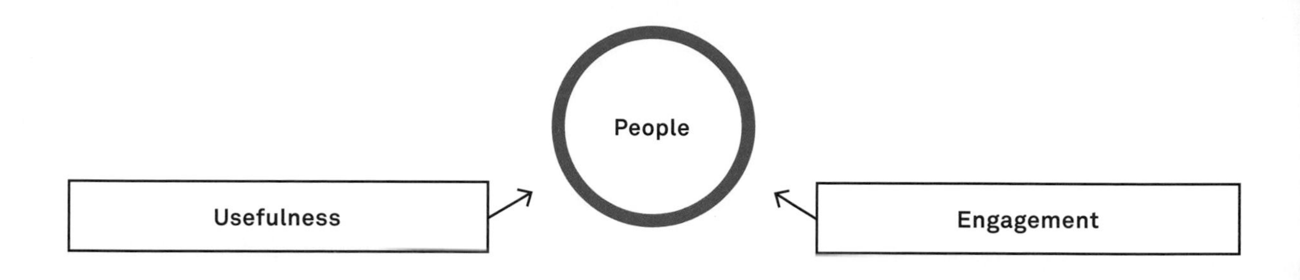

The focus of the people frame is to make the outcomes of the design meaningful to people, driven by two key aims to be achieved:

_ USEFULNESS

making the outcomes of the design process useful in that they help people achieve their goals, addressing the very personal reasons they have to engage in activities or tasks. This involves making things usable, findable, accessible and, useful for the audience.

_ ENGAGEMENT

making the design outcomes compelling and desirable by addressing people on an emotional and subconscious level. Going beyond pure usefulness, engagement involves elements of stimulation, persuasion, and personality, to generate attraction and build trust.

_ EXAMPLE

Imagine the perfect watch you want to give a close friend for her birthday. The basic purpose of looking up the time is quite obvious, so that just about any product that shows hours and minutes fulfills it. To find the perfect watch, however, you will have to take into account a whole range of individual factors and considerations, which depend on your personal relationship and insight into her life. While people with a lot of business travel might like switching between time zones or sophisticated alarms, others might prefer a simple and elegant display. Beyond the functional, a watch might be considered a fashion item, so that brand and appeal come into play. Finding or designing the perfect watch for your friend therefore requires a deep knowledge of her individual lifestyle, daily routines, characteristics, and personality.

Methods traditionally used in Marketing or Product Development, working with standardized models of people based on demographics and quantitative research, are not suited to generate the insights needed to acknowledge and consider people as human beings. Instead, we apply methods used by anthropologists and social sciences, spending time with people to get a personal understanding of their lives, both broad and deep. It requires effort for qualitative research, collaboration, and validation of outcomes, along with a mastery of appropriate techniques of ethnography and shared creative work. In real-life projects, we usually do not have the luxury of designing for close friends or for ourselves. Finding out about the people addressed and involving them in the process therefore is vital to develop the knowledge necessary to determine a good direction for further design work and to identify potential solutions that work and fit.

_ The enterprise as a social space

The people frame perceives the enterprise as a social construct, as a group of persons. It enables a design team to think of the people targeted, involved, and affected by their outcomes. It allows them to immerse themselves in people's daily life, and to develop the knowledge and empathy that are behind good solutions and great ideas. As a strictly human-centered viewpoint, it reduces the challenging problem settings of a complex enterprise environment to the everyday reality of people and the personal and social factors that influence it.

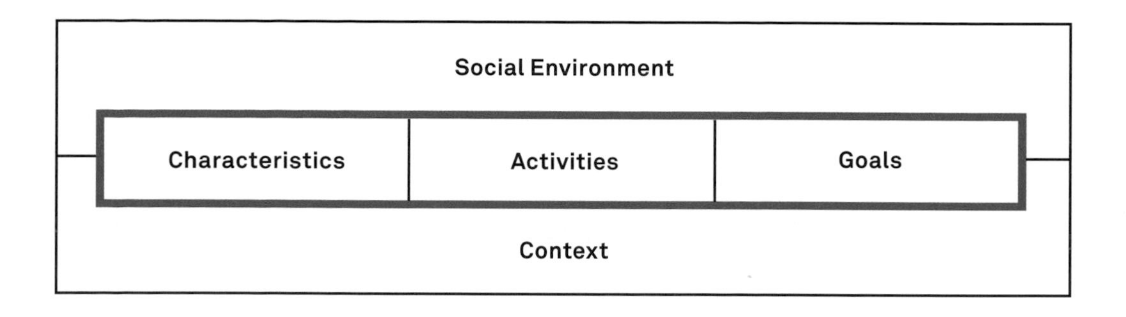

In order to develop knowledge about the people you address with a design initiative and gain deep insight into their lives, take into account the following dimensions in research and modeling:

_ CHARACTERISTICS

demographic data such as their age and profession, but also aptitudes and skills, habits, and attitudes.

_ ACTIVITIES

looking at the nature of tasks, frequency and duration, tools or artifacts used, and interactions with others.

_ GOALS

what people want to achieve, immediate and long-term goals, reflecting rational or practical issues related to feeling happy.

_ SOCIAL ENVIRONMENT

groups, roles and identities, cultural norms and beliefs, local particularities, and the nature of relationships.

_ CONTEXT

both physical environments and states of mind in which people carry out activities and interact with each other.

Beauty and brains, pleasure and usability – they should go hand in hand.

Don Norman

Elements of relationship Anatomy are about the very concrete manifestation of the enterprise in a person's daily life:

_ ACTORS

the roles of a person in social terms. This relates to the individual reflection of the relationships to others in the enterprise space, responsibilities, and the expectations, leading to a certain self-image and behavior. Engaging actors as people means designing for individual aspirations and motivation with regard to their role.

_ TOUCHPOINTS

the situations someone experiences across the journey of interactions with and within the enterprise, and the particular physical and psychological contexts in which they appear. This refers to the environment, time, duration, and intensity of those contacts, and the factors to be considered to address that context.

_ SERVICES

the activities carried out in the enterprise to support people and generate benefits for them. Services appear to people only in the effects they produce, and the communication processes and exchanges happening around them. Looking at services from a people perspective allows exploring their role and how they are perceived.

_ CONTENT

information and messages being exchanged between people in the enterprise, relevant to people only as knowledge that is useful by informing choices and supporting decision-making, or otherwise considered of interest or entertaining. Content is the essence of any exchange, to be designed to fit and make sense to people.

When looking at the Big Picture of the enterprise from a people perspective, it presents itself as a dynamic social space of people interacting with each other in a multitude of different ways:

_ IDENTITY

identities that appear in the enterprise space and the way they are perceived by people, and lead to certain images in their minds. To people, brands and informal identities are meaningful symbols expressing organizational identity and culture, connected to the enterprise as a social and cultural system, having manners and beliefs, language and rituals, and following behavioral norms.

_ ARCHITECTURE

the way the enterprise works for people and the role they play in the way it is structured and executes activities. From a people perspective, architecture refers to organization and formally defined responsibilities, individual tasks that contribute to the running enterprise (not restricted to staff, think of customer tasks), tools and systems used, and the *human factor* in business operations.

_ EXPERIENCE

while the experience quality of the enterprise is inherently people-centric, looking at experience with an individual's eyes is all about their overall personal context and condition. It allows examining the meaning of the enterprise of people in touch with it and important to its success, and the role it plays in their lives.

HUMAN-CENTERED DESIGN

The term Human-Centered Design (HCD)—originally coined by Don Norman, and often also called User-Centered Design—has been around for quite some time now, used in the context of various design disciplines. It usually refers to a general attitude and mindset applied to the design process as a whole, bringing people and their needs and characteristics into focus of all efforts. While there is no consensus in the design industry what such a process must include, there are some approaches and methods which are developed and evangelized by HCD practitioners. The basic assumption is that human reality, in all its diversity and complexity, is somewhat hard to understand, capture, and describe. It is quite a challenge to consider it in the level of detail needed to capture everything relevant to a design process, and to be able to envision the change it will produce for people.

These tools help a design team to develop a deep understanding of the people they consider relevant to the project, because they are being addressed or impacted, or involved in bringing it to life. In enterprise-wide projects with a large and diverse set of stakeholders, they help to keep human reality in focus during the entire process.

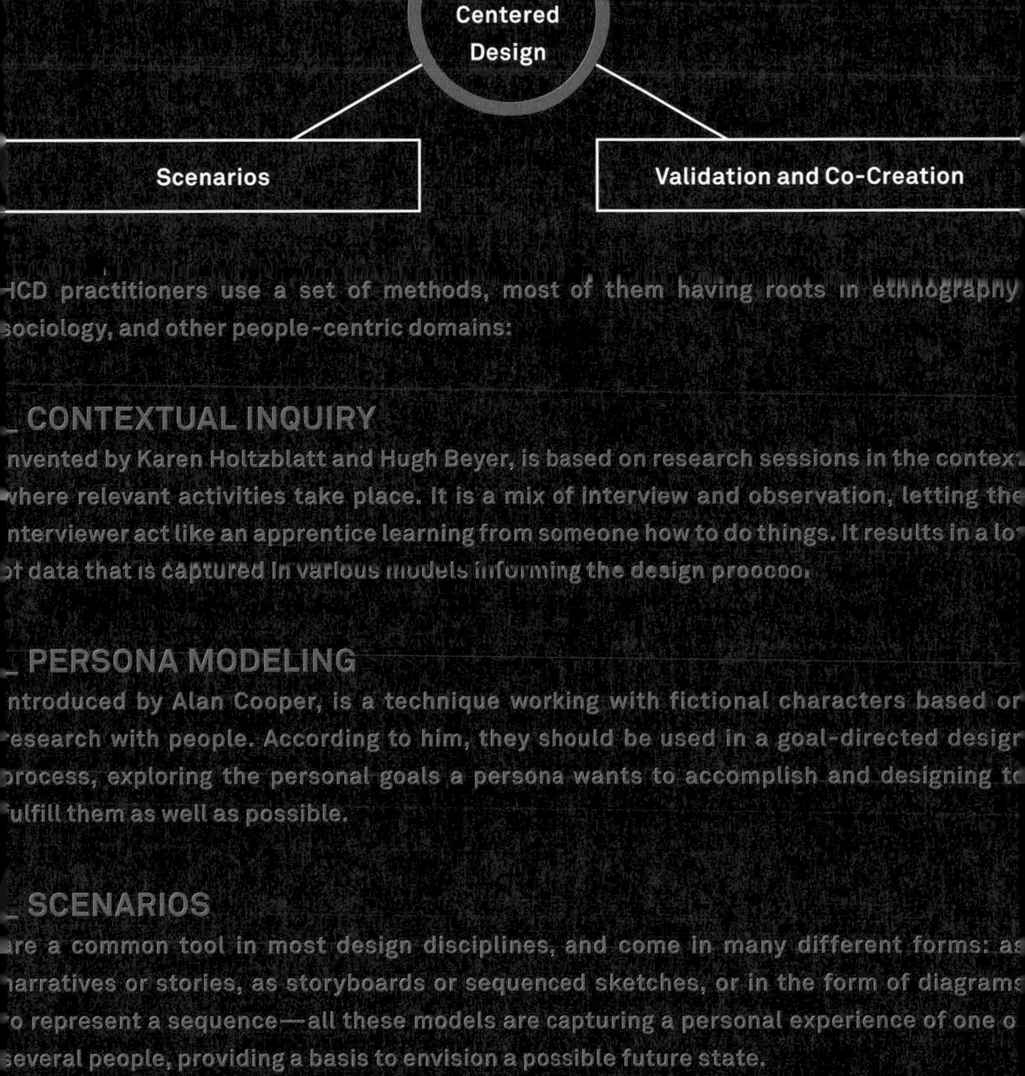

HCD practitioners use a set of methods, most of them having roots in ethnography, sociology, and other people-centric domains:

_ CONTEXTUAL INQUIRY

Invented by Karen Holtzblatt and Hugh Beyer, is based on research sessions in the context where relevant activities take place. It is a mix of interview and observation, letting the interviewer act like an apprentice learning from someone how to do things. It results in a lot of data that is captured in various models informing the design process.

_ PERSONA MODELING

Introduced by Alan Cooper, is a technique working with fictional characters based on research with people. According to him, they should be used in a goal-directed design process, exploring the personal goals a persona wants to accomplish and designing to fulfill them as well as possible.

_ SCENARIOS

are a common tool in most design disciplines, and come in many different forms: as narratives or stories, as storyboards or sequenced sketches, or in the form of diagrams to represent a sequence—all these models are capturing a personal experience of one or several people, providing a basis to envision a possible future state.

_ VALIDATION AND CO-CREATION

are approaches to involve people actively in the design process. Validation techniques allow putting potential outcomes to the test with people, by facilitating test sessions (also known as Usability Tests) or informal reviews. Co-creation goes beyond that by involving

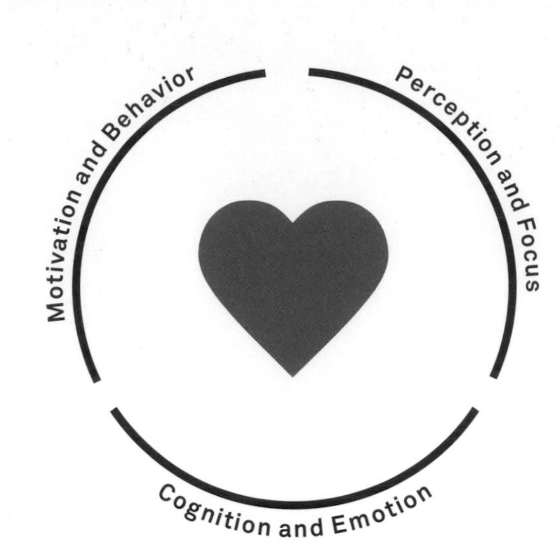

Designing with the people frame is applicable for research, design, and validation activities. It is taking into account multiple distinct but intertwined psychological qualities, and addresses them by applying both factual knowledge about them and subjective conceptualization in an individual context:

_ PERCEPTION AND FOCUS

are about the impressions of the human senses, what a person sees, hears, smells, tastes, or touches. People filter what they perceive to direct their attention on messages and things they focus on, guided by emotion and cognition. Design works with these elements to direct the focus and facilitate comprehension.

_ COGNITION AND EMOTION

refer to thinking processes and feelings, and their complex interrelationships. This includes expectations people have with regards to an activity, associations they apply to develop conceptual models of the world, and questions of memory and learning. Design considers both aspects of the mind, conscious and reflective thinking as well as affection and emotional reactions.

_ MOTIVATION AND BEHAVIOR

are about what makes people take an action, to what kinds of activities and conduct that leads them, and how they adapt to the world around them. Understanding motives and intentions in the context of a design problem provides the basis to discover and address the hidden and unexpressed needs of people, by predicting situational change and motivating new behaviors.

Applying the people frame requires investment in research and collaborative work with the audience being addressed, and with the stakeholders considered important. Beyond just knowing the raw facts, working with actual people enables us to *walk in their shoes*, experiencing the struggles of activities first hand, and to empathize with the people we are designing for. In our design projects, that deep understanding gathered about people has proven to be an important source of ideas, and enables the team to re-frame a given problem in a way that it guides all design decisions. This however requires translating the findings into actionable insights.

The implications of following an HCD principle seem to be interpreted very differently among the various fields where the fundamental paradigm is applied. While practitioners of usability or ergonomics work with analytic approaches driven by metrics, such as task efficiency or cognitive load, methodologies used by design professionals strive to empathize with people, or to engage in co-creative design exercises. Still, all of these different interpretations share a common goal: making people a central consideration in every design decision, and actively involving them in the design work.

In the enterprise context, the people frame connects the design process to the human reality that it seeks to influence—applicable not just to the target group the designers are focusing on, but to the whole range of actors identified as relevant. This helps to expand the human-centric focus beyond an individual outcome. Instead of considering people just as users of a website, customers addressed by a service, or other predefined roles, it allows looking at people first and foremost as human beings with needs and aspirations.

Designing for the enterprise as a social space means considering personalities, individual meanings or roles, and offerings to key stakeholders as persons. This view goes beyond usability and satisfaction, concepts which basically strive to make things less bad—delivering what is expected to not disappoint people, and avoiding needless hassles and efforts. The people frame aims for the big goals: engaging people, helping them to have fun, or making their work a smooth flow. It allows adapting the result of enterprise activities to their lives, their problems, and what makes them happy.

10_FUNCTION

Design is not how it looks like and feels like. Design is how it works.

Steve Jobs

Taking a functional perspective is common in many domains where design is practiced. Because of the difficulties of planning, designing, and implementing any kind of system beyond a certain degree of complexity, it is a natural way to look at things, especially in engineering disciplines, and has also finds its way into Marketing and Product Management. In essence, it is about what a product or another outcome of a design process is actually being made for. It allows specifying, communicating and documenting the scope of a project, resulting in the functions or *features* it should support.

The idea of a function is representing one of the most relevant and universal values in design work. American functionalist architect Louis Sullivan expressed this in his famous quote that *form follows function*. Unlike art that must not serve a clear purpose or goal, the outcomes of design work by definition always have a function. Despite the creative nature of the work itself, drawing on individual experience and inspiration from outside the given problem space, any result of a design process follows the underlying rationale, designed to be consumed, perceived or used by someone. It strives to generate value for both the people commissioning it and people being addressed.

The function frame in design attempts to clarify the intentions behind a project. It helps to define what you want to create, but also—often more importantly—what you do not. It is a driving factor important to any design initiative even if not consciously addressed. Applying it consciously in the design process allows us to clarify the purpose of a project as an explicit and shared idea, and to design the behaviors to fulfill that purpose. A key challenge is to determine and capture function in definite requirements.

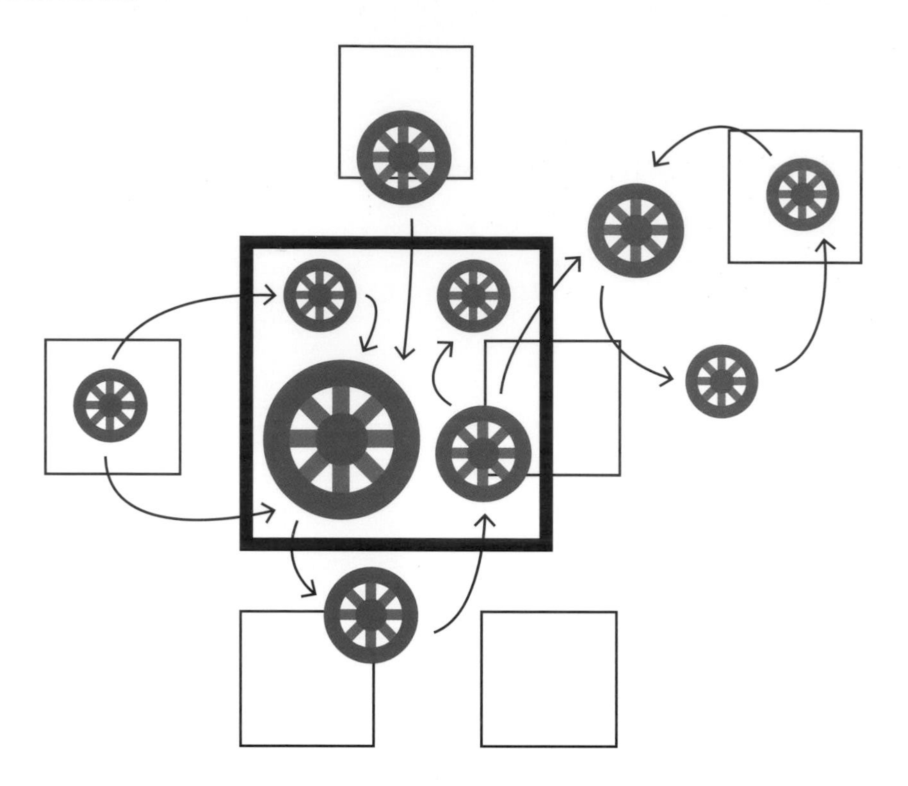

_ Eliciting and modeling functional requirements

In the most general sense, function refers to everything related to the rationale behind a design project. It is about the meaning the outcomes will have for persons and groups, the client or owner commissioning the work, and other potential stakeholders, and about planning the resulting system's behavior in advance in a way that anticipates that meaning. In the end, it is not determined by the designer or developer—it is the user who attributes function. Therefore, predicting function and specifying features usually requires taking into account a large amount of different facts, stakeholder opinions, and observations made in the environment of the project and using this knowledge to elaborate a set of requirements.

In our consulting practice, we found this part of the design process to be particular difficult to get right. While it is necessary to get together with the customer and the people addressed to ask for their needs, in isolation such an approach often leads to failure. It risks producing a large wish list of features and trendy topics, and neglects the potential of hidden needs, ideas available in the project team or among stakeholders, and insights gathered from applying the other perspectives described in this chapter. While having a clear picture of what is wanted is of high value for design decisions, a feature-driven approach bears a great danger, known as *feature creep*: adding more and more to the list under the assumption that more is always better, or under pressure from the client.

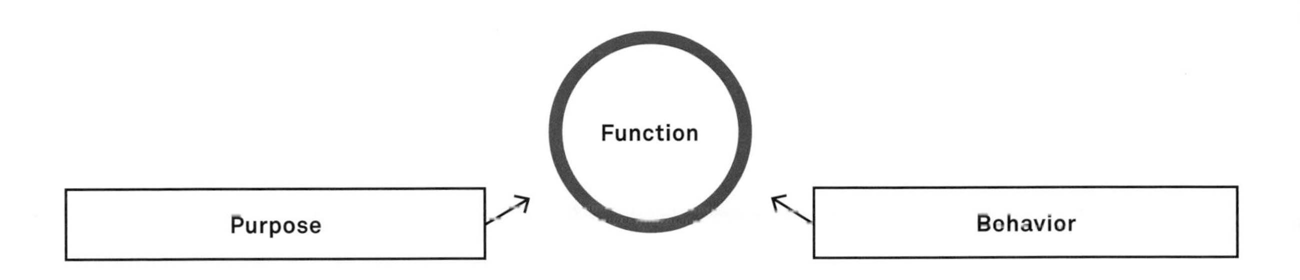

The function frame looks at how something works or should work to serve a defined need, taking into account two main considerations:

_ PURPOSE

what the project is about, making outcomes of the design process respond to actual requirements and expectations, including goals to be supported and qualities or constraints to be addressed.

_ BEHAVIOR

how its outcomes should work as a system to accomplish a goal, supporting activities, tasks or business processes, defined as a behavioral description of interactions with the environment.

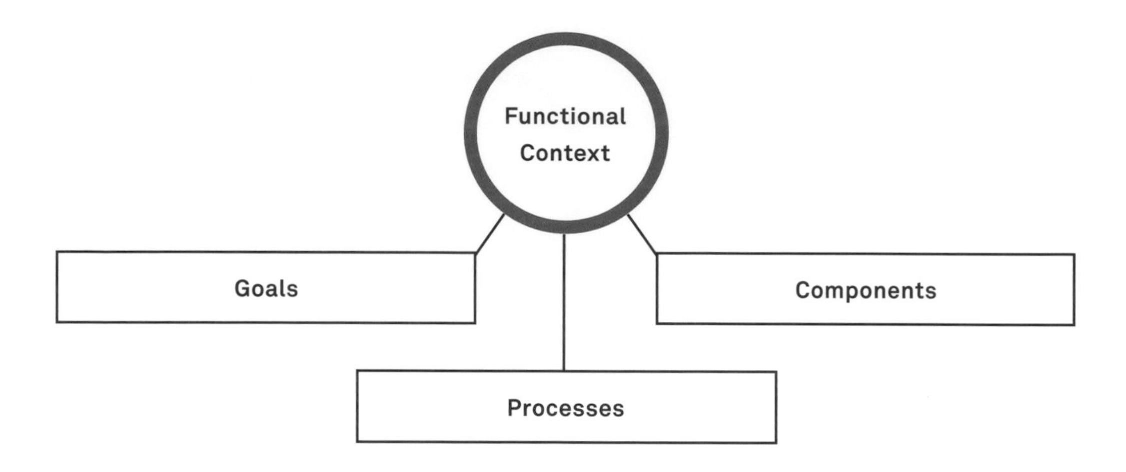

The elicitation and analysis of requirements means understanding and modeling the functional context in which the resulting system will be working, considering a set of distinct elements:

_ GOALS

either explicitly expressed by stakeholders or otherwise becoming apparent, that can be translated into a function the resulting system has to fulfill. Models about goals are informed by conversations with stakeholders or research, but also by ideas that might unveil a hidden need.

_ PROCESSES

representing any form of organized and predictable activity corresponding to the fulfillment of a certain set of goals. A functional process model is a detailed description of what is going on, in order to determine the functions required to support and improve those processes. It might also detail the flows that appear in the system as movement of information, money, value, time, or physical resources, and how inputs are transformed and turned into process outputs.

_ COMPONENTS

are conceptual objects that group a set of related functions as part of the overall system, acting in a process to support the achievement of goals. Components are a mental construct to allow decomposing sets of functionality into systems and subsystems, as well as external and internal parts, helping to determine the working structure and behavior of a system.

Requirements are often just laundry lists: someone makes them up, they are arbitrary, out of context and just plain wrong.

Sally Bean, Enterprise Architecture Consultant

Even if the functionality selected is based on a clear understanding of the stakeholders and the concerns the enterprise needs to address, there is a chance that we will end up with a big list of unnecessary, arbitrary, or conflicting requirements that make the results needlessly complicated and ill-suited to solve the key problems. It is essential to distinguish between the functional context of the goals and processes to be supported, and the components designed to perform a function, again looking at both the current state and a potential future.

A thorough modeling of the functional context lets us capture what is expected of the system resulting from a design process in terms of the purposes it has to fulfill, instead of resorting to subjective and arbitrary lists of features. This in turn is a suitable basis from which to deliberately address function in a design process: once the required functions are known to the design team, it can proceed by elaborating a system of components to perform them. Such a systematic approach to function might lead you to completely different functional requirements than originally envisioned, but helps to identify and solve the actual problems behind vague wishes expressed by stakeholders.

_EXAMPLE

Because any enterprise serves a multitude of stakeholders, a way to look at it from a function perspective is imagining it as a machine that carries out an important task for someone else. In the case of an airline, it is seen by passengers as a transportation machine, while investors see it as a money-making machine. To crew members it appears as a job-making machine, to the supplying airplane manufacturers it looks like a product-buying and consumption machine. Looking at the enterprise holistically involves considering these various functions, and envisioning functional components (or subsystems) that perform them. Components that play a role in this design could be a reservation function, financial performance management, or an HR service for employees, embodied and visible to people as tools or procedures related to those functions.

The function frame views the enterprise as a working system that exists to serve a specific purpose, by performing a certain set of functions:

_ IDENTITY

the purpose and functions associated to identities and brands of the enterprise, and the behavior the system exhibits in executing these functions. In Marketing practice, this is also known as the *brand promise*, and allows designing functional components to deliver on that promise by providing value and endorsing adequate behavior.

_ ARCHITECTURE

the way the enterprise operates by making functional components work together. A functional perspective on architecture ties all formally defined structures and activities to the functional context they are used in—the processes and corresponding goals they serve—and enables designing the enterprise to fit into this context.

_ EXPERIENCE

how the functions of the enterprise are accessible and usable to people, supporting them to meet their personal goals. This refers to the purpose the enterprise has for a person, and the behavior it exhibits. It allows conceiving abstract enterprise functions in a way that makes them useful and valuable for people's activities.

Elements on an Anatomy level can be seen as the context and drivers for functional behaviors:

_ ACTORS

users addressed with a particular function, feature, or behavior. Finding out about the role in the enterprise and individual goals is the basis to elicit requirements and conceive functionality that makes sense and becomes valuable for a certain actor.

_ TOUCHPOINTS

interfaces between users and functions, enabling interactions and providing the input/output channels (or front-end) required for any function. Applying this frame on touch-points enables designing them as interfaces to support certain functions.

_ SERVICES

sets of functions that in combination provide a benefit to service users in the enterprise. Services in functional terms are pre-planned behaviors being used in people's activities. They allow conceiving tools, products, or resources serving functional needs.

_ CONTENT

the input needed for functions to work and the output they produce (also known as functional parameters and results). Content from a functional perspective refers to data needed for a defined function to work, and the data produced after its execution.

_ Functions of the enterprise

Any enterprise is created around a sense of purpose shared by its members, which makes it play a certain role in their lives. Applying the function frame allows us to undersand that role in detail by analyzing processes and corresponding stakeholder goals, and lets us capture and document the functions it must perform to fulfill its purpose. This general purpose the enterprise fulfills for its community of actors drives the decision of which functions to address with a design initiative. It is based on the idea that any function made available as a product or service contributes to that purpose.

Therefore, developing functional models of the enterprise forms the basis to elicit, select and justify the requirements and qualities to be met, and to agree on the functional components and features to be included in the scope of potential outcomes. By applying the functional frame beyond the scope of an individual project, the potential results can be considered part of the larger functional context of the enterprise in order to identify redundancies and integration needs, and to seek opportunities for functional gaps to be filled.

_ Designing with function

In everyday practice, the function frame plays an important role in almost any design project, since it relates to the purpose and goals driving it. In Enterprise Design, it helps to envision the enterprise as a functional system, and to design in terms of the functions it fulfills for its stakeholder community. When looking at function on an enterprise-wide level, designers often face a mass of conflicting, overlapping, and intertwined needs. To address these usually requires a certain degree of complexity in the solution. In order to choose and prioritize the functions to address, they have to systematically manage requirements.

REQUIREMENTS ENGINEERING

The professional practice of selecting and managing requirements comes from engineering-driven approaches to projects. It is particularly in evidence wherever technical systems exhibit complex behaviors, and on projects where the question of feasibility is prominent. Even when not explicitly addressed by the team, it appears in the form of feature lists or briefing documents, requirements specifications, or other informal means of describing what the outcomes are supposed to do.

Requirements Engineering is an approach to systematically manage functional and other types of requirements. It has its origins in technology-driven disciplines such as professional software development or systems engineering. The activities performed in practice encompass the structured elicitation, analysis, and recording of the needs and conditions that the outcome of a project must meet in order to be successful. That outcome can be a machine, a piece of software, or any other type of system to be planned and constructed.

The practice of Requirements Engineering usually puts a particular focus on the functional aspect of what a system's behavior should accomplish, but also looks at non-functional or quality requirements such as maintainability or security. Although establishing requirements is part of any design initiative in some form, applying it deliberately and with some formality helps tremendously to agree on a project's scope.

More than anything else, the process of establishing requirements is a collaborative endeavor of envisioning the outcomes of a project, prioritizing functions and qualities based on that vision, and determining how to implement them as a system. Instead of being carried out in isolation, it has to be an integral part of the design process, with the goal of achieving a common understanding of its outcomes, while the methods applied and level of detail required depend on the particular setting.

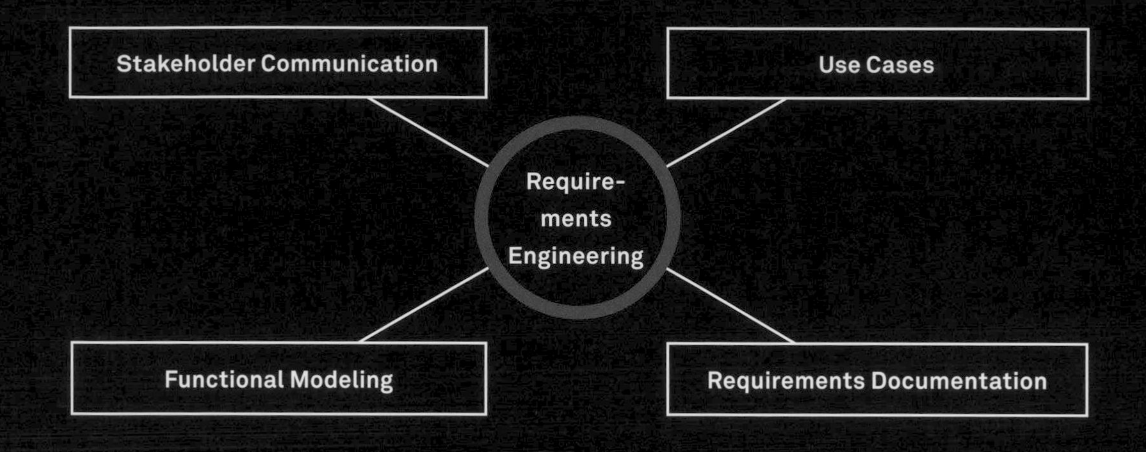

The practice of Requirements Engineering applies a variety of techniques to come to a set of requirements to be addressed:

_ STAKEHOLDER COMMUNICATION

activities to determine potential requirements by collaborating closely with the different parties involved in the project and those impacted or addressed by its outcomes, for example, by conducting stakeholder workshops to inform use cases.

_ USE CASES

written or graphical descriptions of the behavior a future system exhibits, captured in terms of outside interactions with it. This practice treats the system like a magic *black box*, avoiding any technical language and excluding details on how it works on the inside.

_ FUNCTIONAL MODELING

using simplified models to describe how the target system or current system works. This is subject to functional decomposition to divide the functional components until the model captures the behavior on a level of detail suitable for construction.

_ REQUIREMENTS DOCUMENTATION

capturing requirements in a way that is they are specific and measurable to enable implementation. Some methodologies suggest very detailed documentation while others, such as in agile software development, favor writing short user stories.

Modeling function relevant to the design project and to the larger enterprise is a valid basis for establishing requirements, because in the end all requirements — including those describing qualities rather than behavior — pertain to function, to the purpose to be fulfilled by the project's results, and to the anticipated usage and value generation. It allows us to ground the vision of what the outcomes should do in the goals and processes to be supported, and to design components to fit into that functional context. Such a shared definition of functional models and solid requirements in turn is the basis for working with engineers, system designers, and technologists, for mapping out the scope of the solution to be built, and for ensuring that it is actually feasible.

The choice of requirements is informed by many sources of inspiration — findings from applying the other perspectives, competition or similar projects, new technologies and their possibilities, as well as the free flow of ideas that emerge in a multidisciplinary project team. Applying the function frame enables the design team to agree on what the outcomes should be, to reduce the functional scope to support the most relevant goals and processes, and to provide relevant input to engineers planning the technical part of the project.

11_STRUCTURE

The three frames explored earlier illustrate the difficulties of applying design practice to something as complex as an enterprise. In many cases, the designer is not familiar with the particular domain of a design challenge, which requires detailed knowledge of a certain industry, target group, local culture, or particular use context. It involves gathering and making sense of a large amount of data, in order to develop a thorough understanding about the real-world setting where the outcomes of a design process will be applied.

The structure frame supports this endeavor, by focusing on things that matter to the design project, and their interrelationships. It helps when developing conceptual models of the context and the subject matter of an individual design initiative, and of the structural context of the problem being solved — also called the *domain*. These conceptual models represent the mental models of different people involved in the design process, and of those addressed by or in touch with its outcome. As with the other perspectives, it is used to actively generate knowledge by modeling and exploring the current state as the context of a design challenge, as well as by envisioning the possible change enacted by introducing a new system of structured elements into that setting.

Structural models have their roots in Linguistics and Systems Thinking, and are embraced in engineering approaches such as object-oriented programming and business analysis. Known as object or domain models, and sometimes called ontologies, they describe a set of objects (also called referents or entities) and their relationships. In design projects, working with structural models helps us to learn about the things relevant to a certain design problem. Concrete techniques include concept maps, class diagrams, and entity-relationship models. In the Enterprise Design framework, the structure frame is used to map the things that are important to the enterprise in one way or another, to explore their roles and relationships, and to think about how the results of the design process might transform that structure.

Structures are seen differently by people involved in the project or impacted by its results, reflecting the view of different roles such as business owner, customer, staff, or planner. In the course of a design project, it becomes clearer what parts of the structure have to change to respond to people and business objectives.

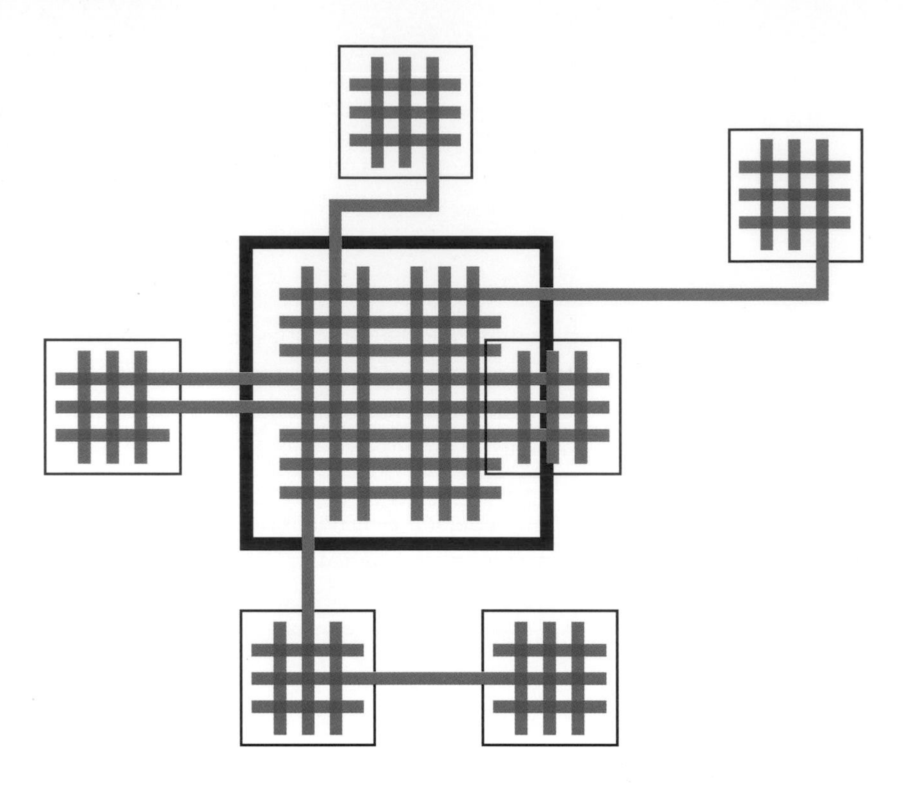

The emphasis of this aspect is on developing a holistic understanding of the structural context a design process seeks to improve and envisioning the role its outcomes will play in the wider environment of interrelated objects.

Working with structural models is another way to frame a complex design challenge, focusing on an understanding of the problem domain, and exploring the objects to consider in a design. A conceptual model describing a structure can be expressed in different ways and using different techniques, but usually contains a description of entities and how they relate to each other. It always reflects subjective conceptual models of the people creating it, and can therefore be seen as creative and generative as a business or mental model. Change introduced by a design process is never neutral in such a conceptual model, but becomes a part of it, enhancing or modifying things and resulting in an evolved structure when applied in real life. By introducing new objects, making others obsolete and rearranging relationships, design always attempts to restructure the world.

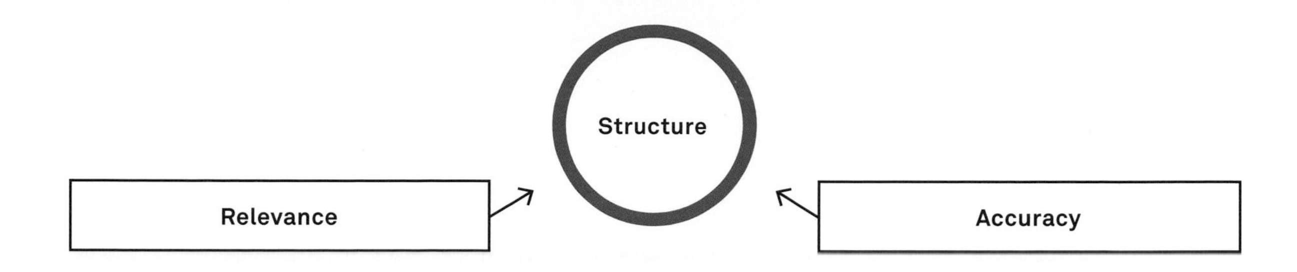

Based on insights from stakeholder research and expert knowledge, the list of things to consider in such a structural model quickly becomes quite large. The structure viewpoint focuses on two main concerns:

_ RELEVANCE

placing a “spotlight” on a subset of objects which are important to the enterprise with regard to the business objectives, the people being addressed, and the structural context.

_ ACCURACY

capturing the relevant parts of problem space, domain, or potential future state in a way that corresponds as much as possible to reality, as well as to common conceptual models.

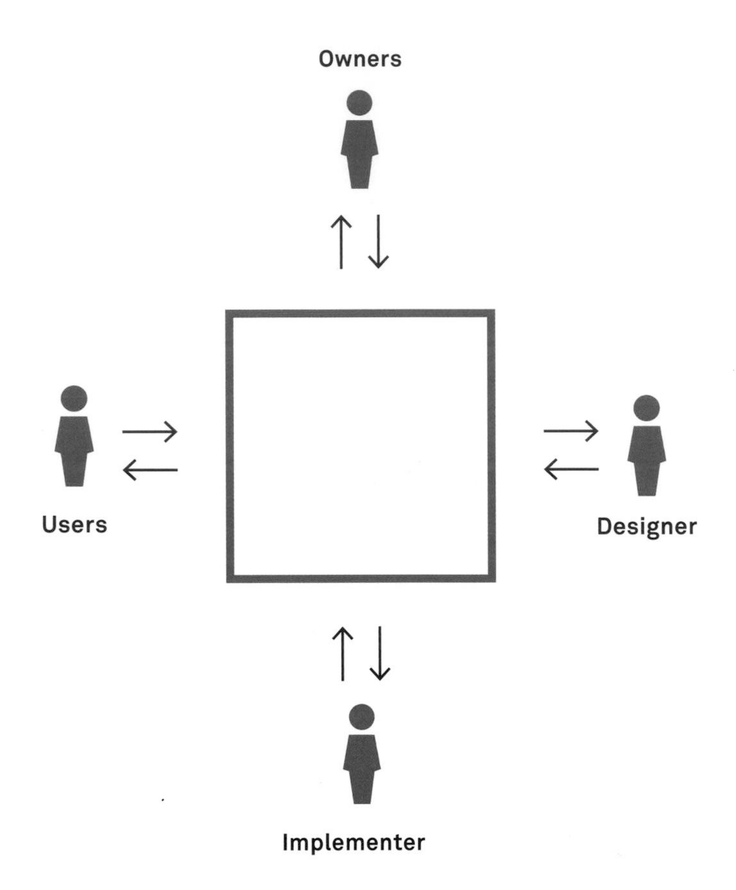

_ Modeling the structures to be designed

One of the particular strengths of the structure viewpoint, despite its subjectivity, is its ability to align and integrate different conceptual models into a coherent representation. The interrelated objects expressed are shared between many different conceptual models of a system, and allow for comparison, prioritization, and alignment.

For the creation of structural models, virtually any noun that people use might qualify as an object to be included. These could be physical and virtual things, people or organizations, representations and symbols, activities or processes, as well as mental constructs or categories—the important question is not the type, but the relevance and accuracy of the model for the design process. It can also help to deepen the analysis by exploring different relationship types such as generalization, instantiation, cardinalities, or by including attributes of objects and the tasks associated with them. Working with structural models enables the designer to externalize and study the different models based on research, prior knowledge and experience, as well as logical implications of the respective model.

Expanding on an early definition from Don Norman and Stephen Draper in the context of designing user interfaces for software products, there are a number of conceptual views on a system that can be subject to structural modeling in an Enterprise Design context. They can be mapped to the actors addressed and the roles active in the project, and aligned with findings from other perspectives:

_ THE USERS' MODEL

describes the enterprise's structure the way they are understood by its target groups. In most cases this type of model understands the enterprise as a black box, with only a certain parts of the overall structure being exposed to the people it has been created for, embedded in an existing context. The focus is to capture the mental model of people, so that such models often are incomplete and incorrect, being constructed in a person's mind based on observations, experience, assumptions, and superstitions, and embedding the system into structures from the individual environment it is used in.

_ THE OWNERS' MODEL

contains the structures important to the enterprise as a system, based on the views of business stakeholders—actors involved in making the enterprise work. This might include various concepts such as units, offerings, activities, events, competencies, or key partners, focusing on assets, resources, and infrastructure that ensure the viability of the business. It often has a significant overlap with the corresponding users' models, but includes also background structures invisible to the outside world of customers.

_ THE DESIGNER'S MODEL

is a conceptual representation of a system as it is designed by a designer, someone envisioning and planning its structure. It describes the system, its components, and their relationship as the designer intends them to appear and act. The focus of a designer's model is to structure things in a way that they make sense for users and owners as a coherent whole.

_ THE IMPLEMENTER'S MODEL

captures a system in the way it is built or developed, detailing the technical, operational, and organizational structure in sufficient scope and detail to plan its implementation. The focus being feasibility of a new or evolved system, an implementer's structural model describes the technical components and operational details put in place to bring the design to life, usually widely hidden from the system's users.

_ EXAMPLE

As an example for the structure frame applied, imagine a design project for a hotel. Some essential structural objects playing a role in the business come to mind quickly, such as rooms, beds, or keys. Thorough design research will result in many objects not quite so obvious — think of abstract objects like reservations or requests, elements outside the realm of the core service like a hairdresser, or even things outside the hotel but relevant to guests such as a bar nearby that serves good drinks, but also produces noise. Other structures could be invisible to guests but equally important to the business and personnel, like staff rooms or the cleaning plan. All these objects are potentially an important part of the design.

Regardless of the focus, creating a good structural model depends on working with domain experts, people who have a profound knowledge and expertise about the domain targeted with a design initiative. Depending on the individual project, those experts can be users of the system being designed, stakeholders, or external actors. Their role is to enable an understanding of the problem setting, making the models relevant and close to reality. With their help, a designer can iteratively refine the structural models, closing the gap between the different conceptual models and the designer's model, and preparing the subsequent implementer's model.

_ The enterprise as a structure

In an Enterprise Design project, the diversity of stakeholders makes an overarching structural representation very valuable when seeking alignment of their various concerns and ideas. The structure frame captures the enterprise as a system to be understood and transformed by rearranging and modifying its constituent parts. Any outcome of a design project is an evolved structure, introducing or removing elements and changing their associations. Such a redesigned structure can be described explicitly in a designer's model, including references to the way the enterprise is structured as a whole. This is again where the ubiquitous qualities and elements described earlier come into play.

The enterprise features structures and substructures related to the three ubiquitous qualities, which can be described and mapped out in structural models to inform design decisions:

_ IDENTITY

elements in the enterprise that constitute identities relevant to the design initiative. Such elements are often named in the users' or owners' conceptual models, formally introduced as brands or just there, such as organizational groups, products, trademarks, or tools.

_ ARCHITECTURE

elements that make the enterprise work as a system and are related to the design challenge, such as essential services or processes, organizational entities, technical systems, or infrastructure elements, either visible to actors addressed or hidden from view.

_ EXPERIENCE

objects introduced by the enterprise that appear in the user's lives and minds, playing a role to make a design achieve objectives toward its target group. This includes physical products and environments, but also brands, services, or messages as *things*.

Relationship elements on an anatomy level can be mapped to objects in structural models. Including those elements permits capturing relevant structures in individual human-enterprise interactions:

_ ACTORS

people interacting with objects in the context of a particular relationship to the enterprise, to perform a task or activity.

_ TOUCHPOINTS

instances of objects appearing in a certain situation, enabling exploration of the level of object visibility and access required.

_ SERVICES

objects introduced by the enterprise that make available a benefit to actors, such as a website, software tool, or physical store.

_ CONTENT

the way objects are represented in documents, communications, or other types of descriptions at a certain touchpoint.

DOMAIN-DRIVEN DESIGN

In the area of software development, a common practice is the construction of structural models as a basis for the implementation of a software solution, known as object-oriented analysis and design, or domain modeling. The goal of such activities, done either before coding or in parallel, is to capture the problem setting in a level of detail that permits an accurate representation in an information system, implemented separately from the other development concerns such as data storage or the presentation of information on a screen.

A key concept in that regard is the domain where that software will be applied. Any context of application can be a domain that applies to a piece of software: the structures, logic, and rules behind its function and behavior. The domain is driven by the concerns of business stakeholders and users, and the problem they want to solve with the software — examples for domains include industries such as banking and healthcare, but also personal settings such as managing a music collection, or the fictional world behind a role-play game.

In his book *Domain-Driven Design*, development expert Eric Evans introduced a framework of the same name. The core idea of DDD is that the domain should drive the design process, since it represents both the real-world environment where the outcomes will be applied, and the conceptual models in the minds of people. This paradigm is used to achieve an alignment between the problem setting and the proposed solution, on a fundamental level that is neutral to any technology or technical implementation paradigm.

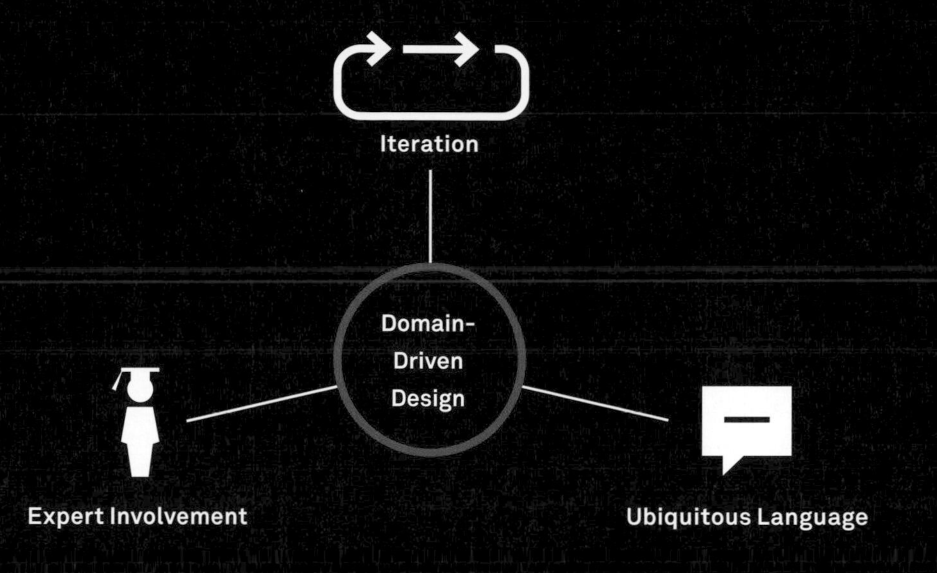

The DDD methodology is used for domain modeling in the context of complex software development projects, applying 3 key principles:

_ EXPERT INVOLVEMENT

any model is based on an active exchange between the designers of the solution, and experts in the domain. By discussing the details of problems and sketching out models together, they collaboratively generate knowledge about the things important to the design, and develop structural models to capture those findings. These models form the basis for developing a designer's model that fits.

_ UBIQUITOUS LANGUAGE

one of the key challenges of such interdisciplinary work is one of language. Experts and designers know about different things, have different concepts in mind and use different words. One goal of domain-modeling as used in DDD is to develop a common language based on the model, where every term has a single and clear meaning. This shared language is used in communication, in documents, and as a universal naming convention.

_ ITERATION AND REFINEMENT

there is awareness in DDD that no model can actually represent a domain as a whole, so that the choices made during modeling are in part analytic and in part subjective statements of how things should be. Instead of a single modeling phase, the domain definition is continuously improved and refined during the design process. This involves discussions and co-creation with experts, application in models and prototypes, and ongoing adaption of the model to reflect the latest thinking.

_ EXAMPLE

Although originally conceived for software development, Domain-Driven Design is applicable to any project that deals with the transformation or representation of things. In a project for the BBC, Mike Atherton and his team applied the same technique to structure a vast amount of content, reorganizing many of the website's main sections. Abandoning the tree-like hierarchy of many websites, it takes the connections between things from a user's model as the basis for navigation and page composition, creating pathways between related items. This way, the navigation model applied fits much better with the user's mental model, and allows for dynamically pulling together content into useful chunks. By starting from a domain model, the design process at the BBC were able to explore topics as diverse as food, music, and films, guiding the design process independent from the representation of content. Instead of pages and screens, the design team contemplated the actual role these objects play in their business and the lives of their audience.

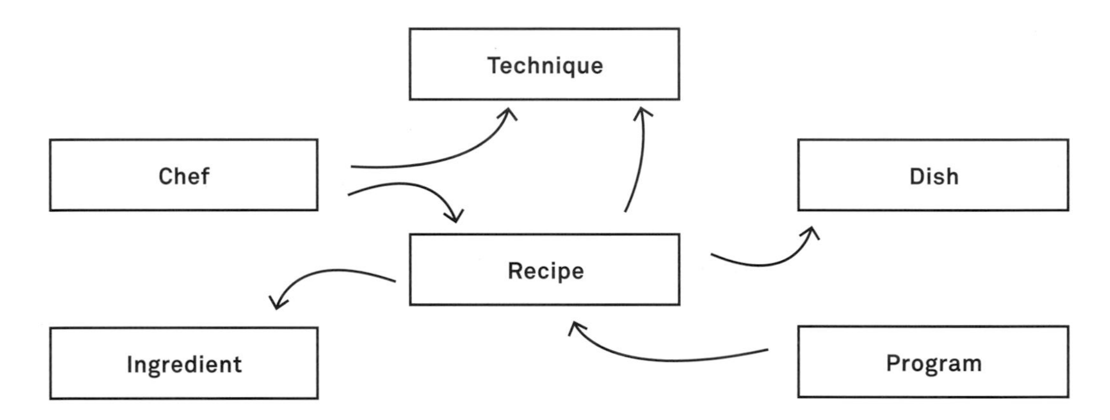

People think about the things that matter to them, not the documents or records that describe those things.

Mike Atherton, User Experience Designer

Designing with structure _

Modeling the objects relevant to a design challenge provides a powerful foundation for a design project. It allows us to document every relevant object and its relationship to other objects, and even to inlcude the potential results of the design process as a part of that structure — exploring how the world is represented in the design, and how its outcomes integrate into the world.

Applying design on an enterprise level means dealing with a problem setting that shares many characteristics with that of software development in a specific domain. Therefore, the core principles of DDD described by Eric Evans also apply to any difficult design challenge in the enterprise context. The enterprise can be seen as a special case of a domain — especially in large organizations, the jargon used, the constraints imposed and the hidden structures to be considered form a core area of interest for the designer. The creation of structural models in close collaboration with experts helps us find out about the issues, rules, relationships, and concerns.

The outcomes of a design project, all of the *things* that have been envisioned and produced, have to blend with the real-life context where they are introduced. The structure frame in Enterprise Design helps us explore that context and make structures visible, and to consciously decide on the objects introduced, their designation and attributes. This forms the basis for defining a new state that fits with the conceptual and mental models of people addressed with the results and other stakeholders.

_ DESIGNING WITH FRAMES

A key objective of the Enterprise Design framework described in this book is to apply a holistic and global view to the complex environment that design projects face in the enterprise. As opposed to purely user-centered, business-driven, service-oriented, or similar design paradigms, it aims to provide a vocabulary and approach without imposing a particular focus or main area of interest. That combination of non-focus and a wide range of aspects covered is based on the central idea of taking into account everything that matters, of overcoming the silos, biases, and preconceptions that appear naturally in any multidisciplinary team or diversified stakeholder community.

The four aspects described in this chapter are applicable as universal perspectives, to frame the problem to be solved and reach a global view. They appear in any design project at least in an implicit way, as the ideas and mental models of the project team, and normally they are at least partially made explicit in research findings, concepts, requirement documents, graphical models, or other types of deliverables.

The complexity of the enterprise context makes it worthwhile for us to consider each of them in a more systematic fashion. In essence, the frames help us model the enterprise as a system. Such models can then be used for exploration, synthesis, and concept development to transform that system during the design process.

Although one of the aspects is called people, in fact all of them are intrinsically human, relating to different but related concerns of the people addressed by or involved in the activities of the enterprise. They describe the same system from different points of view, exploring the enterprise as design context, envisioning the way outcomes might transform it, as well as measuring results against the expectations of different actors. They help us deal with the ambiguity of the design challenge by providing frames for both exploration and conceptualization.

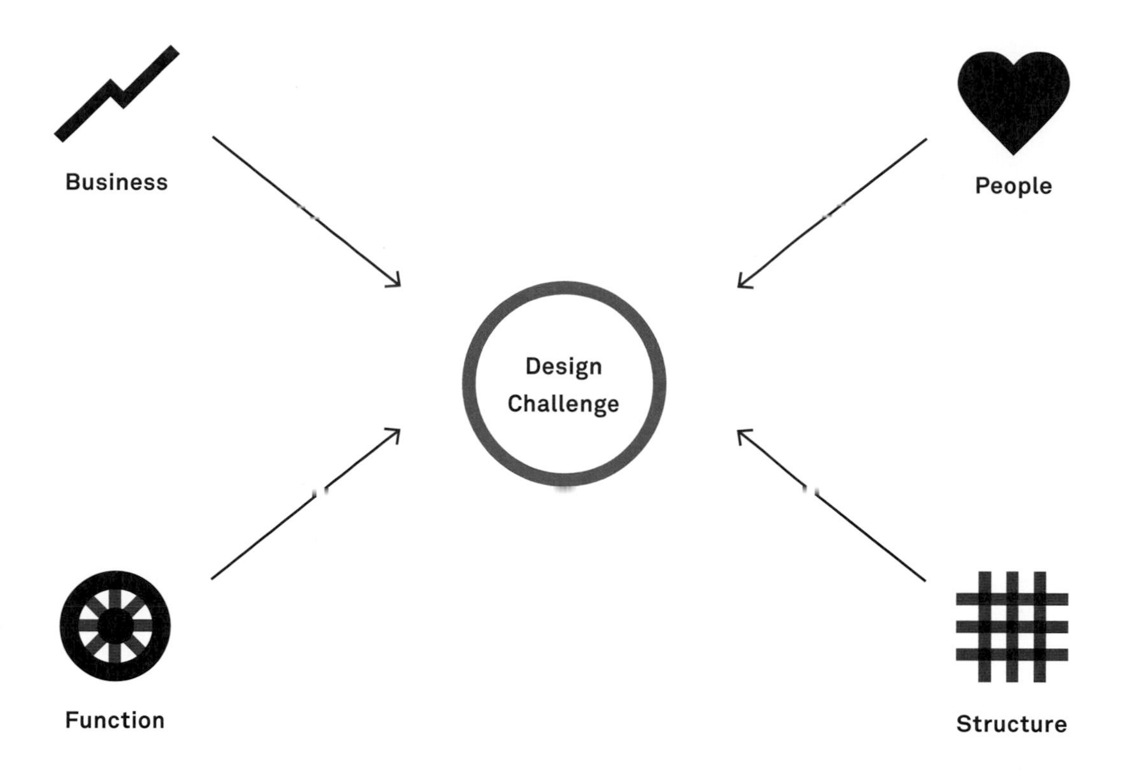

The four aspects capture the enterprise as a system, looking in detail at the different aspects relevant to a strategic design process:

_ THE BUSINESS FRAME

results in models of the enterprise as an economic system, which envision change in terms of business models, markets to address, and opportunities to seize.

_ THE PEOPLE FRAME

helps to create models of the enterprise as a social system, in order to develop insight into the lives of people impacted by a transformation and adapting outcomes to them.

_ THE FUNCTION FRAME

sees the enterprise as a procedural system, and allows modeling goals and processes, how the enterprise fulfills its purpose today and how it could work better tomorrow.

_ THE STRUCTURE FRAME

helps to model the enterprise as an organized system in terms of objects and their relationships, to envision an evolved system and make it fit into the context around it.

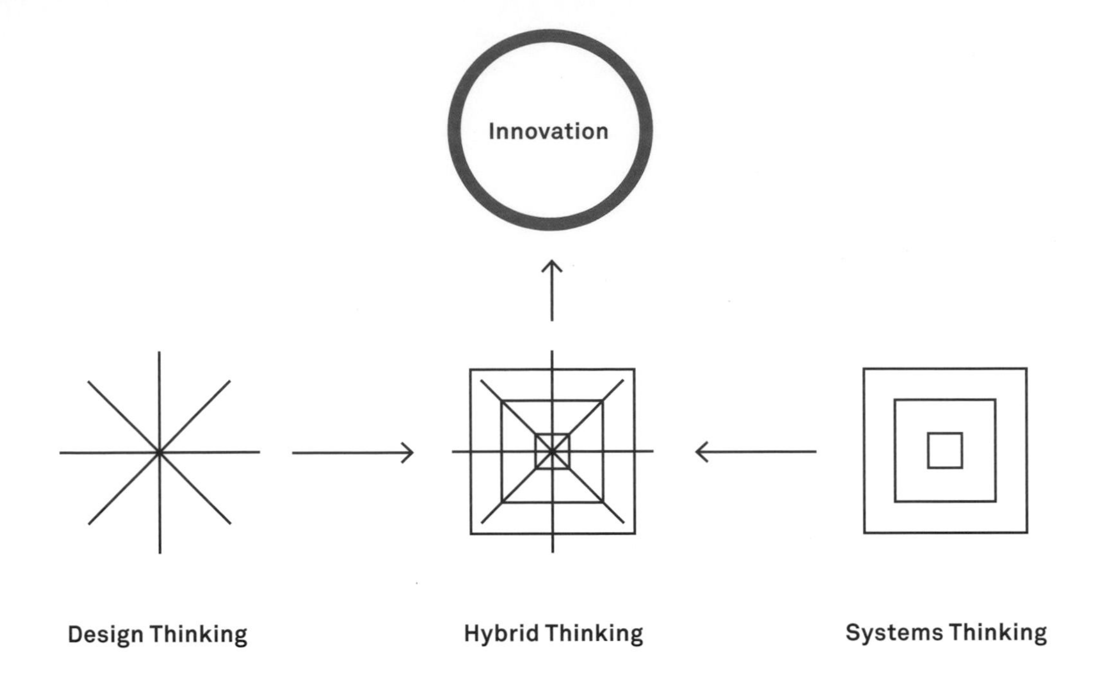

The aspects portrayed in this chapter are thinking devices for dealing with complex design challenges in the enterprise. Applying them requires constantly switching between different modes of thinking:

_ SYSTEMS THINKING

to explore constraints and dynamics, looking at a problem as part of a larger system and consisting of subsystems rather than in isolation, and defining the context and conditions of the enterprise as a system.

_ DESIGN THINKING

imagining the future of the enterprise based on various sources of inspiration and creative exploration inside and outside the given system, applying iterative refinement and collaboration to generate and validate ideas.

_ HYBRID THINKING

bridging both states of mind to synthesize opportunities for innovation in the enterprise.

Closing the gaps _

A strategic design project in the enterprise context rarely starts with a blank page. There is always some context and a range of external factors to take into account even when defining the problem to be solved. Applying a frame to a design challenge, such as one of the aspects portrayed in this chapter, is a common way to formulate the problem setting that is to be subjected to a design process. Frames are means to reduce the scope of an exploration or solution definition phase by consciously leaving out other aspects. A frame makes it possible to work on a potential future state, as a redesign applied to just that limited subset of elements and factors.

Because all frames consider the same elements, just from different viewpoints, any change in one model also implies changes in other views. Rather than building detailed individual models in isolation, an exploration of the the relationships, transitions, and dependencies between the different aspects provide the most interesting inputs to a design process. They enable the project team to push the boundaries of their work beyond a single frame, to encompass the greater enterprise context.

To close the gaps between the different models, aspects have to be approached by considering their relationship to each other, in a process of iterative inquiry and constant adaptation. This is applicable in all phases of the design process—framing allows us to explore the concerns to be addressed and the tradeoffs between them in research activities, generating ideas for potential solutions to address them while dealing with divergent views and needs, and validating them against the different concerns identified.

_ Design themes

Regardless of the frame applied, the choice of issues to be tackled and possible ways to go forward is a difficult one. In order to make progress, a key task is the identification of suitable design themes to guide the direction being taken. Themes are present in any design activity, in the ideas followed and the decisions made by the team, often based on a rather intuitive selection informed by client demand, personal priorities and research results. If made explicit, such a theme is usually expressed as a broad statement that conveys a very basic idea to drive a design process.

All design work is deeply concerned with choice and priorities. Working with the four aspects as thinking devices enables a design team to identify the major factors that drive the enterprise, being also the drivers behind a design initiative. Business concerns and market developments, issues, and opportunities discovered by immersion in people's lives, potential structural and new or evolved functions envisioned—these framed insights provide the necessary input to choose and articulate a set of suitable design themes for further steps.

Themes guide the activities carried out in a design project by imposing a strong core idea to the thought processes, creative work, and decisions being made. They respond to the major topics identified and prioritized during modeling—instead of attempting to deal with everything with equal measure, designing with themes means reducing the design choice according to priorities and central goals. It means addressing these issues with consistent principles and individual, dedicated solutions, while allowing compromises and generic solutions for other topics.

This is the basis for the six aspects introduced in the next chapter, a set of cross-cutting areas of choice in the enterprise which span the space for all conceptual design decisions.

AT A GLANCE

Designing at the enterprise level requires working in a complex space of underdetermined problems. It involves working with people with varying views and ideas of the future, understanding different conceptual worlds, framing the problem from different angles, and translating findings into principles that guide the design work.

Recommendations

_ Understand your client's business model and define how investing into a design process and subsequent transformation significantly contributes to business success

_ Understand the people addressed, their lives and what makes them tick, and develop empathy to make their lives better with your design

_ Create models of the functional context to be supported and conceptual structures that will play a role in your design, using these to map the enterprise now and in its desired target state

_ Work with models that apply different frames on the enterprise environment, to define the challenge to be addressed and to identify the central themes that span the design space

CASE STUDY _JEPPESEN
(May be Test Only)
SAN FERNANDO
114.5 SAN
NORTH SECTOR
MANILA
119.30
128.30
RP-IMTA 1
RP-IMTA 2
CLARK TMA (E)
SUBIC
113.5 SBA
CLARK
113.1 CIA
CABANATUAN
112.7 CAB
MANILA
113.8 MIA
JOMALIG
116.7 JOM
LIPA
115.1 LIP
LUBANG
117.5 LBG
MANILA TMA (E)
KALAYAAN I CORRIDOR
Ninoy Aquino Intl
RPLL 75-112
Baguio
RPUB 4251-55
TAREM
BIGBI
MALIB
ESCAN
ALPAS
NABAL
TADEL
VILAR
ZULUR
SILAG
KARAG
RINAL
OLIVA
NATIB
MINDA
SIRCA
RADEG
VOTAS
DILAN
BORGA
EXORA
KEERO
SUSIE
BETEL
INDAN
BALAY
TAALA
BANOS
TALIM
CAMBA
KABAN
PABLO
BUCAL
OLRAX
VERDE
GALER
TAPAP
LAIYA
CONDE
IPATA
TIMON
LABAY
ALABAT

JEPPESEN'S ENTERPRISE

Providing accurate, comprehensive and timely navigational information for safe and efficient routes in the air, on the water, and by land.

Jeppesen, part of The Boeing Company, provides navigational charts, real-time information, and operational services for the aviation, marine, and railroad logistics industries. The company was created over 80 years ago in Colorado, USA. Founder Elrey Borge Jeppesen was an American pilot, and the first to produce charts for in-flight use, still called *Jepps* by many pilots today.

As electronic navigation and planning systems are being adopted in virtually all areas of transportation, Jeppesen started a series of research activities to come up with digital solutions, reinventing and expanding their business model and defining new offerings. The company wanted to find new ways to process and adequately deliver data in navigational solutions, optimized for its context of use. The Advanced Research department commissioned eda.c to contribute to a series of research projects. Jeppesen wanted to adopt a design-led approach to work on new offerings, looking at new potential target states of their enterprise beyond technical shifts and industry trends. Because of the mission-critical nature of such systems, this involved a deep dive into complex system architectures, industry regulations, and international standards.

The task required the design team to work with a complex problem space of constraints and opportunities, both operational and technical in nature. In the course of our work on multiple strategic design themes, we made extensive use of modeling, shifting viewpoints to address different concerns. Such models informed the design process by capturing both the conditions of the enterprise environment and the conceptual thinking behind new products and services, and led to innovative design solutions. Today, Jeppesen is making design and User Experience practice an important part of their working culture.

_ BUSINESS

Applying a Business frame was essential for the design team to understand the value proposition behind the offerings. Jeppesen's services are characterized by their ability to gather, process, combine, and deliver data relevant to their customers' operations. Although targeted at very different markets, all offerings are fundamentally based on data collection from various sources, processing and enriching it to make it part of the company's high quality content.

In order to envision tomorrow's most compelling market proposition, the design team had to develop a full understanding of the business context. We reached out to various stakeholders in the company and carried out extensive desk research, looking at both existing offerings and external market developments. Major trends provided the background for some fundamental shifts in the market Jeppesen is addressing, such as digitization of data and systems, increased safety and quality requirements, and more coordination efforts between different regulatory authorities.

In parallel, Jeppesen's own research teams significantly developed their capabilities to make sense of more data, to bring it together more flexibly, and to use automated processing to enrich it with semantic layers and live information — while maintaining superior standards of data quality and process optimization. These techniques enabled them to combine data with layered information, tying together different navigational aspects for a range of channels, from onboard equipment to the iPad.

Jeppesen recognized that any business model renovation needed to put their unique data pool at its heart. It would connect airlines, shipping companies, and other customers to that consolidated base of high quality digital data, flexibly tailoring its delivery to the context of use. Instead of a cluttered portfolio of software products and digital documents, the technical components needed to get out of the way, becoming merely channels for custom data delivery services.

The considerations from a Business perspective led to the universal design theme of consolidated service interfaces, delivering just the information needed in a way that related to both business need and user task. It fueled a whole set of detailed questions, such as collaboration scenarios between the actors, service integration, or financial licensing models, to be tackled in the course of the projects.

In today's software world, highest quality ease-of-use and brand identity are key product differentiators. World-class User Experience and Interaction Design have turned into key competencies to be able to compete in this market.

Jens Schiefele, Director, Jeppesen Advanced Research

PEOPLE _

The navigational information from Jeppesen is used in highly critical environments—by crews on the flight deck of airplanes and onboard large ships, and also by staff on the ground and on the shore to facilitate the operations. All these use contexts require pilots, air-traffic controllers, navigators, and other users to obtain the information needed to make the right decisions. Consequently, most of the people we were designing for are highly trained professionals. They are executing a set of tasks within a well-defined operational setting, and in close collaboration with other actors in the ecosystem.

The fundamental premise of all design work at Jeppesen is therefore to adapt information display and functionality to the task at hand and tailoring information to the human condition. Only by looking closely at the way these people worked, at their preferences, habits, and routines, could the design team develop the understanding necessary to render the systems we designed useful and engaging. Jeppesen had adopted a formal Human Factors approach typical of the aviation industry, and possessed very detailed information about the operational environment and procedures. Through some intensive user research, both conversational and in context, the design team expanded this base with first-hand knowledge about the tasks, user characteristics, and the physical and social contexts of the activities.

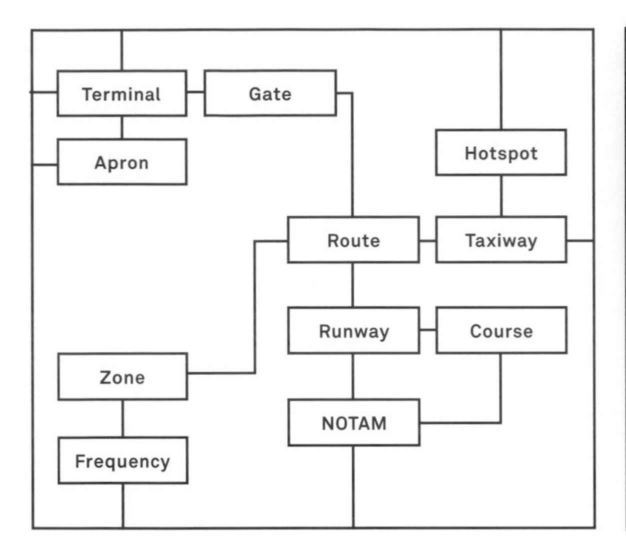

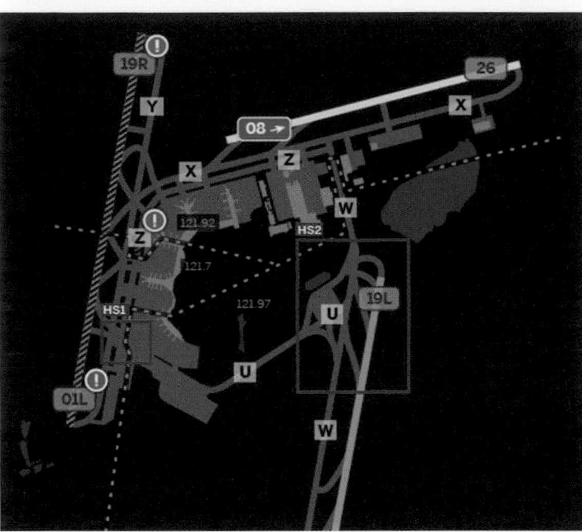

Structural model on airport layouts, informing different design outcomes such as the graphical representation in a map view

This immersion enabled the team to develop models from a people perspective. Although the check-lists and procedural descriptions of the tasks were quite accurate, research results showed a different picture. The models captured a multitude of interrelated tasks and tools, ad hoc decisions, information overflow, communication, and collaboration—aspects of a complex human reality, simply impossible to formalize in prescribed procedures.

They led the way to a set of key design challenges to be tackled in the subsequent design process phases. Issues of exception handling, de-cluttering information, notifications, and transitions between touchpoints were identified as the most relevant problems.

We designed for an environment where the center of attention must constantly shift between different parts of a complex system landscape. People adapt to this complexity by developing working habits, mental scripts, and strategies to deal with exceptions. To make their systems play a key role in that setting, design must aim beyond usability and usefulness. While these goals remain relevant, the goal shifts to designing solutions that people actually like to use.

FUNCTION _

As a technology-driven company in a highly standardized market, Jeppesen has a long track record of capturing requirements and specifications related to the functional aspect of complex systems. All research projects involved the creation of a functional specification based on input from the business stakeholders, human factors specialists, and technologists. Such a document describes the purpose and functional context of goals and processes, and external factors relevant to the project.

In the course of the projects, the design team was able to make the functional goals a key part of the design process. The design work also incorporated different types of behaviors into a single model, mapping user activities, operational procedures, and business processes. This consolidated model, applying a function frame, resulted in a continuous alignment between the functional specification and the further modeling and design activities.

The key purpose of Jeppesen's navigational solutions is to support situational awareness. This goal led to the principle of data-driven functionality, automating everything non-critical as much as possible based on the system's knowledge of the current situation. Instead of more features and functions that add to the complexity, the systems' behavior should be driven by external variables, and only trigger user interaction when needed. The background information about mission-critical tasks provided the input needed to incorporate goals of error prevention, system redundancy, and exception handling.

The extensive collaborative work on the functional scope of each project enabled the design team to agree on an adequate set of functionality, and to generate a Design Space to work in. Every vision concept and prototype elaborated could be tied to a specified functional context of behaviors and the goals that it supported, and informing the specification of functional components, their interplay, and the modeling of user interactions. This process was based on the joint creation of models, describing both potential behaviors of the systems to be designed and the functional constraints of the environment they would have to fit into. By aligning the functional scope with the input from the other frames applied to modeling, we could reduce the functional scope to give users the information they need instead of making them deal with features.

Manage inventory & updating	Plan a voyage	Watchkeeping (reports, observations)	Deal with emergency, contingency, exceptions
Update charts	Check available charts	Look up/verify position	Man over board
Mark corrections	Find/add a port	Transition position	Trigger emergency state
Register chart	Plan a route	Calculate gyro error	Adjust to bad weather
Update weather	Edit/split/join route	Calculate magnetic course	Break off voyage
Check credit	Plot a detailed route	Make observation	Adjust to piracy activity
Update piracy	Plan reporting	Make log book entry	Rescue mission
Check inventory/portfolio	Perform calculations	View current chart	Contact security officer
Browse inventory/portfolio	Create/edit voyage plan	Look up chart coverage	Contact emergency officer
Replace charts	Calculate route	Trigger new phase	Security

Functional model diagram repartitioning different levels of activity as a basis for requirements elicitation

_ STRUCTURE

The design work for Jeppesen required the design team to get very familiar with the world of aviation and marine. In our quest to develop a holistic understanding of the things that matter for our projects, taking a Structure viewpoint has proven particularly valuable. Capturing the domain in structural models serves several purposes in parallel.

To design a system for navigational information, it is imperative to understand how that system will fit into its context. This includes other systems — in our case software components, communication and control devices, background systems, and similar technical parts of the overall systems architecture. But it also, and perhaps more importantly, includes the objects that make up the mental objects users care about, and that have to be represented as information to be conveyed.

Because of the mass of objects to be considered, we started with simple lists of things that we found to be described in documentation, used in existing solutions, or included in technical data models. To develop domain models as the basis for further design work, we chunked the domain based on task models derived from the other perspectives, and reduced the selection of objects to those relevant to that task. We collaborated closely with domain experts on different navigational and technical topics, such as weather, routing, or traffic control. In parallel, we validated the models with users to ensure they represent their users' models, and iterated them alongside business stakeholders and technology experts.

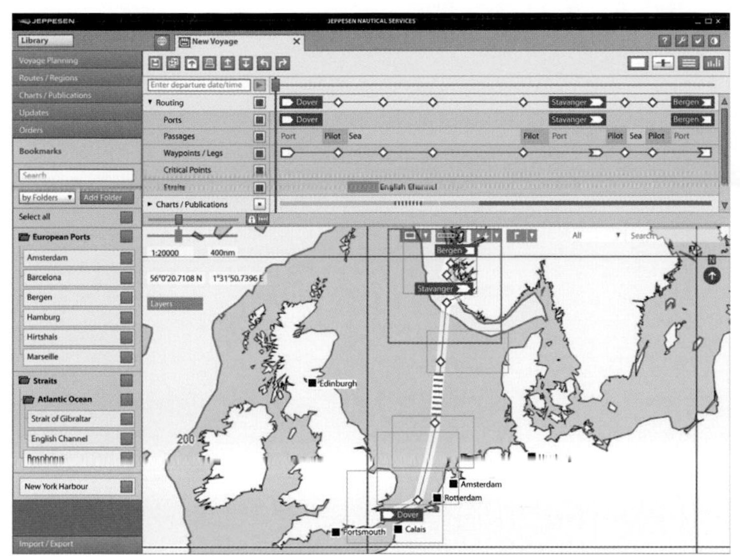

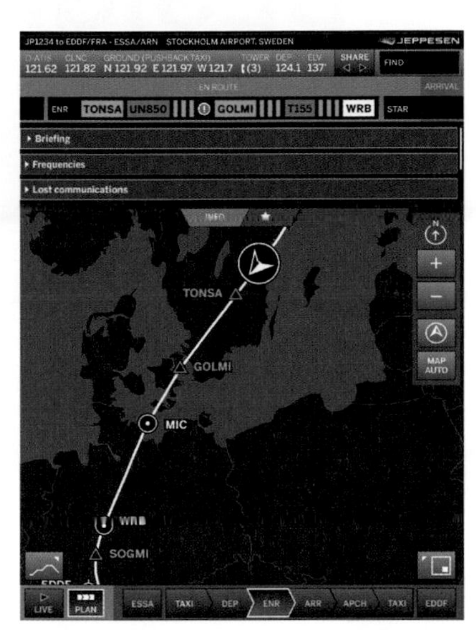

Digital solutions to support route planning and navigation for aviation and marine

By exploring the relationships between the different objects, the design team was able to develop a consistent picture of the domain, limited to those parts to be incorporated in the design. We developed a set of designers' models, regrouping and re-relating objects to make them fit to the tasks and information needs in different use contexts.

These models became the basis for choosing domain objects to be shown on screens or using other media, for representing them across geographical, time-based, or task-based depictions. By taking a structure frame to look at the domain elements isolated from existing solutions or abstract functional requirements, we were able to shift the thinking away from today's practice. The design team found new ways to combine and connect navigational information, and to arrange it in a way that makes sense adapted to individual situations.

7_ DESIGN SPACE

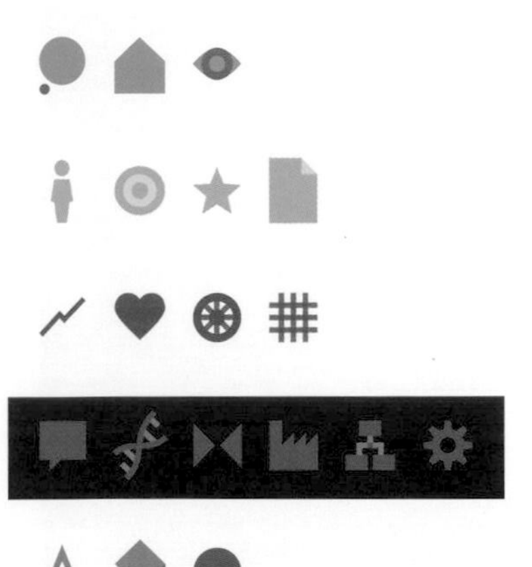

Any strategic design project in today's enterprise environments has to deal with the challenges of a dynamic, uncertain environment. Shifts of an economic, political, technical, or social nature are forcing organizations to reinvent their business models in ever faster cycles of change, to transform themselves into new configurations, learn new capabilities, and address new markets. The ecosystem they address and thrive in is constantly changing, forcing the enterprise to adapt. One of the major hurdles to coming up with a strategy that responds well to new conditions and opportunities is the issue of coherence — aligning the internal activities with external market reality, and thinking beyond individual departments, professional disciplines, or competences.

Design applied to strategic purposes can be seen as a practice of developing just such a coherent a plan for the active part of change, of deliberately introducing new things into a given situation with the intention of transforming it to a desired state. In general, design is about arranging and reshaping the world in a way that it makes sense to people and stakeholders, as business models, as products and experiences, as units of structure and function. Instead of attempting to analyze and optimize the existing, it promotes combining creative leaps with informed synthesis to a discover and make visible the desired target state.

A deep understanding of the environment and the driving factors behind a strategic design project from different perspectives, as described in the previous chapter, is the basis to get a clear picture of the challenge to be solved. It enables us to choose which themes to address, and usually also initiates a number of ideas for potential outcomes. At this point, the focus shifts from inquiry to conceptual thinking.

© 2013 Elsevier Inc. All rights reserved.

In essence, those framed models deliver the elements to be included in a desirable future state of the enterprise, defining it as a vision to be executed. What is missing from that vision is a definition of the way those elements work together as a system. To come to such a coherent vision, design teams have to bridge multiple divergent views and cross-cutting concerns, by aligning the different frames and mapping their intersections. The following six conceptual aspects help us to achieve such an alignment, to pave the way to defining the outcomes of a design initiative.

This chapter describes six interrelated aspects which together represent the space of conceptual decisions about a potential future state of the enterprise. Those aspects are information, interaction, communication, organization, operations, and technology, representing significant fields of innovation in strategic design. Each of them corresponds to an area of strategic thinking and alignment with the overall strategy for sustaining and developing the enterprise, and helps us respond to important external trends and identify internal opportunities.

Any transformation of the enterprise means triggering a shift in the way these aspects are addressed, envisioning new ways of doing things and dealing with change. They act as bridges between the different frames applied, to align diverging views and stakeholder concerns and bring their interests together in a common conceptual vision. Depending on your background and experience, there are probably many considerations described in this chapter which you already address as part of your professional practice — whether or not you would call what you do "design work." Others are perhaps less familiar or generally seen as outside the scope of your work.

When aiming for a strategic impact on the enterprise level, the interrelationships and dependencies between these aspects make them all relevant to design work. In this framework, all aspects are based on connecting and aligning at least two framing aspects from the previous chapter. This means that the conceptual vision must be achieved by framing the problem, deciding on how to design according to different aspects, then applying these decisions on a system of interrelated outcomes.

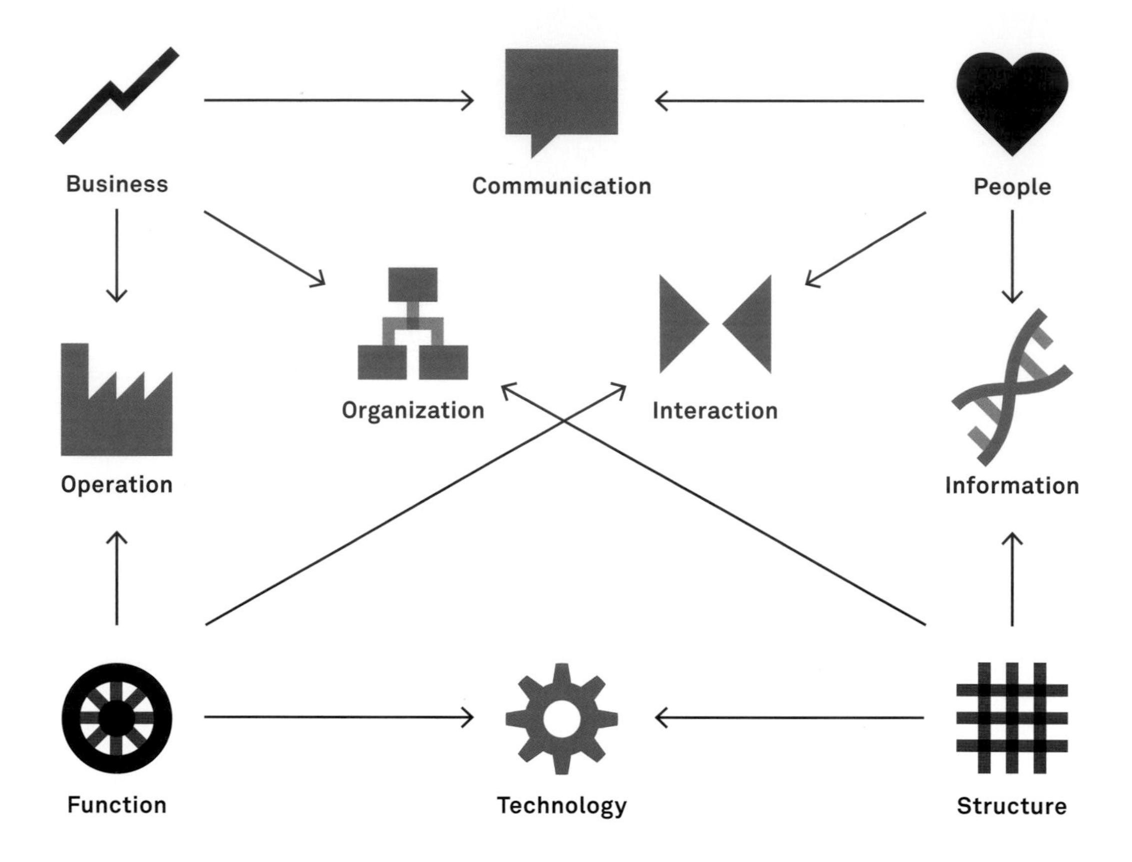

The conceptual aspects to be addressed in the Enterprise Design framework are the following:

_ COMMUNICATION

is about the ongoing exchange between people in a business context. In today's design work, designing communication — in terms of both messages and media — is heavily influenced by the interconnectedness of digital channels, with Internet and social media having turned markets into spaces for rapid and dynamic conversations going on in the enterprise.

_ INFORMATION

is being made available to people about the things that matter to the enterprise (its structure). Designing information is about providing the right things to the right people at the right time, also influenced by the development of digitization and the difficulties of drawing out meaning and gaining knowledge from ever more quickly growing mass of data.

_ INTERACTION

is about connecting people to functions they are using in the enterprise realm. Designed interactions are omnipresent in the enterprise, where virtually no activity is carried out without the help of digital technology and tools. Shaping interactions as behaviors therefore is a basis to define and design useful tools, which are supposed to facilitate them.

_ OPERATION

refers to the way the enterprise works and carries out its activities, making use of the functions as capabilities playing a role in ongoing business processes. This aspect is about designing both automated procedures and human work, striving to rethink the way work is being done in terms of both internal efficiency and customer value, defining the flows and drivers behind work processes.

_ ORGANIZATION

is about the structure of a business, put in place to allocate responsibilities and tasks across organizational units and roles, and building and managing teams. Strategic design projects require us to take into account the shape of formal organizational structures, skills and competencies, and their influence on emerging team culture and working habits.

_ TECHNOLOGY

is a means to enable the delivery of enterprise functions, facilitated by the structures put in place to support them. Nowadays any design project depends on a creative usage of technology, playing a vital role in most human activities to be addressed. Designing technology means leveraging technical possibilities for human usage, and shaping it in a way it fits.

12_COMMUNICATION

Markets are conversations.

The Cluetrain Manifesto

The arrival of the internet and new forms of digitally supported conversations has transformed and amplified our communication habits. Formerly separate channels of mass communication such as TV and print are converging with instant one-to-one channels such as phone calls or email. New communication platforms appear for specific forms of private or work-related, instant or continuous communication, for various purposes from team collaboration or instant messaging to online dating, auctions, or sharing pictures.

While some communication platforms already play a vital role in the lives of millions, there is a large number of much smaller and more specialized communities and tools appearing. The Internet and its various communication tools provide the user with an enormous choice of platforms and media, and hundreds of competing modes of communication with other people. These communication spaces are closely related and intertwined, with messages transcending the boundaries between them by repetition, syndication, and transformation. People select from those spaces for a particular communication activity based on goals and habits, heavily influenced by local culture and specific group conventions.

Today, we see organizations of all kinds struggle to adapt to this new reality. Traditional Corporate Communications, Advertising, and PR practice has focused on unidirectional mass communication and its indirect influence on reputation and sales success, and has considered the dialogues among individuals as out of scope. The advent of dynamic digitally supported communication spaces removed the boundaries between mass and personal communication, resulting in connected and converging media. In a way, modern enterprises can be seen as a communication space much like communities and platforms on the web—a dynamic community of participants exchanging with each other, across multiple contexts, tools, and communication modes, and with varying reach.

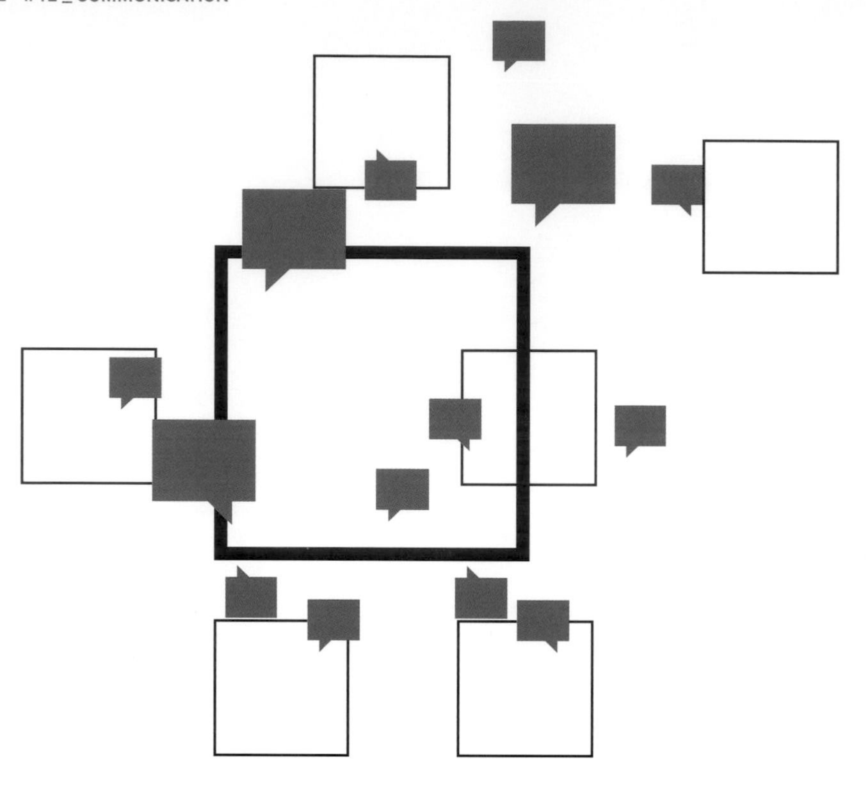

This aspect is about providing the right communications using adequate and useful channels. In the enterprise context, this translates to the fundamental challenges of appearing and participating in these exchanges, as well as facilitating communication with adequate channels and modes. Any business success is the result of communication processes, these being the prerequisite for actual transactions to take place. This requires a dialogue involving customer communication and Marketing, internal collaboration as well as coordination with all suppliers or business partners involved, making the case to invest in modeling and designing communication in the enterprise. It also requires constantly monitoring what is going on in that space, and actively and coherently participating in conversations when they are happening.

_ EXAMPLE

To send a message to a friend, people have the choice of a large variety of communication channels and modes, from giving a call or sending a short message or email, to private or public messages via Twitter or more special forms such as a greeting card, or even sending a letter on paper. The channels available and considered appropriate vary depending on the context. For example, many people in Japan never had to distinguish between mobile short messaging (SMS) and email, since sending and receiving email was supported by most phones much earlier than in other countries. Building on a service similar to email, Facebook allows invitations for events, supporting tailored interactions such as RSVP. However, the complexity of this function made people accidentally invite the general public, turning private birthday parties into overcrowded mass events and causing the German government to consider prohibiting Facebook events altogether.

Modeling channels and messages _

Based on a large body of research in semiotics, social sciences, and communication theory, models that capture communication processes describe both the nature of the dialogue and the community of participants engaged in it. They help designers for developing an understanding of the ongoing or potential conversations, topics and trends, social interaction modes, and communication needs. Communication models usually describe cycles of social interaction and feedback loops. These models provide a basis to develop a communication strategy by envisioning a future dialogue, and formulating the messages to convey. They capture the enterprise as the business context for an ongoing dialogue between people. In such a model, the social dimension is not regarded as separate from the purchasing decisions and transactional business activities, but as the driver behind them. Designing effective communication channels and good messaging requires us to understand the potential social structures and communication processes, based on insights about people and how they relate and interact.

The communication aspect in design by nature deals with a complex environment where effects are hard to predict. With the myriad of communication modes available today, the design challenge moved from developing and conveying a message to leveraging and redesigning media that effectively facilitate social interaction, while at the same time supporting the goals behind a design initiative. In essence, models about the communication process should not only illustrate how to convey a message to a target audience, but also define the channels and media to facilitate a dialogue and engage in it.

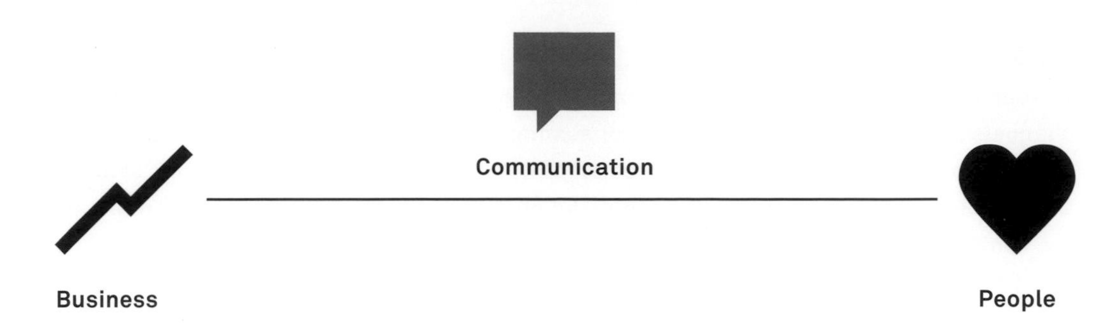

The communication aspect in the Enterprise Design framework is driven by two frames portrayed in the previous Chapter:

_ PEOPLE

the foundation of designing for communication is learning about the intended audience to engage, inform, persuade, attract or inspire. Knowledge about identity, interests and motivations, relationships and groups allows tailoring channels and contents. Further, a deep study of the participants of the ongoing dialogue reveals insights about the cultural background that applies in terms of social norms and habits or the repartition of power and influence.

_BUSINESS

for all activities going on in the enterprise as a functioning ecosystem, communication processes are the basis. Applying the business frame allows mapping that ecosystem as a market space and capturing the dialogue behind the business transactions. It enables us to understand the economic motivation of market players, and to envision a communication strategy that connects the right actors, positions offerings to the market and facilitates relationships and trust.

In 1948, political scientist Harold Dwight Lasswell formulated his famous model of communication, serving as the archetype for many models to follow:

Who says what to whom in what channel and with what effect?

_ COMMUNITIES / WHO OR TO WHOM?

modeling community structures describes groups addressed with and involved in communication activities, and apparent social roles. Design projects have to identify and address leaders, followers, moderators, considering interests and motivations, sub-groups, barriers, and relationships.

_ MESSAGES / WHAT?

the messages communicated, encoded in signs and their meaning (semantics), relationships (syntax) and pragmatics (interpretation). Design is about facilitating reception and reaction, considering media and message format, style, metaphor, and narrative.

_ MEDIA / IN WHAT CHANNEL?

media are the basis for a communication process by providing means for senders and receivers to engage in an exchange of messages and feedback. In that context, design helps to support the choice of media by identifying and providing media platforms to engage in, and applying suitable interaction modes.

_ CONVERSATIONS / WITH WHAT EFFECT?

communication is always aiming to produce an effect on the intended audience. Modeling conversations allows describing how media are used to receive information and again participate in an ongoing process. Such a model captures how social interactions among participants are happening over time, including triggers for activity and feedback loops.

	Real time	Asynchronous
Same Place	Face to face	Continuous interaction
Distant	Remote interaction	Communication and coordination

To describe communication media, it is useful to distinguish between different fundamental modes. A model developed by Robert Johansen in 1988 to describe the circumstances of Computer-Supported Cooperative Work (CSCW) captures the different contexts of communication in time and space:

_ FACE-TO-FACE INTERACTION

participants are in the same place and interact directly — think of a team using a white board installed in a meeting room.

_ CONTINUOUS INTERACTION

participants share a place but frequent it at different times — for example a war room or a shared blackboard.

_ REMOTE INTERACTION

participants are in different locations but interact in real time — such as phone calls, virtual meetings or other live channels.

_ COMMUNICATION AND COORDINATION

participants are in different locations and interact asynchronously — think of web-based media such as discussion boards or feeds.

Built and sustained by communication, "social" emerges because people like and get along with each other. That's what the water cooler is for, and why the water cooler is no sooner going away than is the oasis that punctuates the barren passage.

Adrian Chan

Designing conversations _

As communication is a fundamental part of any human activity in the enterprise, strategic design projects depend on getting that aspect right. Regardless of whether the intended outcome is a product or service, internal change or other types of transformation, success depends on engaging and involving people. Therefore, both the channels used and messages conveyed are subject to conceptual design, keeping the intended audience and their conversations in mind. In that context, designing communication involves mapping the enterprise as communication space, identifying existing platforms and adequate modes of interaction, and extending these platforms with media tailored to particular communication needs.

The term Communication Design is used in different ways and with various meanings across the design community, from being just a synonym for graphic design or visual communication, to a systemic initiative of defining the entirety of media and messages in a given social environment. At the core of the disciplines are the encoding and conveying of messages, such as in advertising campaigns, adapted to the now scarce resource of human attention and comprehension.

Designing communication applies in large-scale product launches as well as on the level of projects or teams. Both approaches, the traditional and the social model, are valid for designing communications. Instead of being applied in isolation, they are increasingly converging, with conversation topics being triggered by large-scale communications requiring organizations to be prepared to follow up in individual dialogues. This comes with a certain loss of control and the need to take into account the reality of social dynamics in communication, but also contains the opportunity to establish real links to people important to the business.

While the classic practice of Communication Design is based on a model of mass communication, evolved communication strategies take into account the new reality of hyper-connected people and markets.

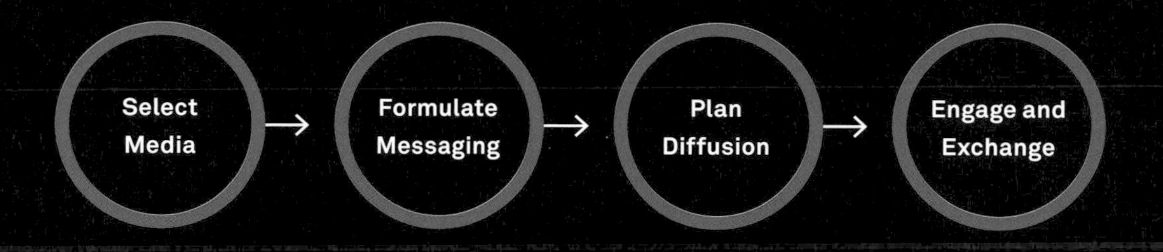

Depending on the project, the steps of a Communication Design process may involve different activities:

_ SELECT MEDIA

selecting and designing adequate media and modes as the environment for the communication process to happen. This may involve mass or personal media, visual or auditory communication, print or interactive, classic or new tools such as annual reports or social network platforms. It might also involve inventing and developing entirely new platforms for social interaction.

_ FORMULATE MESSAGING

designing and producing the content to convey to the audience across the chosen range of media, including a narrative basis, defining the key messages, and elements of the communication that vary according to different audiences. This messaging concept is the starting point for all further conversations.

_ PLAN DIFFUSION

publishing the content and bringing it to the intended audience, defining the timing, frequency, and reach of the communication activities. Depending on the individual strategy, this plan might be centrally organized or building on network effects, be planned long in advance, or be adapted based on opportunities.

_ ENGAGE AND EXCHANGE

preparing for the impact of a communications initiative. This might involve collecting metrics or feedback and adapting communications accordingly, or engaging with participants in personal exchanges according to a set of rules and incentive mechanisms.

_ COMMUNICATION IN THE ENTERPRISE

_ IDENTITY

how to encode the brand message in communications in a way that it supports organizational identity, and building trust and reputation in the marketplace.

_ ARCHITECTURE

how to support communications with adequate and useful media and technologies, responsibilities, and operations.

_ EXPERIENCE

how to design media and messages to reach and engage people, and to adapt to their communication needs and interests.

_ ACTORS

who to address with communications as audience for directed messages and participants contributing to an ongoing dialogue.

_ TOUCHPOINTS

where and when to deliver, tailor, and format messages to participants or recipients, and how provide adequate ways to respond.

_ SERVICES

what communication processes and platforms are needed to support the ongoing service provision via the touchpoints.

_ CONTENT

what are the messages to be exchanged in communication processes from the different actors.

The enterprise as a bazaar _

Designing for communication in the enterprise requires developing an integrated approach for all messages and conversations, including formal and informal communication, small and large scale, short and long term. This applies to the exchanges with market players as well as to the internal communication processes of organizations involved.

In some ways, the enterprise is like an oriental bazaar. Different market players promote their offerings and interact with each other to agree on doing business, with communication being the mediating force between them. The mechanics of such an environment are complex and interwoven—price, placement, and decisions are all dependent on to the dialogue between customer and vendor.

Dealing with the communications aspect in a holistic fashion means designing how the enterprise talks to itself—people involved in its activities—and about itself in the markets addressed. The established forms of advertising campaigns and classic corporate communications will prevail, but merely as a means for reaching a large audience to get the actual conversation going. With today's large and hyperconnected communication platforms, these conversations are subject to complex social dynamics and network effects that are hard to predict. There have been numerous cases where companies were surprised by the speed, reach, and consequence of social interaction. Designing communication enables organizations to participate and have a say in ongoing conversations in the enterprise.

#13_INFORMATION

Information is constantly becoming more relevant in our daily lives. With the shift from print media to digital information and the Internet, we got used to being able to learn just about everything in an instant. But beyond the classic range of digital media, information is now available on a large variety of devices across all our activities and contexts of use, from the smartphone we use while on the move, to the projected interfaces and augmented reality used in professional settings. Essentially, information is everywhere, with tools like Google pulling it out of a hidden space whenever we want it. While this development in many cases makes life a lot easier, the sheer mass of information and its ubiquitous availability across touchpoints multiplies the challenges of organizing it in a way that it is useful to people.

In the enterprise, virtually everything being done requires the acquisition and processing of information. Management decisions, Marketing communication, customer purchases, collaboration between various actors and daily business operations: all activities obtain, use, and produce information, and both human and automated decision making depend on its availability and quality. The acquisition, processing, management, and consumption of information form the basis to run and develop the enterprise.

The challenge of dealing with information is particularly present in today's business environment, as information and data can prove quite difficult to tame. In many cases, it is scattered across the organization, managed in divisional silos, and reproduced or reorganized on local levels. At the same time, information is floating outside of organizational boundaries and controlled systems, being shared with customers and business partners in the wider enterprise environment.

In their book *Information First*, Roger and Elaine Evernden argue that in order to address that challenge, organizations must treat information as a business resource, much like capital or labor. They require expertise and strategic thinking to use that resource as part of a business strategy, and to leverage its potential. More than just data to be used in operative processes, information must be seen as the essence of all decision-making and knowledge-building efforts in the enterprise, something that must be adapted to the people using it and interacting with it.

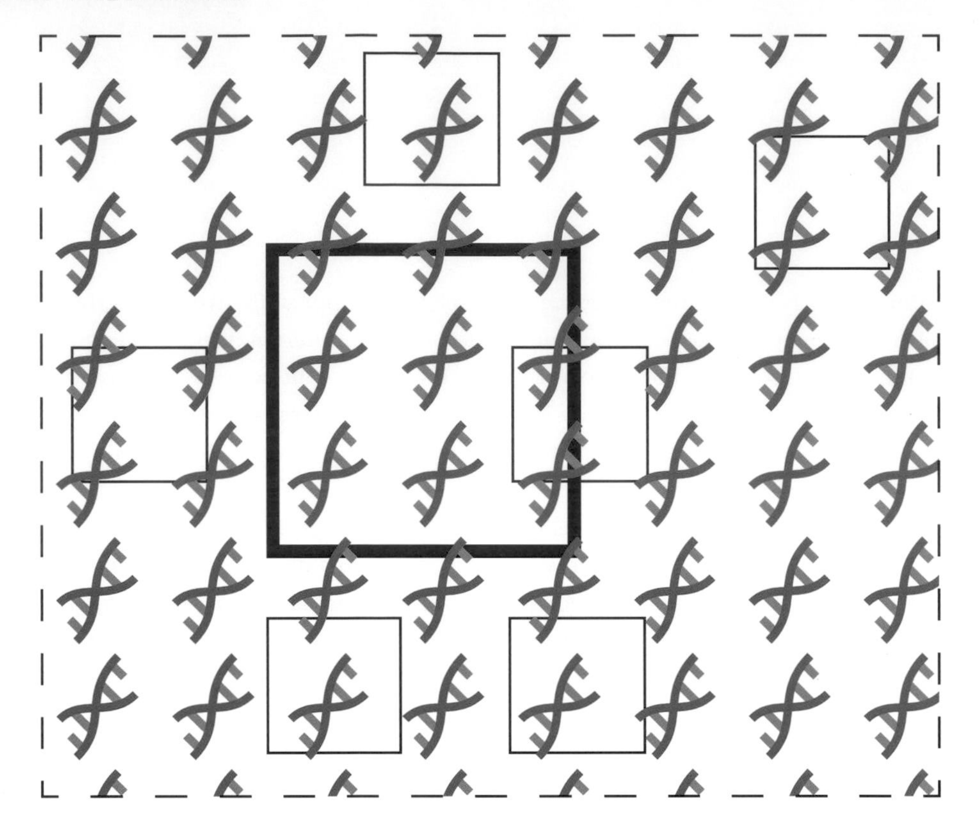

In a world where digital information systems capture data about everything that happens and store an overwhelming amount of content, it is clear that just adding more information to the pile not only provides little value, but can actually make things worse by contributing to the mass that is already there. Already, people are struggling to find and filter the relevant bits out of the general noise, and the knowledge needed is often hidden like a needle in a haystack. Therefore, all structures and systems have to be based on real information needs, retrieval behaviors, and usage patterns.

The information aspect in the Enterprise Design framework is meant to support designers dealing with that challenge — the problem of making available the right information to the right people, in an appropriate way and format, and at the time they actually need and want it. Instead of further building up the haystack, the design challenge with regards to information is to help people find the things they are looking for, and to provide information that is relevant in context.

Modeling information as design material _

In the broadest sense, information represents the world around us, helps us making sense out of it and develop an understanding to inform the choices we make. The words, labels, categories, and attributes we use to convey messages in conversations reveal the large diversity of organization systems underlying our information usage. Information is chunked into groups of genres, industries, quantities, time, or geography, all shared concepts and categories used to facilitate access and enable comprehension.

Applied to a design project, the information challenge translates to a problem of organizing it in a way that makes items findable, comprehensible, and useful to different people. This requires defining, structuring, and classifying bits of information, turning them into nodes in a network of connected elements. As Jesse James Garret points out in his book *The Elements of User Experience*, such a node can be as small as a single number, or as large as an entire library, depending on the individual context of the design project. If done well, such a classification system answers people's real questions, shows the way to the next item of interest, and makes information accessible according to actual needs, adapted to the respective context of use.

Based on a deep knowledge and models about both the audience to be reached and the structures to be represented, information can be designed to fit. Organizing principles applied to information are agnostic to particular choices of technologies or media, and can be applied regardless of the individual medium used to convey it. The system can be used for signage in a store, a navigation menu on the web, chapters of a printed catalogue, or something different. In order to create define how information nodes can be accessed, designers have to externalize that structure in models, describing the logic of the underlying classification system.

Modeling information is the basis for designing classification systems that actually work. Depending on the individual activity, goals, and context, the strategies people use to seek information vary widely, making the case for a flexible and adaptive classification system. The organizing principles applied therefore have to facilitate different modes of access and usage to support active and passive, directed and undirected behaviors with adequate means of information access.

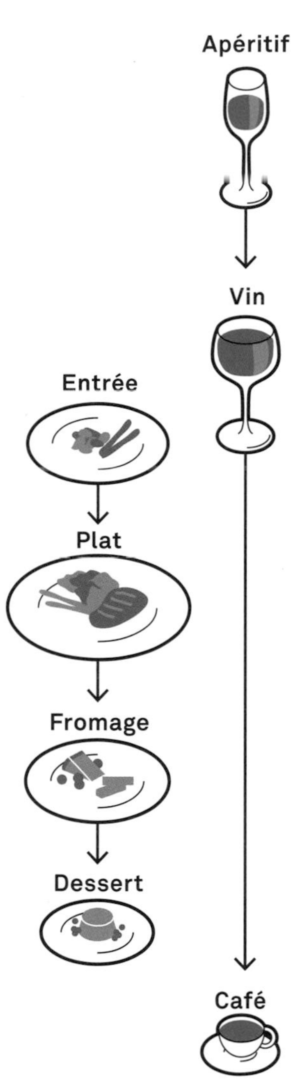

The French dinner classification system

_ EXAMPLE

Systems organizing information are all around us. Some are subject to entire fields of science, such as biological or chemical systems, while others are highly personal. As an example, think about a large music collection. The way to label, sort, tag, and group items in that collection will depend largely on the way the person doing it will listen to music. A fan of classical music might use periods, while someone into jazz or heavy metal could dive into the subgenres. Some people prefer to listen to entire albums; others just shuffle through the songs. Some might pull out songs for specific occasions, moods, or any combination of the above. Whatever system is used, the organized information about the actual items—the songs or albums in the collection—supports its creator in accessing and consuming his or her music. This challenge multiplies when designing an organization system for a larger audience, such as for a store or online music service.

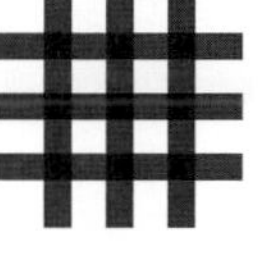

The information aspect in the Enterprise Design framework is driven by combining insights from two aspects portrayed in the previous chapter:

_ PEOPLE

following the objective of providing information that fits to people's needs and questions, the insights from applying the people aspect are the starting point to classify and design information. Design uses information to educate, enlighten, or advise people, requiring designers to learn about their information acquisition and consumption behaviors, and to classify and organize information nodes accordingly.

_ STRUCTURE

the structure aspect captures the things that matter to the enterprise and people in touch with it, and their interrelationships, based on expert involvement and knowledge. Any piece of information is representing a subset of entities which are part of that structure. Therefore, structural models are delivering the basis for organizing information nodes, capturing the underlying domain in terms of entities, language and labels, and relationships and rules to apply.

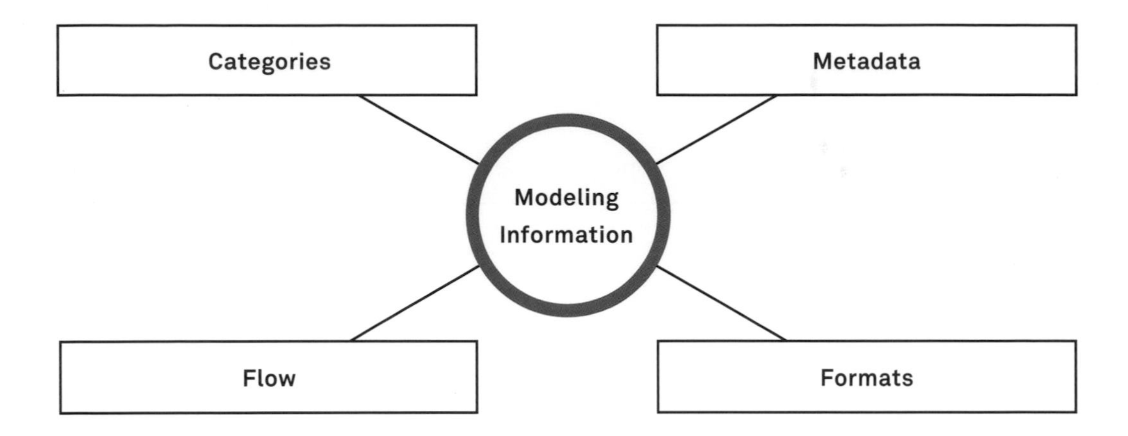

There are a number of different methods to model information, all aiming to formulate structures guiding its organization:

_ CATEGORIES

help distinguish one object represented in a model from the others, and provide the basis for consistently grouping and labeling information nodes according to their topic. Used to guide people to containers of information nodes, a single node might be assigned to a single (hierarchical) category or multiple (faceted) categories.

_ METADATA

is information about information—the attributes describing it according to different dimensions of classification, such as author, time of creation, genre, size, or tags. Modeling metadata is the basis for making information searchable, or to pull out related nodes in context.

_ FLOW

models capture how information moves around in a system of some kind and permit to describe their creation, modification, and usage. They are used to study how information is acquired and processed to make decisions, and how it evolves over time.

_ FORMATS

describe the different kinds of information in a system, the type of media, and the level of structural formalism applied. Models of formats enable us to define the way content is made available, and applying this to the definition of structures, styles, and templates.

INFORMATION ARCHITECTURE

Certainly, an Information Architect will define the meaning of every element on a website. Because the question of meaning is at the very heart of Information Architecture.

Sylvie Daumal

As a professional area of knowledge and practice, Information Architecture deals with structuring of information, and the systems and environments that are its containers. Coined by Richard Saul Wurman, the term has been taken up by Louis Rosenfeld and Peter Morville in their book *Information Architecture for the World Wide Web* from 2001. Beyond a practice applied to websites or business information systems, there is a wide range of schools influencing information architecture practice, ranging from computer science or informatics to library science, linguistics and visualization. In the end, all these different approaches share the goal of leveraging the potential of information and data, turning it into a useful resource, accessible and findable for those who want or need it.

While classic Information Architecture deals with systems such as websites or databases, the professional practice had to move beyond controlled information environments. The new reality of social media and a ubiquitous Internet makes the distinction between new and old media, between providers and users of information obsolete. The constantly growing mass of user-generated contributions makes the issues of information classification and organization even more significant, calling for dynamic and adaptive Information Architecture solutions. This is especially the case for the dynamics and uncertainty of complex enterprise environments.

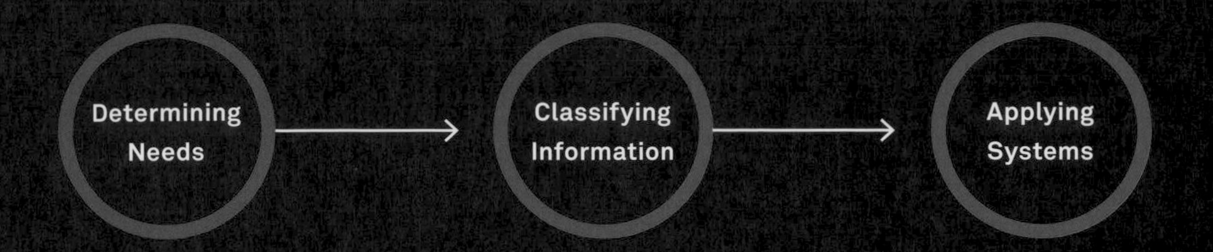

Applied in concrete projects, Information Architecture follows a process of exploration and synthesis, often combining top-down, bottom-up, and opportunistic approaches:

_ DETERMINING NEEDS

gathering requirements and information needs from potential information users and other stakeholders, analyzing how information is currently structured and used. This activity is supported by techniques such as card sorting, working with domain experts and structural models, or looking at existing information repositories.

_ CLASSIFYING INFORMATION

developing conceptual systems to organize information items. This activity involves developing information and data models to capture the semantics, informing the creation of vocabularies, taxonomies, labeling systems, or logical data schemes. It is informed by information needs derived from the top down, or classifying existing items.

_ APPLYING SYSTEMS

designing means to find, access, and retrieve information that support different behaviors based on the classification systems. These systems are designed to enable navigation, orientation, and wayfinding, providing relevant information in context. They can be based on a rigorous application of the first two steps, or around information needs as they emerge.

	Directed	Undirected
Active	Searching	Browsing
Passive	Monitoring	Being aware

According to American information scientist Marcia Bates, information is only rarely actively sought for, when there is a concrete need justifying the effort. In 2002, she developed a scientific model describing how we find and use information. Her scientific model uses two axes of different behaviors to distinguish four essential modes of information retrieval:

_ SEARCHING

actively searching for information to answer a specific question, supported by search functionality, indexes, or maps.

_ MONITORING

passively being alert to notice relevant and interesting information when it appears, supported by alerts or status displays.

_ BROWSING

actively looking around for information, guided by interests and loosely defined goals, and supported by menus, news, or guides.

_ BEING AWARE

passively absorbing information during other activities, supported by contextual shortcuts or lists of new or hot items.

_ INFORMATION IN THE ENTERPRISE

_ IDENTITY

how to design information to encode brand symbols, language and semantics into categories, labels and structures.

_ ARCHITECTURE

how to design information to support the enterprise in performing its activities and operations, and how to protect information.

_ EXPERIENCE

how to design information in a way that enables perception, understanding, use, and generating knowledge.

_ ACTORS

who are the consumers and generators of information in the enterprise, and how to accommodate their journey.

_ TOUCHPOINTS

where are points of information exchange between actors and the enterprise, and how to create paths between them.

_ SERVICES

what information feeds are needed to support service delivery and operations.

_ CONTENT

how to design information to chunk, group, and connect enterprise content as information nodes.

> The value in Information Architecture's structuring the information in an enterprise is not in attaining some abstract goal of imposing order on disarray but in enabling the provisioning of the right information in the appropriate context to the stakeholders who need it.
>
> **Gene Leganza, Forrester Research**

The enterprise as information space _

Information only comes to life when applied to a given use context—someone accessing and using it to achieve a certain goal. Therefore, any model describing an abstract organization system is meant to be applied to the design of accessible information spaces, such as electronic archives, web communities, signage systems, or other media. Good models about information provide the blueprints for structuring and connecting information, and ways to organize it in a classification system. To make use of such a system, it has to be turned into perceivable artifacts, delivering information so that it helps us solve the real-world issues of making decisions and developing understanding.

To be applied strategically in an enterprise context, information architectures must be consistent, but at the same time flexible to support varying needs. In practice, this often requires a mix of top-down, bottom-up, and pragmatic approaches to come to an adequate organization scheme. This allows us to balance the role of information systems as a secure and reliable *single source of truth*, a dynamic and evolving space for social exchange.

The way information is shared, processed, and used requires architectures that span multiple contexts, channels, and media. They have to connect information nodes across the entirety of people's journeys and business processes, connecting search engines, social media, information systems, and documents. Such a system then is the basis to make offerings findable on the web, inform management decisions with business intelligence, or enable navigation in physical spaces. This makes the underlying organization system similar to the DNA of the enterprise—a consolidated structure for relevant information, a strategic resource when applied in various contexts.

14_INTERACTION

From simple tasks like making a call or shopping for an item online to complex problem-solving and creative work, sophisticated tools and devices are a part of everyday life. In fact, they always have been, as using tools to support our activities and achieve our goals is a major aspect of human civilization. In the last few decades, we can see a major shift from the industrial manufacturing of comparatively simple physical tools to a new generation of technology, driven by software and data, embedded in virtually all the products we use.

Such products, both physical and virtual ones, are part of our social and professional lives, supporting us in working, playing, communicating, and learning. Even tools which used to be comparatively simple mechanical or electrical devices now include enough functions, states, and information to make them behave like computers. Compared to the tools and machines of the industrial age or before, these new products are exponentially more complex in nature, as probably earlier using software or electronic devices will have experienced. Software, and the things incorporating it, is famous for not working as expected, being hard to learn and difficult to use.

Interestingly, the modern enterprise as a whole is facing a similar challenge. Businesses have a long history of using software to support and automate their activities—consider the role of machines, data, and automated processes in the enterprise. Within this trend of ever greater digitization, the tools made available to people only got little attention. The same problems of complexity and impracticality are appearing in interactions between people and tools in the enterprise space, hampering business activities and market success.

Digital tools or media are the prevalent mode for interactions in the enterprise, be it customer interactions, collaborative work, or executing operational processes. But still, most of us encounter frustrations in our interactions with even the most advanced organizations, quite similar to our experience with the software-enabled tools we use for our professional and private activities.

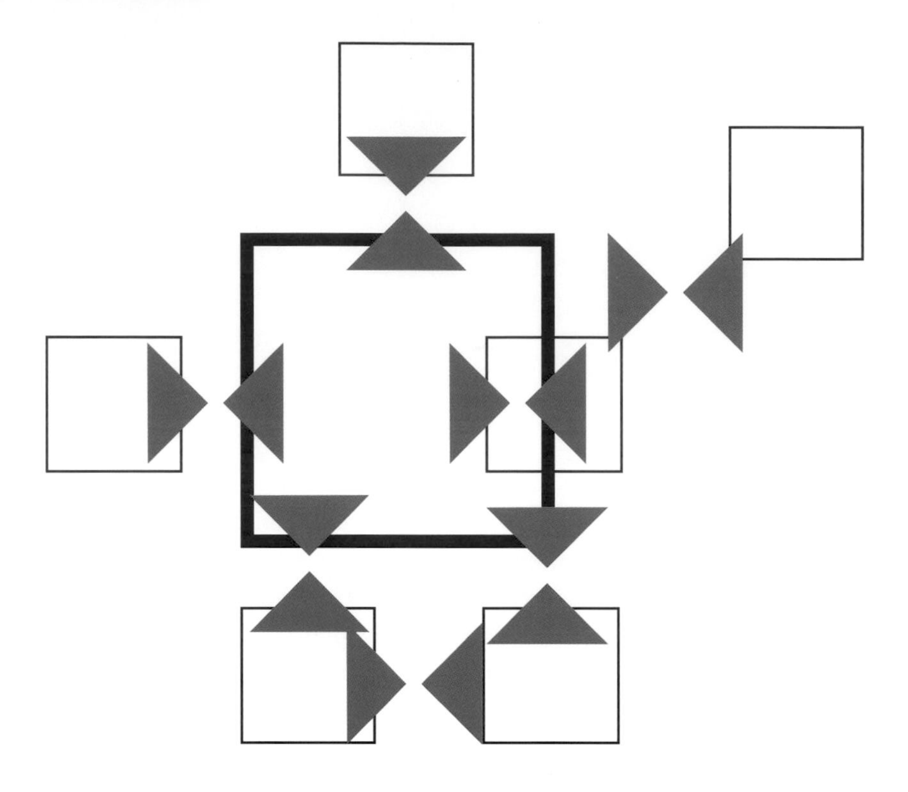

The interaction aspect is about consciously and proactively considering the exchanges between people and organizations in the enterprise, and supporting these interactions with automation and tools. It is about shaping behavior as it occurs, by designing the automated part of the dialogue, while anticipating the human part. While it is important to know what's technically feasible, that is not the driving factor behind modeling and shaping good interactions. Instead, it is the quality of the interactions that make the use of technology a success or a failure.

_ Modeling interaction and behavior

Learning about the interactions people are engaging in enables designers to consider how the results they produce will be used and to base their design on actual usage patterns. To best support interactions, the design of any tools and systems needs to be based on a solid definition of the dialogue envisioned. Therefore, it is useful to create models of the interactions to be supported as a basis for the design of any tools, by describing the dialogue between tools and their users over time.

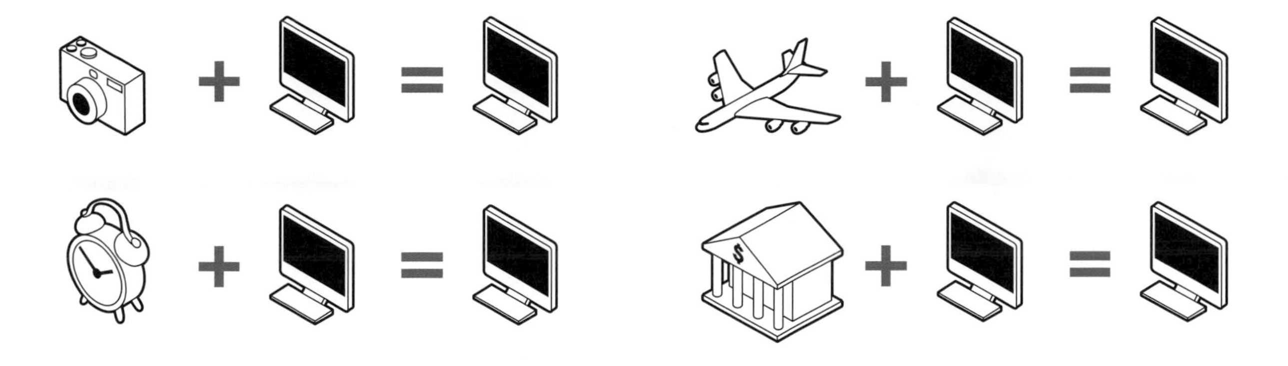

In his inaugural book, *The Inmates Are Running the Asylum,* interaction design pioneer Alan Cooper describes the effects of introducing computers as part of everyday life—a whole new world of possibilities that comes at the price of a dramatic rise in complexity, resulting in just about everything behaving like computers. Consider these examples from his book:

_ CAMERAS

that used to be relatively simple personal devices with just about three buttons start to offer hundreds of options and functions, are required to *boot* in order to work, just as telephones or clocks do.

_ AIRPLANES

are by now fused with computing technology, and a large part of the emergencies and accidents that have happened in the recent past can be tied to a chain of technical failures, usually in combination with *human error* from dealing with the mass of data, options, and automation.

_ BANKS

interact with their customers using online banking and ATMs, making simple transactions such as initiating a standing order to update your address a complex matter to deal with. Instead of a human operator, you are dealing with a computer and following its set of rules.

Interaction

People

Function

Interactions in the context of the Enterprise Design framework can be seen as exchanges between people and the enterprise, supported by tools or devices that make enterprise functions accessible.

_ PEOPLE

knowing about who you design for allows insights into the cognitive, physical, and emotional aspects of interactions, and lets us design the dialogue between a tool and a user accordingly. This involves researching, understanding, and modeling people's activities and motivations. Interaction models are based on these factors to determine the tasks people will attempt to solve with an interactive tool, the strategies, methods, and habits they apply, and scripts applying to the dialogue to be created.

_ FUNCTION

interactions in the enterprise space are happening in the context of a wider system of actors involved in business processes. Applying the function frame permits us to determine the *behavioral topics* behind an interaction. Interactions have to support the functional context of goals and processes by meeting defined requirements. They make any functional components in the enterprise accessible to people, supporting the wider purpose to be fulfilled.

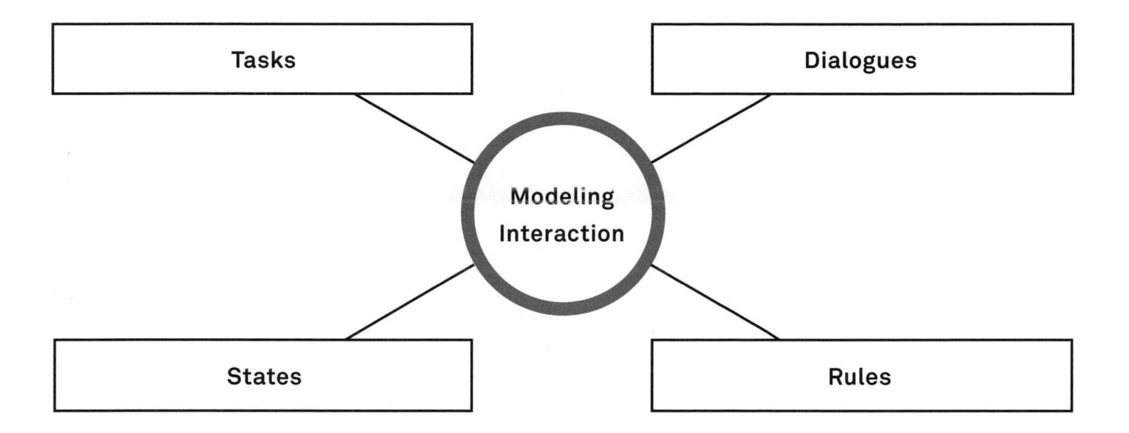

Techniques for modeling interactions have their roots in the areas of Human-Computer Interaction (HCI) as well as applied psychology, cognitive science, and ergonomics:

_ TASKS

are models about activities users perform to reach their goals. They are usually represented as steps to be taken in sequence. Additionally, they might contain different variants of a task flow, points where decisions are being made, information about the complexity or repetitiveness of an activity, or typical scripts and routines.

_ DIALOGUES

represent the interaction between a user and a system in the form of questions and responses. A common notation of dialogues is as a swim lane diagram, capturing who does what and when in the flow. Such a model describes the roles of people and automated components involved, the triggers of user action, and system feedback.

_ STATES

describe the different conditions presented by the system to be designed to its user, and the transitions between them. Such a model us to create a map of all states that have to be considered as part of the design activity, guiding a user from an initial state to a desired end of the dialogue.

_ RULES

are collections of behavioral instructions to the system, often in the form of if-then statements, in decision tables or trees. Modeling the conceptual rules behind a system's design captures how it reacts to whatever user input and the automated behavior it exhibits according to a set of environmental variables to be defined.

_ EXAMPLE

As an example for your interactions with digital systems, think of the way you interact with your bank to manage your money and your accounts. Depending on the offerings and your choices, interactions are supported by an online banking environment, ATMs, and similar machines, as well as by physical interaction with employees in their local branches or via phone. Each interaction largely depends on the tools made available, either to you as a customer or to the staff you are dealing with. While most institutions by now realize the importance of making these tools useful and valuable to their users, they are usually designed in isolation, looking only at people-tool interaction instead of the bigger picture of customer-bank interaction. Problems such as call center staff unaware of requests made online are due to the missing links and transitions between the interactions.

Models of interactions allow us to describe the dialogue between a system and its users, in order to anticipate the behavior on both ends. They capture the interactions in terms of the purpose to be fulfilled, activities and goals to be supported, and the individual contexts where they take place. This enables us to define and describe the desired interactions prior to designing the tools to support them, or even planning to implement them on a technical level. Instead, interaction models give a design team a clear idea of how interactive systems will appear in people's experiences. In order to create persuasive, useful, and desirable tools, they have to be seen as automated participants in a dialogue between human and machine, designed to shape the interaction as it occurs.

_ Shaping interactions and interactivity

Just like the other conceptual approaches portrayed in this chapter, designed interactions are invisible and intangible to the people they are made for. In order to actually work, they need to be bound with interactive artifacts that facilitate the interaction envisioned. Based on modeling of anticipated and desired interactions, such artifacts can be designed in a way that the choice of technologies, user interface paradigms, and metaphors enable the dialogue to be supported over time, working together to form a coherent whole.

INTERACTION DESIGN

It is impossible to design interaction per se, but what interaction designers do is to create conditions for interaction.

Jonas Löwgren, Malmö University (interaction-design.org)

As a professional practice, Interaction Design is a comparatively young discipline concerned with the design of dialogues between interactive systems and their users, or more generally between humans and the things they use as a means to achieve their goals. Compared with the more classic approaches of usability engineering or HCI, Interaction Design puts an emphasis on designing behavior over time as a creative activity. The challenge is understood to be about triggering dialogues and influencing behavior by creating interactive things. Such artifacts are made to accommodate dynamic exchanges with users, taking into account a deep understanding of users and their goals and use contexts, and the possibilities and constraints of technology.

While the outcome of Interaction Design usually is some kind of user interface, Interaction Design defines it rather as an implementation of a more abstract behavioral definition than an independent artifact. Mastering the practice requires a deep knowledge of technology and its promise to enrich people's lives or solve business problems. Its core area of concern lies in the behavior of the human beings that the technology is being made for. Therefore, turning Interaction Design into interactive artifacts often starts by pretending it works by magic or by people pulling the strings, in order not to get caught up in purely technical considerations too early in the process. The aim behind shaping the behavior of technology-supported products and services is to have an impact on the way their users behave.

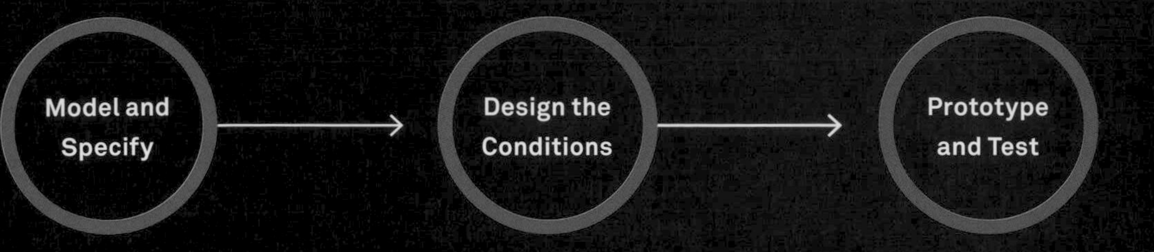

An Interaction Design process goes from researching and modeling a certain set of interaction qualities to a specification of the interactive outcomes of a project:

_ MODEL AND SPECIFY

modeling the interactions to be supported and envisioning ways to facilitate them with tools and systems, based on insights from research and the larger conceptual vision.

_ DESIGN THE CONDITIONS

design the tools themselves as responsive systems, to be used to support the interactions. This requires a cognitive leap to translate dialogues into interfaces and systems.

_ PROTOTYPE AND TEST

more than most other design disciplines, Interaction Design relies on a trial-and-error approach that emphasizes prototyping and iteration based on validation with future users.

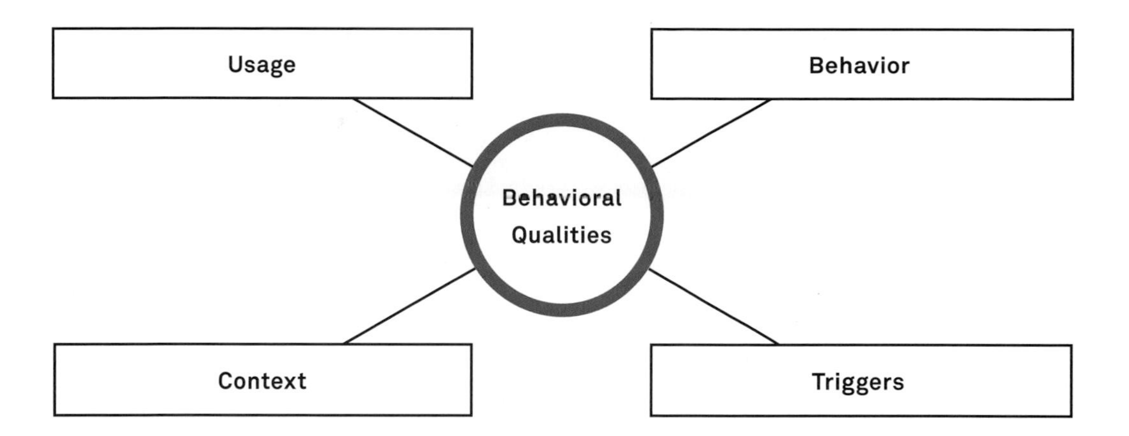

Designing interactions touches a set of fundamental behavioral qualities of interactive systems:

_ USAGE

learning about people addressed, their activities and usage patterns is a good starting point. It enables the designers to determine different groups such as beginners, intermediaries, and experts, to get a detailed understanding of how the same goals are achieved today and what an interactive tool as an outcome has to do to improve things.

_ CONTEXT

the individual context of use is the basis for choosing the right means to support the interaction — devices, media, and interface paradigms, for example, audio, websites, or a mobile app.

_ TRIGGERS

triggers are the concrete means of input and output to be used as part of the dialogue. This includes choosing affordances and metaphors, and turning them into interactive elements.

_ BEHAVIOR

a definition of the behavior of the system to be designed. This can be done in different forms, but usually includes a procedural description and a set of rules.

_ INTERACTION IN THE ENTERPRISE

_ IDENTITY

how to turn desired and adequate behavior into designed interactions and dialogues, encoding brand identity and values.

_ ARCHITECTURE

how to support the activities of the enterprise with interactions, supporting their effective and reliable performance.

_ EXPERIENCE

how to make interactions with the enterprises or within its realm, making it useful to people and great to work with.

_ ACTORS

who is interacting with the enterprise in what role context to participate in the dialogue to be mediated with interactive tools.

_ TOUCHPOINTS

what events are triggering interactions—when and in what kind of context do they appear, and how to accommodate these conditions.

_ SERVICES

what are the services made available to people in interactions, and what type of dialogue they require.

_ CONTENT

what content is being exchanged in interactions and how to support them with instructions, feedback, and contextual information.

_ EXAMPLE

As an example, think of the way your company enables people to apply for a job. The interaction starts probably outside the realm of company-owned tools, for example, via a job site, and continues eventually in the career section of your own website with an online application. Both systems can be considered a good subject for Interaction Design. Seen as a whole process, the interaction spans well beyond using these tools to find a job—it involves getting initial feedback by mail, scheduling an interview, and going through a quite long and extensive recruiting process.

Companies are notoriously bad at this, often failing at the first step of sending feedback to everyone who applied within a reasonable timespan. Applying Interaction Design holistically to the wider dialogue enables us to accommodate for this in the design—for example by supporting HR consultants' communication with applicants, introducing a tool that reminds them in time and provides the status and next steps of the recruiting process. Although this is a purely internally used system, the impact transforms the enterprises' behavior towards an external stakeholder group.

The enterprise as a dialogue _

Applied on the enterprise level, Interaction Design expands its scope beyond the design of isolated interactive artifacts—it means designing a space for interactions between actors in the ecosystem. Every tool or medium used to facilitate these interactions has to be seen in the context of its wider meaning, as a means to shape the interaction between people and the enterprise and as a way to affect human behavior. This applies to automated components of products and services offered to the market, as well as to internally used tools or systems.

Some leading organizations have tapped into the possibilities of investing in the quality of their interactions, which enables them to deliver a great Customer Experience by providing products and services embedded in a wider system. Instead of isolated tools, customers and other actors are supported by a system of functional elements, designed with their users in mind and facilitating transactions, conversations, relationships, and flows of work. Applying Interaction Design on the enterprise level, deeply rooted in human behavior, unlocks the potential to design more than just usable products and tools—considering the larger dialogue allows us to design enterprise interaction holistically and makes a difference in terms of behavior.

15_OPERATION

What from the outside looks like a well-oiled machine is, from the inside, lots of people working really hard to keep up that impression and make it truer each day.

Mei Lin Fung, Institute of Service Organization Excellence

The idea of designing business operations has its roots in the idea of operational excellence. It has been developed as part of the lean manufacturing approaches at Toyota and other Japanese organizations, which aim to maximize production quality while minimizing effort and costs and are applicable equally to industrial products and service provision. Efforts in operational design define and plan in detail what is being done, and how it is done in an organization, to produce and deliver its products to the market, and to turn this design into a standardized, repeatable process.

Most people have to deal with large organizations of various kinds, such as phone companies, energy suppliers, or public administration. Airlines transporting your baggage to the wrong destination, after-sales support failing to address your specific problem, mail orders that never arrive, departments unaware of what their colleagues are saying or doing, inability to communicate important information, waiting in long queues—the list of stories of operational failures people have to face is long and frustrating to say the least. Such failures can get downright dangerous in critical areas such as aviation or healthcare.

Despite the long tradition of operational optimization and the advent of digitally supported automation, this is still a particularly error-prone area in the enterprise context. Across all industries, employees feel trapped in narrowly defined processes rather than empowered by them, so they attempt to break out of the system instead of working with it and improving it.

The operations aspect in the Enterprise Design framework deals with the design of all activities executed in the enterprise, tackling the challenge of defining collaboration practice and managed procedures that maximize performance and quality of delivering what the enterprise has been created for. It is the basis for creation or transformation of an operating model of the enterprise, in alignment with the potential outcomes of a strategic design initiative.

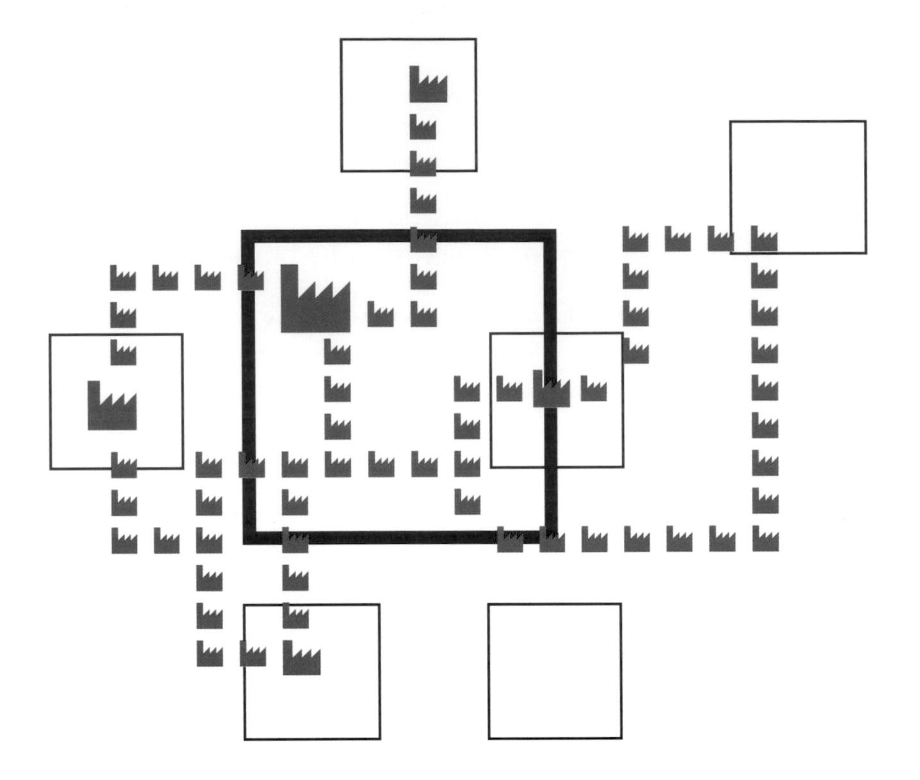

Operational design aims to support design teams taking the operational factors into account in projects, to analyze and redefine how business activities are being done to support the larger goal behind the design initiative. To do so, operational designs work towards improving performance across the value chain, identifying and removing unnecessary steps, phases of inactivity, over- or underproduction, defects, or other types of waste. But to make an operating model actually work in real life, this also means making processes transparent to the people carrying them out. It requires empowering them to unlock the flows of work by resolving problems and eliminating errors when they appear.

_ Modeling operational processes

A model of operations is a definition of the activities an enterprise undertakes to compete in the target markets and deliver products. It is collaborating with suppliers and partners, managing employee relationships, and sustaining its ongoing business. In general, it describes the flow of people's work and decisions, materials, information, and money through the enterprise's value chain, starting at a given initial state and aiming to reach an intended satisfactory end state. It reaches across technology, work practice, ad hoc collaboration, and management incentives, and therefore should address everything considered relevant to the achievement of operational goals.

_ EXAMPLE

In 2010, one of the largest national railway companies in Europe decided to reward loyal customers with a special promotion, one of them a free upgrade to first class. The upgrade was printed as an icon on the frequent travelers' cards, entitling the holder to claim the right to pay reduced ticket fares. It was also added as a section to a long public document describing the terms of service.

However, the promotion was not communicated to the personnel effectively, leaving them unprepared to meet customers' expectations. Conductors were unaware of the special offer and did not accept the upgrade card. They humiliated customers by sending them to the lower class compartments or (even worse) attempting to charge a full first class fee. This is a typical example of designing great offerings to [illegible] in the Marketing [illegible], but turning them into a nightmare by failing to taking the operational aspect into account to make it actually work as a process.

Such a model captures the business processes being executed as a series of activities arranged in time and space, and enables design teams to incorporate operational elements in their concepts. Design decisions about operations involve the order and sequence of things being done, and requires us to account for dependencies between activities. Furthermore an operational model specifies ways to deal with any exceptions and deviations to the planned process, allowing people to step in if something goes wrong.

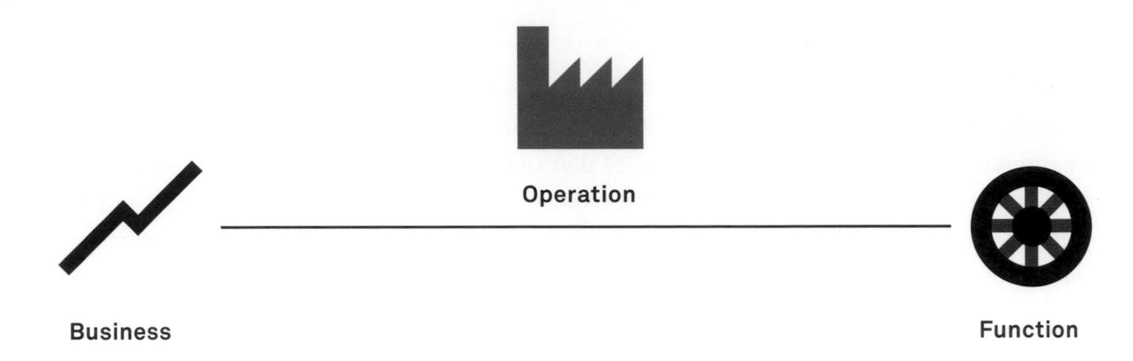

This aspect is informed by insights from the business and the function frames, providing the context for operational design:

_ BUSINESS

the business aspect reflects the reasoning behind an operational model. Insights from the business model to be implemented allow us to draw conclusions on what to prioritize when designing a subsequent operating model. It clarifies which tasks provide value to the customers, which areas have the highest financial leverage and contribute to market differentiation. It allows us to define metrics as operational objectives, and identifying resources required for execution and exception handling.

_ FUNCTION

the function aspect is the other main driver for designing operating models. In a way, operation is the coordination of activities directed at supporting the purpose of the business. It is about connecting and orchestrating business functions — or capabilities — helping the enterprise to achieve its objectives, generate value, and work properly. Well-defined functional requirements provide the input needed to identify and connect operational capabilities, detailing what exactly the enterprise needs to be able to do.

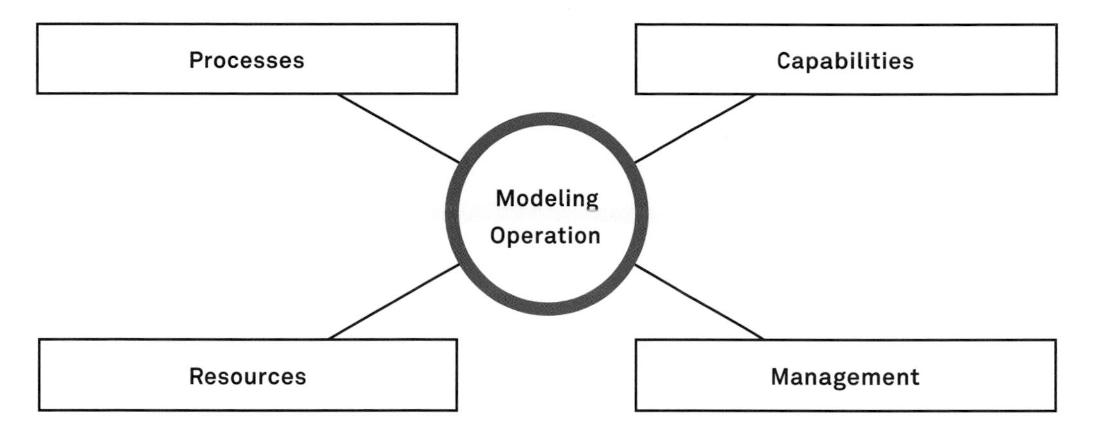

Depending on the industry and nature of the activities, items to capture in an operating model vary. In general, elements fall in these categories:

_ PROCESSES

the order, sequence, and business of activities and important events, from a starting condition to an end state. The scale depends on the modeling purpose, but can reach from single processes to entire industries from supplier to consumer. It captures how things are done, how to process work, capital, and information along the chain.

_ CAPABILITIES

the organizational skills and instruments needed to achieve an operational target state. Modeling capabilities allows identifying what an organization has to be able to do in an operating model. Based on a map of capabilities, operational steps can be repartitioned across organizational units. This is also the foundation for local optimization of a capability, and supporting it with tools and technology.

_ RESOURCES

the finite inputs needed for the operating model to work, such as raw materials, manpower, and capital, as well as the capacities needed to process them to produce the desired outputs. Modeling the kinds of resources that are needed where and when is the basis of addressing questions of planning, transportation and logistics, queuing and scheduling.

_ MANAGEMENT

the systems put in place to manage the activities, including monitoring and support, initiatives and projects to evolve the operating model, and ongoing controlling and measurement. Including the management system in the operating model allows mapping the design to organizational responsibilities, and operationalizing activities such as quality assurance, communication, and security.

Because operations has such a large scope, it is vital to determine what operational factors and details are important to consider in the context of an individual project. This depends on the repetitiveness and predictability of the activities to be done, the degree of possible automation with the help of technology, the expectations of customers and internal stakeholders, and the operational goals. An operational model might cover an entire enterprise ecosystem with processes between a network of intertwined organizations, or just a small subset of activities considered important for the problem at hand. The elements to take into account are consistent independent of the level of detail covered, equally applicable to a big picture model and a detailed process.

_ Designing work

Many of the disciplines dealing with operational design concentrate on improving the effectiveness of existing operating models using a large variety of approaches. Most approaches focus on achieving better performance, while leaving the existing mode of operation unchanged. They aim at improving the model by wiping out inefficiencies and delays, reducing cost, optimizing resource usage, and generally working towards continuous improvement.

Making the task of developing a potential future operating model part of a strategic design project allows project teams to move beyond just incrementally improving what there is today, letting them look for opportunities to dramatically shift the operational process. As a starting point, such a project usually deliberately sets difficult or unrealistic performance goals, opening the designers' minds to radical ideas. Key to this is overcoming the paradigms applied in a current operating model—applying design thinking to operations means to radically rethink how work is done and processes are handled.

BUSINESS ARCHITECTURE

Business Architecture enables stakeholders to make key business decisions by taking an integrated view of the business and aligning all the various moving parts in an adaptive 360° model.

Mike Clark

As a professional discipline, Business Architecture focuses on translating strategic decisions and initiatives into structures and systems to be implemented in an organization's evolved operating model and business functions. Practitioners perceive the enterprise as a dynamic structure of interrelated elements that contribute to maintaining, sustaining, and developing business activities. Applying a Business Architecture approach, this structure is subject to proactive and informed design decisions, providing the link between a defined business model and its execution as a transformation or extension of the operating model.

A Business Architecture approach enables design teams to map out the overall execution of a solution as envisioned in a strategic design — its operating model — as an abstract representation on paper or in a spreadsheet. Sophisticated design paradigms such as Lean or Six Sigma can be applied and tested in such a model as a simulation, trying out different approaches and iteratively validating and refining them in the course of a design process. Once a target operating model has been identified, the team proceeds to encode it in systems, instructions, or other kinds of documentation, as a plan for execution.

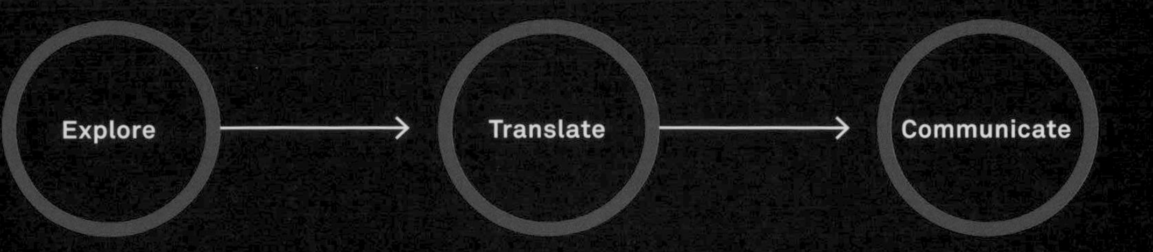

The work of Business Architecture professionals consists of a set of interrelated activities, informing the development of business strategy and models, and turning these elements into actionable *blueprints* of the business:

_ EXPLORE

the enterprise environment to discover developments and changing conditions or trends that might affect the business model applied, and identify threats and opportunities to inform strategy.

_ TRANSLATE

strategic initiatives into concrete redesign and improvement projects, defining the evolved business and operating models and communicating changes to other parts of the business.

_ COMMUNICATE

and socialize with stakeholders the changes envisioned with a new architecture, and advise during its implementation.

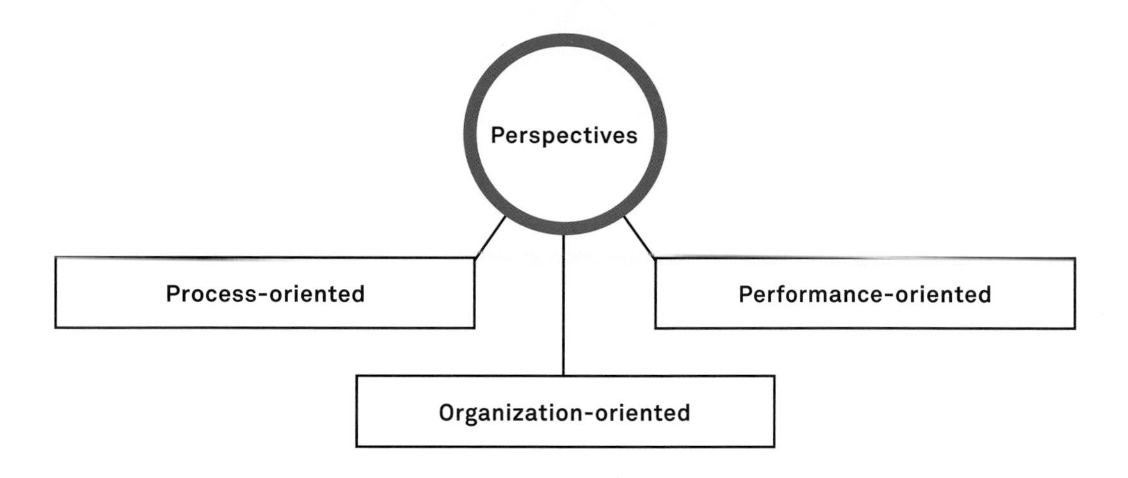

While there is a multitude of domains potentially addressed by initiatives defining the operating model of an enterprise, there is a set of common perspectives used as starting points to such an endeavor:

_ PROCESS-ORIENTED

approaches see a business as a set of interdependent activities to be executed. They apply techniques from Business Process Management and Reengineering to model operations as repeated procedures, sometimes employing Operations Research techniques, mathematical models to optimize operations.

_ PERFORMANCE-ORIENTED

models emphasize on business performance as captured in metrics and goals. Such an approach is focused on measuring and improving business performance in numeric terms, using financial models, scorecards, and Business Intelligence technology.

_ ORGANIZATION-ORIENTED

methods model a business as a system of areas, responsibilities, and formal relationships between them. They aim at describing and designing effective systems for governance and work practice.

A design approach to operating models finally aims to reshape human work, supported by systems and tools, rules and incentives, defined procedures and flexible arrangements, making these elements contribute to achieving the goals defined. It also involves specifying and implementing the model into systems and procedures, and also communicating it to people to make it part of their work practice.

An important part of this is a human-centric approach, starting with customers, staff, and other actors, and modeling operations according to the activities and needs to be supported. To be able to envision radical innovation, designers need to explore the activities of people involved in and addressed by the processes to be designed.

_ The enterprise as a process

In his business novel *RecrEAtion*, corporate strategist Chris Potts makes the point that although many organizations like to see themselves as the owners of their operational processes, in the end it is the customer who actually owns the process. So instead of designing discrete processes to deliver value to the customers, organizations have to decide how and where they want to appear in their customer's activities. This then is the basis for designing an operating model that fits.

For the enterprise, an operational design approach can help to rethink its operational systems and subsystems. As the usage of the terms *architecture* or *design* in Business Architecture circles suggests, such an approach is actually conceptual design based on insights about people, which makes it very relevant to the goals of any strategic design initiative. It lets us choose an operational mode and to design operations around the role an organization assumes for its customers and other stakeholders.

OPERATION IN THE ENTERPRISE _

_ IDENTITY

how can our operating model encode and support the values and behavioral goals embedded in our brand identity.

_ ARCHITECTURE

how to structure the formal systems in the enterprise to best support the operating model and the ongoing processes.

_ EXPERIENCE

how to design operations so that we appear in our customers' and other stakeholders' lives, support their goals, and provide real value.

_ ACTORS

who is involved in our operations in what way—what do staff, partners, customers, or other stakeholders contribute.

_ TOUCHPOINTS

where do we interact with people as part of our operations, and how to turn these exchanges into rewarding opportunities.

_ SERVICES

how to support service provision to customers and what internal services do we need to provide as part of our operations.

_ CONTENT

what content has to be communicated as part of operational processes in order to make them work and useful.

#16_ORGANIZATION

Bringing about organizational change has been a hot topic in management circles for quite a while. The converging effects of group culture and identity, operational processes and planning and organizational politics, and authority make designing and implementing organizational change challenging in the enterprise. This diversity of viewpoints on organizational transformation has produced a wide range of practices dealing with their design and implementation, such as Change Management, Organizational Development, and applied methods like training and coaching.

Any major endeavor in the enterprise requires its organization to evolve and adapt to a new state — introducing new products or services, addressing a new market, or revamping internal operations — developing the business always means destroying existing structures and replacing them with something new. Therefore, the success of a design initiative with strategic impact depends on its ability to influence the organization behind it. In effect, this makes larger design initiatives into organizational change programs, because they have to reach the people involved in delivering the outcomes they intend to shape.

Driven by Marketing initiatives, many companies make unclear promises in their communication to customers. The things that appear in those communications are usually subject to design practice, such as products, interiors, or visual brand identity, driven by a customer-oriented view from the outside. More often than not, there are wide gaps between promise and reality, especially when there is a problem and the standardized operations break down. Equally, many design projects fail because of unclear responsibilities, resistance to change, and ill-adapted organizational structures, especially in the long run after a design project has ended.

In order to make an intended transformation happen in a sustainable way, design projects are dependent on support from many executives, departments, and staff. Any substantial change usually triggers changes in the way work is being done, requiring evolved organizational structures and a new mindset to actually make it work. Transforming organizational structures is by nature a co-creative endeavor, requiring collaboration and support from people involved in the activities and impacted by a change. These projects have to affect the habits and beliefs people are used to in order to change the way things are getting done. Their manifestation in business practice is driven by a mix of formally defined structures such as organizational hierarchies and job definitions, and of informal social norms, habits, and assumptions leading to certain attitudes and behaviors.

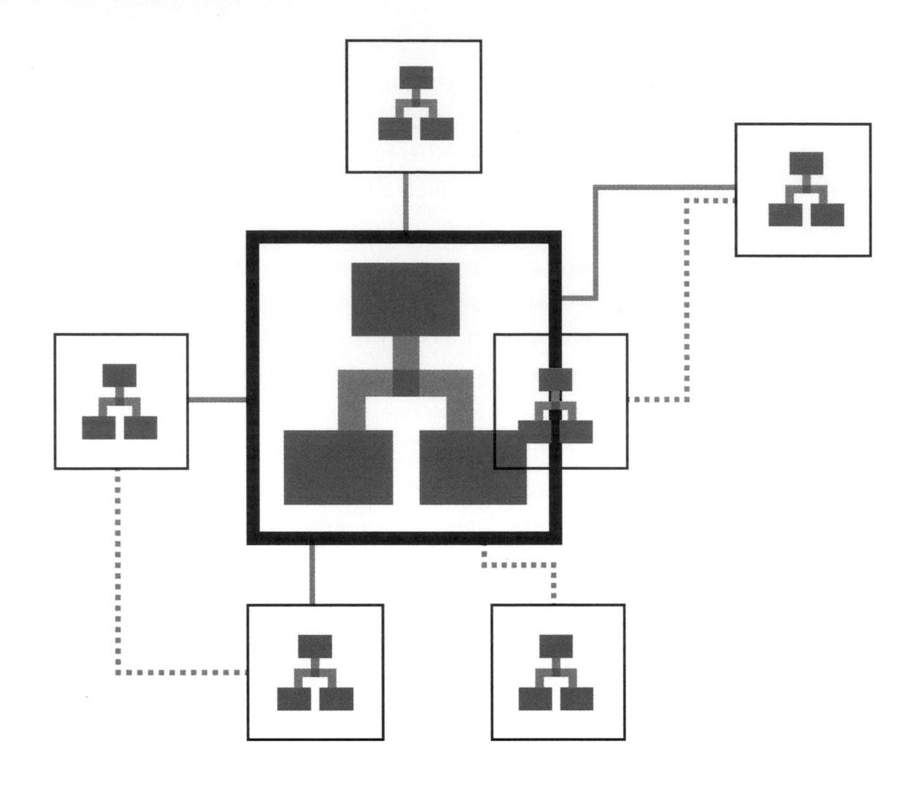

Addressing these issues requires us to take into account the organizational reality around the processes and products that come out of a strategic design initiative. The Enterprise Design framework therefore makes the organization aspect an individual conceptual topic independent of any accompanying project or change management. It assumes organization is an aspect that is as worthy of a design initiative as the other conceptual aspects in this chapter, and meant to be modeled and designed to support the objectives of the project.

_ Modeling organization

Design practice never happens in a vacuum, but is always embedded in an existing or emerging organization, impacting both structure and culture. Classic approaches usually start at the periphery of organizational structures and cultures, focusing on the products, services, or communications to the customer. In reality, their success is highly dependent on the adaptation of human behaviors and attitudes in order to really deliver on their promises. Without aligning results to the people intended to work with them, the benefits envisioned from new customer interactions, tools, and technologies, or redefined operations cannot be realized.

While organizations of all sizes and types need to adapt and reinvent themselves in ever shorter cycles, the traditional tools and methods for understanding them are still based on comparatively static metaphors, that neglect the complexity and dynamics that influences their development. They attempt to shape organizations as though they were analogous to buildings, machines, or living organisms, but fail to acknowledge the fact that they are made of people — determined not by fixed parts or processes, but by the mindsets of individuals and the dynamics of groups. So in contrast to those other man-made systems, designing organizations involves dealing with social systems or communities, resistance to change, and politics.

Based on a substantial background in social research and theory, organizational modeling captures the essential factors that determine how people work together to contribute to the shared goal of the enterprise. In the context of a design program, such models are about the organization as an environment and enabler to its outcomes. They let us map both formal and informal organizational factors that would influence the success of the program, to address these factors in subsequent design decisions, outcomes, and transformation activities.

Bringing design to life _

The success of a strategic design initiative largely depends on the collaboration with the people who implement it in daily business. In order to affect the organizational structures and cultures, designers need to transform the rules and the environment in terms of formal systems, relationships, and incentives, but at the same time inform and educate the people involved in order to make organizational designs actually work. Such a design cannot stop at envisioning how people are supposed to work together according to the target state model, but also to connect, communicate, build trust, argue, and convince. It is about affecting the life in an organization by combining formal transformation and informal change.

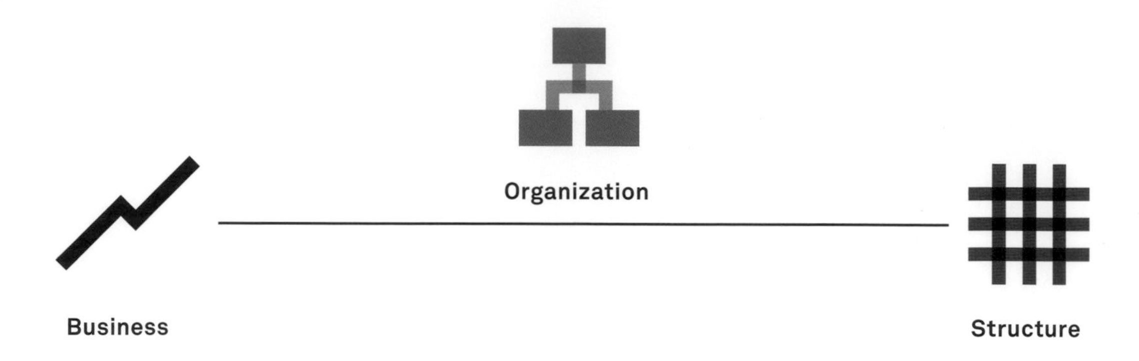

The organization aspect is based on applying insights from the business and structure perspectives, setting the context of goals to be achieved and the structure to be changed in the course of the project:

_ BUSINESS

organizations are designed and managed systems that support a business in implementing its strategy, business model and achieving defined goals. Therefore the business aspect is one of the main drivers behind designing organizational systems, delivering the background that drives conceptual design decisions with regards to organizational structures and choosing actions to transform the way people work together. It also captures the formal environment of legal entities involved, their offerings to the market, and their history.

_ STRUCTURE

organizational entities and their different manifestations are part of the overall structures of the enterprise, visible to actors as companies, services, jobs, or departments and their relationships and interactions. These elements are part of the reality in the enterprise described in structural models, and their redesign has to fit into that reality. Domain elements such as products, brands, tools, or regions are reflected in the design of organizational structures as roles and responsibilities, assigned to units or departments.

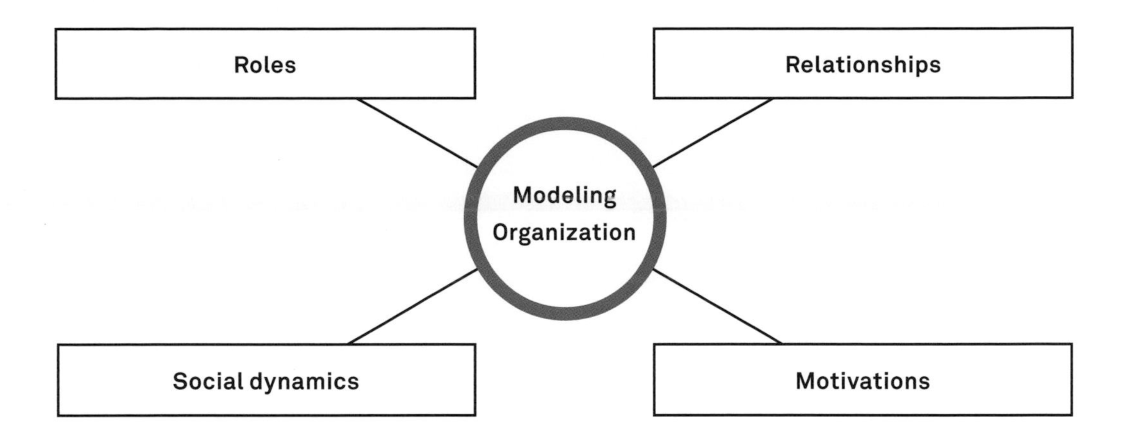

Organizations are complex and dynamic, and the elements to consider in a model will certainly vary depending on the individual background of a project. In general, the following categories might be considered:

_ ROLES

units of activity and responsibility according to the capabilities needed. Models of roles are the basis to define jobs and identify or develop talent profiles, reflecting skills and management levels.

_ RELATIONSHIPS

formal reporting structures such as hierarchical or matrix systems, describing who needs to work together in the context of operations and service delivery.

_ SOCIAL DYNAMICS

informal social structures such as leaders and influencers, established communities, values, and cultural themes or local variations.

_ MOTIVATIONS

shared goals and objectives and how they are supported or affected by incentive systems and cultural values.

ORGANIZATIONAL DESIGN

The soft stuff is the hard stuff!

Alec Sharp

The topic of designing organizational structures and systems is addressed by a large variety of different professional practices and disciplines, from classic management to Organizational Development, Business Architecture and Human Resources. As many practitioners stress, designing organizations holistically and effectively requires taking into account the entire range of formal and informal organizational factors. Most importantly, it means working together with the people whose lives are impacted by the change envisioned.

An organization design initiative is a way to execute the business strategy and the business model by making people work together on a common course. The steps described here reflect such an integrated design approach as it is applied by some practitioners of the various disciplines mentioned above.

The formal way organizations are designed mostly reflects the chosen business and operating models, visible in reporting lines and responsibilities. This leads to paradigms such as products, regions, market segments or other dimensions leading the way for departments and teams to emerge. Depending on the nature of the enterprise, organizational design can either integrate or diversify, with many models between these extremes like shared service units. The way people work together only partially reflects this formalism, so that organizational design practice must always address the human side of things, by educating, engaging, and connecting people.

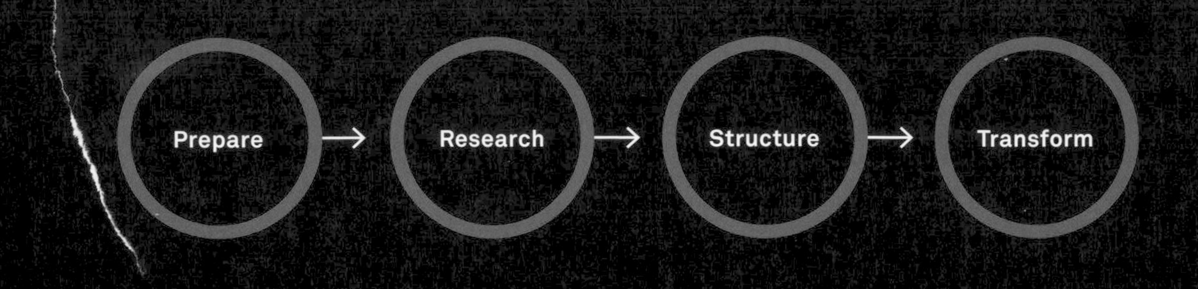

Because of the importance of both formal structures and informal culture, work on organizational design usually has to be approached as a change management initiative, where working closely with stakeholders is equally important as the conceptual design itself:

_ PREPARE

engaging with stakeholders, assess and discuss the strategic drivers making a change worthwhile pursuing.

_ RESEARCH

collecting information about the way people are currently working together, identifying roles, gaps, and cultural factors.

_ STRUCTURE

designing an evolved organizational structure including roles and responsibilities, skill profiles, and formal relationships.

_ TRANSFORM

introducing the changed organization to the team and providing training and leadership, and making pragmatic adjustments.

Design could offer a new way to understand and practice management, leading to more human-centered organizations.

Richard Buchanan

_ ORGANIZATION IN THE ENTERPRISE

_ IDENTITY

how to support an organizational culture that helps reinforce our brand values and goals.

_ ARCHITECTURE

how to structure the organizational structure, relationships, and formal rules to help the enterprise work.

_ EXPERIENCE

how to support great experiences for staff and customers by defining clear responsibilities, and creating a great place to work.

_ ACTORS

who needs to be addressed in the organizational design to make it work as a whole.

_ TOUCHPOINTS

who talks to who in the course of business activities, where and when alignment is needed.

_ SERVICES

how to support services with organizational communication and collaboration and vice versa.

_ CONTENT

what content is being exchanged between members of the organization in alignment and communication processes.

_ EXAMPLE

In 2007, the city of Paris introduced Velib', a public bicycle sharing system for local transport. This large-scale enterprise includes over 1 200 stations and 20 000 bicycles, and a sophisticated kiosk and payment infrastructure. While most of the operational challenges were addressed through technical means, a particular problem is the need to redistribute vehicles between full and empty stations, which requires a fleet of 23 transporters continuously moving bicycles around. In order to address this problem, the owning company decided to offer their customers an incentive to help with the redistribution: returning a vehicle to a station that is repeatedly empty—marked with a specific symbol and usually on higher grounds or toward the outskirts of the city—credits the user 15 minutes of additional free time. This is an example of an organizational incentive on the enterprise level, where customers contribute to solving the issues of internal operations.

Proactively designing organizational structures and addressing change proposed by a design project can dramatically improve its chances for success. In the end, it is always people who have to work with the outcomes of a design initiative, while the designers usually move on to their next project. Beyond finding ways to make a strategic design initiative sustainable by allocating responsibilities and changing relationships in the enterprise, the co-creative nature of organizational design is a good way to profoundly engage stakeholders. It enables design teams to embed the process in the enterprise and to align their results with the people affected — they are perceived as something originating from within rather than from the outside. This makes actors external to the commissioning organization feel involved and taken seriously. Applying the organization aspect reflects the team spirit of a design project and is the prerequisite for sustainable transformations.

17_TECHNOLOGY

Design, in its broadest sense, is the enabler of the digital era—it's a process that creates order out of chaos, that renders technology usable to business.

Clement Mok in *Designing Business*

Technology drives the transformation processes in the world. Automation, digitization, and technical progress are behind the most fundamental shifts in society and economy, such as the advent of the Internet and mobile computation, as well as advances in medicine, construction, and transportation. Entire industries are driven by the prospects of ever greater automation and technical possibilities, resulting in a business model centered on science and technical innovation. Others are using software to drive every bit of their daily operations, be it manufacturing, service provision, or public administration.

This view of technology as the backbone of the enterprise, as systems processing information in the basement, dominates especially the area of information technology, and led to the idea of using it like electric energy or similar commodities. This view is in contrast to the innovation role attributed to technical development, visible in the disruptive effects of certain technologies.

In recent times, technology has become much more present in our daily lives, changing the way we behave and interact with each other. Products and services supported by digital technology play a vital role in almost every activity people undertake today, and the number of microchips silently doing their work around us keeps growing. Beyond the different variants of computers and electronic devices, they now turn up in unexpected places, from kitchen equipment to light switches and clothing. Moreover, the arrival of the Internet and online offerings such as Google or Facebook are examples of virtual products, working for us in a distant cloud of networked servers, and making available a myriad of functions and possibilities whenever we demand them.

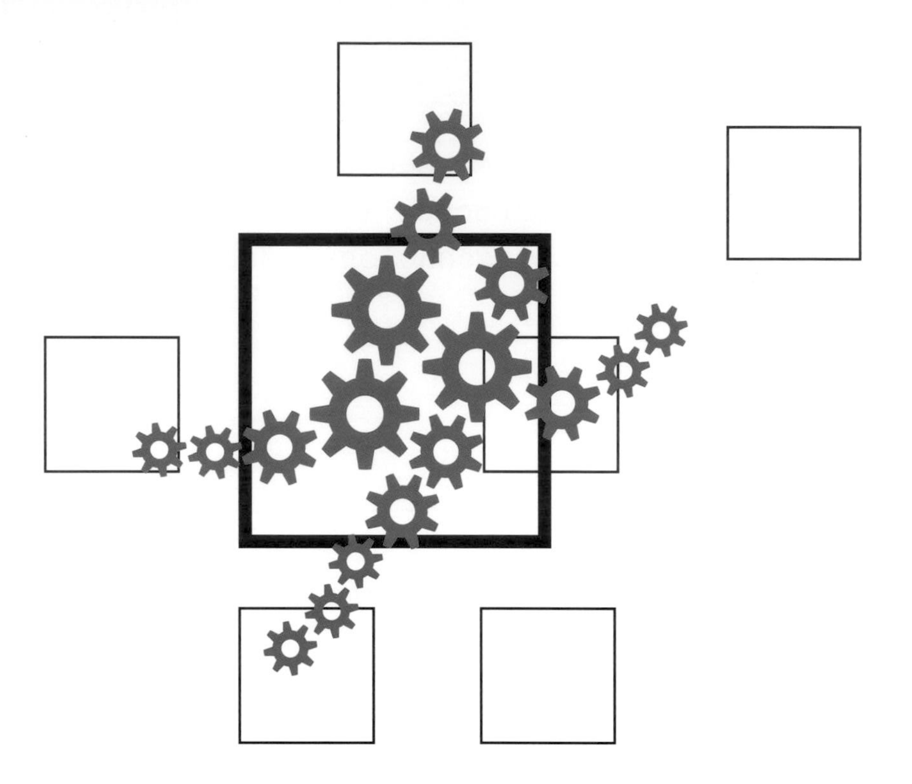

It is this ubiquity of capabilities that makes technology a major conceptual dimension in any design project. With the new role of technology, technical design decisions shifted from studying materials and manufacturing to exploring the possibilities of using technology as part of a product or service, realizing benefits for the people addressed by it. The potential of technology lies in creatively leveraging technical ideas to design adaptive and intelligent offerings to the markets, and making them available to actors in the enterprise.

The Enterprise Design framework approaches technology as a space for conceptual design decisions and as an enabler for producing great outcomes through design initiative, similar to the other aspects portrayed in this chapter. Technology is useless in itself—it has to be turned into meaningful proposals to actors in the enterprise in order to provide any benefit to the organization exploiting it. Therefore, strategic design initiatives have to identify the possible uses and purposes of a technical option, and to design it in a way that it can assume its role. This work has to happen in two ways—by taking all technical developments seriously and putting effort into exploring possible use scenarios, and by informing efforts of technical innovation, researching demand, and designing a possible future state.

Designing with technical objectives and options _

In the context of design projects, technology can be seen as any tool, technique, or machine that supports a task relevant to the design challenge at hand. Models centered on technology are driven by considerations of feasibility and technical possibility. They have the goal of mapping the space of technical options, asserting the limitations and possibilities, and weighing their advantages and disadvantages.

A technical model unambiguously specifies the role a tool or automated service plays in the overall concept. Unlike models to inform engineering processes, modeling technology for design purposes aims to describe the implications of a particular choice for implementation in terms of usage and behavior.

Approaching technology as part of a design project holds great benefits compared to taking a purely engineering-driven approach, as is still common in many areas today. It permits organizations to pair technical innovation and sophisticated applications with an active research of the uses associated with those technologies, and can be used with either a push or a pull approach to technology strategy. Based on modeling and understanding capabilities and possibilities as well as their potential meaning, applying design to shape technology provides the missing link to people and markets. Key to this is combining design and engineering in a creative fashion, being open to unconventional and surprising outcomes.

Technology

Function

Structure

The technology aspect is driven by aligning the functional with the structural perspective on the enterprise:

_ FUNCTION

the function perspective provides insights into the purpose fulfilled by applying technology in the enterprise. The functional context reveals goals to be met and processes to be supported or automated, and delivers the reason behind applications and other functional units of technology that are envisioned as part of a design project. It enables a design team to establish requirements that leverage technology to provide actual value to actors, and to drive design decisions.

_ STRUCTURE

insights from structural modeling and exploring the domain behind a project allow us to envision the possible integration of a technical solution into the real-world context where it will be applied. The structure of the domain, the objects and their relationships, form the basis for modeling technical components, infrastructure, platforms, and representation in data models and code. The other way around, technical components will be objects introduced into that structure, visible to users as tools or devices.

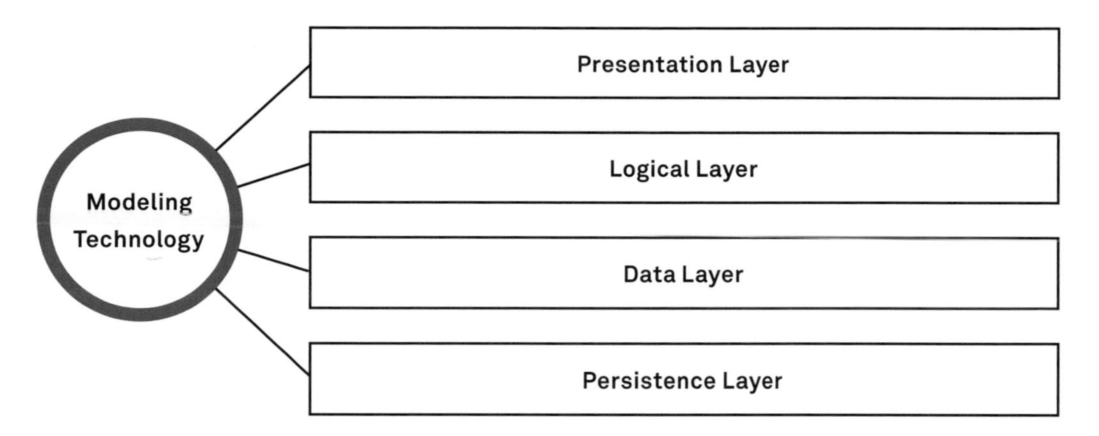

Modeling technology usually describes system architectures of interrelated components in multiple layers or tiers, capturing their dependencies and interfaces in structural or procedural diagrams and separating different concerns. A typical layered architecture consists of about four layers, but there are model variants combining or further dividing these layers, and much more complex models:

_ PRESENTATION LAYER

How does the system connect to its human users, present its outputs, and states and capture input? This includes components like target devices and their capabilities, interaction modality such as touch input, and media channels to be supported.

_ LOGICAL LAYER

What processes and decisions can be made automatically? This layer, also called the business logic or application layer, controls the system's behaviors and functionality, performs calculations, and represents a system's intelligence. It is the basis for designing adaptive systems, often labeled *smart* technology.

_ DATA LAYER

What does the system know, and what does it need to know? This layer is about the information processing and storing of a system as a basis for automated decision making. It describes what data should be captured and represented in order to make the system useful, based on a structural or domain model.

_ PERSISTENCE LAYER

What platforms, devices, machines, networks, and data stores are being used? The bottom layer describes the available infrastructure and resources and how they are used by the system to perform its functions. Decisions made about this layer relate to concerns of performance, security, and availability.

Technology is outpacing our ability to use it. And it's the job of designers to restore balance to this equation.

John Maeda

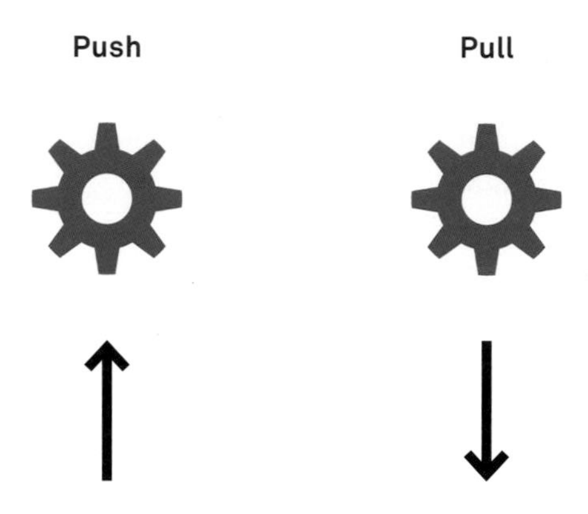

In general, there are two ways of approaching technology from a design perspective, triggered by or triggering technical innovation:

_ PUSH APPROACH

a technical innovation appears through developments in natural science or experimental engineering, without having a definite market need, use, or purpose in mind. Many organizations strive to create technology pushes in their Research and Development departments.

_ PULL APPROACH

an organization makes a directed investment in technical innovation based on a defined need or requirement from the market or another source. Research is conducted with that need in mind, steering the efforts made into a certain direction.

TECHNOLOGY DESIGN

Because of the size and diversity of the field of engineering, there is an overwhelming number of disciplines that are relevant to Technology Design. Although technology plays a significant role in nearly all design disciplines, there is no established discipline or body of knowledge for applying design to technology. However, there are many subfields and professional practices that are relevant for making technology fit for people in the enterprise.

This portrayal of Technology Design does not reflect a homogeneous field, even less so than the other young design disciplines portrayed earlier. Instead, there is a broad range of design-related areas that have mastery of technology at their core.

Applying technology in an Enterprise Design initiative requires developing a holistic view of technology usage in the enterprise, an overarching view of all systems and how they work together, reflected in the work of IT and Solution Architects. In order to be part of the outcomes of concrete design projects, it also needs to go into relevant details and cover the parts of technology that are visible and tangible for people.

In large enterprise settings, specialists in Technology Design are not necessarily developers by training but highly skilled technologists. They enable us to make decisions on what technology to use and what technical paradigms and standards to adhere to, inspired by a shared vision of the desired target state to be achieved with a design initiative.

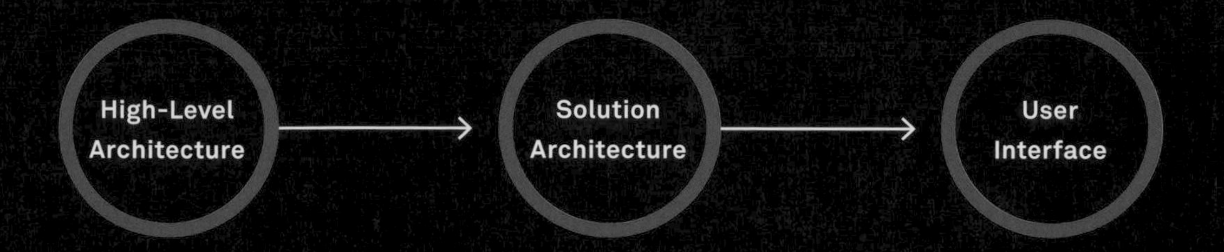

To come up with technical solutions to design challenges that fit, designers need to apply technical thinking and align with engineering, and to model technical constraints and possibilities:

_ HIGH-LEVEL ARCHITECTURE

arranging high-level technical components used and infrastructures available in the enterprise, aligned with overall business requirements and opportunities. This includes fundamental decisions on data management, process support, applications and software, IT systems and networks.

_ SOLUTION ARCHITECTURE

defining technical components such as software, data stores, servers, and devices with regards to a particular problem setting. It involves understanding the business problem and applicable technical possibilities and limitations, to come up with a technical configuration that fits. This activity is also referred to as business analysis or solution design, and there are many related fields such as security or network architecture.

_ USER INTERFACE

often forgotten or neglected in the enterprise context, this involves working with digital media and front-end technology to convey information and make interaction accessible to people. As is the *last mile* of technology, this involves specifying the channel used such as web or print, as well as technical components like pieces of software, sensors, devices, or projections.

The enterprise as an instrument _

While advances in technology are without doubt a key driver in economic development, the experiences people have in the enterprise context are often influenced by the extensive amount of automation and reliance on tools in large organizations. In a way, these tools are the organization using them, since they are what people see and use when doing their activities. This makes it worthwhile thinking about their influence on life and business in the enterprise.

American experience design professional Andrew Meier wrote about the idea of Corporate Interaction Design in a blogpost from early 2010. One of his thoughts was to tie the intended behavior of corporations to the *Three Laws of Robotics* postulated 1942 by Isaac Asimov. These laws are fundamental rules, technically enforced in a way they cannot be bypassed in any way by a robot:

_ *A robot may not injure a human being or, through inaction, allow a human being to come to harm.*
_ *A robot must obey the orders given to it by human beings, except where such orders would conflict with the First Law.*
_ *A robot must protect its own existence as long as such protection does not conflict with the First or Second Laws.*

In a way, organizations can be seen as instruments — like robots — in people's lives, represented by the products and services, and the jobs and resources they make available. They have to be designed and implemented so that they are easy to approach and deal with. While surprising new functions and proposals are welcomed by their users, in daily business they are expected to behave professionally, in a predictable and controlled fashion. And above all, they have to avoid all negative effects on the people related to them.

In practice, addressing technology requires an ongoing dialogue between designers and engineers to determine what is possible, what has been done, and ultimately what could be done.

_TECHNOLOGY IN THE ENTERPRISE

_ IDENTITY

how can technology support our brand goals, communicate the brand message and reach its audience, and encode desired behavior using automation.

_ ARCHITECTURE

how to use technology to help the enterprise perform its activities, to automate what can be automated, and to provide operative intelligence.

_ EXPERIENCE

how to make people's experience better with the use technology, using technical capabilities and intelligence to make information and interactions available.

_ ACTORS

who is using what technology resource in the enterprise, and who depends on it.

_ TOUCHPOINTS

what types of devices, media, or interfaces are used to enable interactions with technical systems.

_ SERVICES

what services are supported by technical components or infrastructure, how to ensure availability and continuity.

_ CONTENT

what content is being stored and processed in technical systems, and how to turn it into usable data structures.

Modern business challenges are wicked problems. All too often, organizations unknowingly pass this complexity on to customers, resulting in negative user experiences.

James Kalbach

DESIGNING THE ENTERPRISE AS A SYSTEM _

Most of the aspects portrayed in this chapter are taken care of in large organizations. In many cases, they are treated as isolated areas of practice — projects redesigning customer interactions, operational processes or deploying new information system are carried out independently by expert groups.

This kind of separation of concerns is important for interdisciplinary work, making experts work on the things they are actually experts in. The way it is practiced in many organizations is one of the major challenges enterprises face today. With regards to strategic objectives, it is a lack of alignment that prevents such initiatives from achieving the results they were aiming for, impacting the design and performance of the enterprise at a whole.

_ EXAMPLE

Most of us have to deal with very large organizations, such as energy suppliers or telecom firms. Many also have already experienced the problem of incoherence between different parts of the same company. They make us shift back and forth between different departments and contacts, establishing walls between media and people. The website is done by Marketing, so Customer Service owns just the phone — your email will not be answered. The website works fine when everything is in order, but operational delays are not accounted for. There seems to be nobody responsible for your problem. All these examples result from a missing alignment, and the resulting blindness for issues outside the own field of responsibility or expertise.

_ Somebody Else's Problems

The designs in place and decisions made in large organizations are often made without alignment, leading to incoherent and unhelpful behavior towards their actors. Using a concept from science fiction author Douglas Adams, you could say that our interactions with organizations suffer because they are putting too many *Somebody Else's Problems* in place.

In the third part of his famous series *The Hitchhiker's Guide to the Galaxy*, the technique is described in detail:

An SEP is something we can't see, or don't see, or our brain doesn't let us see, because we think that it's somebody else's problem. (...) The brain just edits it out, it's like a blind spot. If you look at it directly you won't see it unless you know precisely what it is. Your only hope is to catch it by surprise out of the corner of your eye. (...) It relies on people's natural predisposition not to see anything they don't want to, weren't expecting, or can't explain.

In order to overcome this issue, designing the future of enterprises has to start on a conceptual level, by taking into account all aspects deemed relevant with regards to the design challenge to be tackled.

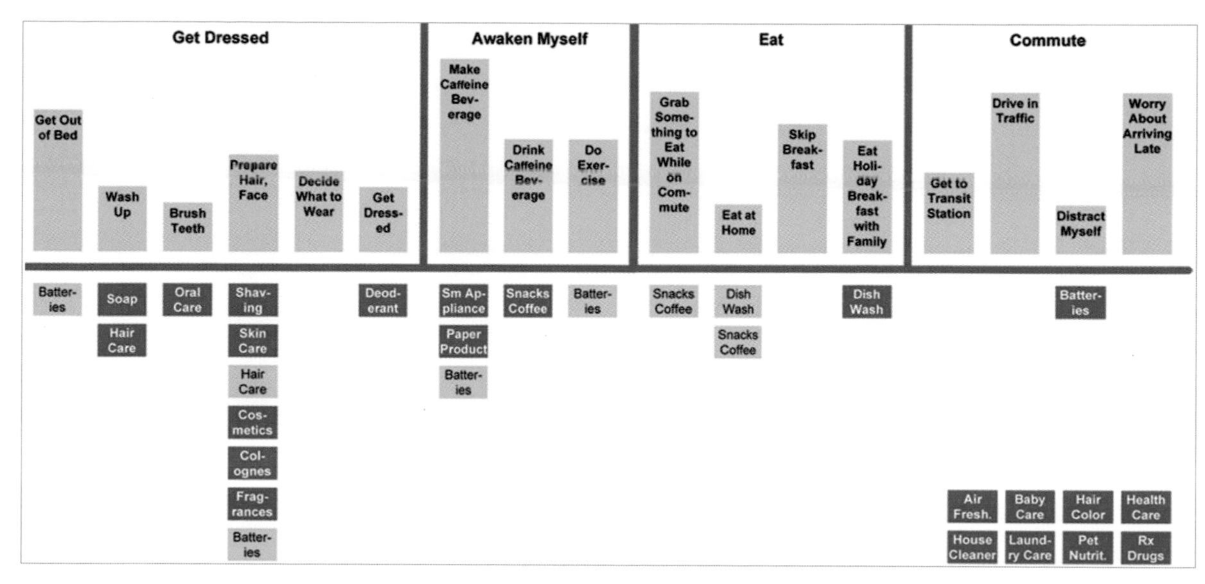

A Mental Model diagram (Source: boxesandarrows.com, © 2008 by Indi Young)

Achieving alignment _

The conceptual aspects described in this chapter lay the groundwork for the transformational goals of a strategic design initiative, spanning the space of design decisions and ideas, and forming the basis for designing concrete outcomes as the result of the design process. Decisions made with regard to these aspects determine the outcomes of a design initiative, both for people addressed, and organizations commissioning it.

The point of the Design Space part of the Enterprise Design framework is to show the relationships between those aspects, illustrating clearly that conceptual decisions and ideas with regards to a particular aspect have to be seen holistically, in terms of their effects on the entire system. As mappings between different perspectives on the enterprise, such as business- and people-centric thinking, applying the conceptual aspects in design work can be seen as an alignment activity, with the goal of synthesizing solutions based on divergent viewpoints and needs. This requires us to capture the proposed solution in models that express how those views work together. In an article for the *Parsons Journal of Information Mapping*, James Kalbach and Paul Kahn propose the creation of visual models to illustrate how different aspects of a conceptual design align. They use the term *Alignment Diagram* to describe graphical mappings between two related aspects, such as the Mental Model diagram above maps people's activities to functions made available to them by a tool or service.

The term fits equally well the examples used throughout this chapter, connecting different perspectives. Such mappings allow us to specify how different elements of a conceptual design are supposed to work together, but more importantly provide graphical notations to make alignment efforts visible. This permits a design team to actively fight the SEP phenomenon, showing stakeholders how their areas intersect.

Depending on the particular challenges and themes behind a project, there will be aspects which are more important than others, and there is a need to set priorities according to that context. In the enterprise context, all of them are relevant. Models of cross-cutting concerns make the thoughts and goals that underlie a certain design explicit, acting as a language for the dialogue between strategy and design, and forming a space to define the enterprise, its qualities and features.

To come to life, conceptual models have to be turned into visible outcomes that implement the concepts. By putting the tangible outcomes of a design project on a strong conceptual basis, design teams can ensure that they are not just making seemingly creative decisions which in the end prove to be arbitrary. Instead, results are the logical consequence of the design decisions made, with regard to the concerns to be addressed.

Translating conceptual designs into visible outcomes is what the following final layer of the Enterprise Design framework is about.

AT A GLANCE

Designing the enterprise means actively engaging in change, and achieving coherence across different domains relevant to its endeavors. This requires aligning different viewpoints to unveil opportunities and constraints and ultimately coming to a clear vision of a desired future state. The conceptual aspects provide a map of the potential design decisions to be made in order to get there.

Recommendations

_ Base conceptual design decisions on insights gathered from applying the frames described in Chapter 6, and choose suitable design themes to be addressed in a coherent fashion

_ Design information architectures, communication channels, and interactive tools for people to have conversations, access information, and interact with or within the enterprise

_ Make operational processes, organizational structures, and technical components part of that vision, consciously shaping their characteristics to support the way the enterprise works

_ Blend virtual and physical, social, technical, and economic aspects of the enterprise in your considerations, constantly redefining the Design Space to take into account all relevant aspects

CASE STUDY _ SAP

SAP'S ENTERPRISE

Helping companies of all sizes and industries run better.

SAP is one of the world's largest independent software providers. Founded in Germany in 1972, the company developed rapidly to its current position as a market leader in enterprise application software for operational business processes, analytics and Information Management. Originally focusing on on-premise installations for resource planning and relationship management, SAP has moved into the areas of mobile apps and on-demand solutions delivered via the cloud.

SAP focuses strongly on supporting how people work, reaching beyond technical enablement through process automation and Business Intelligence to focus on how the user experiences work tasks. Building on its promise to make enterprises run better, the company employs a design-led approach to developing its offerings and practices for customers and other stakeholders. This focus reflects the influence of its founder Hasso Plattner, best known as the founder of the famous d.school (The Hasso Plattner Institute of Design) at Stanford University in Palo Alto, California and a similar institution in Potsdam, Germany. SAP promotes a Design Thinking approach throughout all areas of its business. SAP has a worldwide team of User Experience design and research professionals adopting a human-centric perspective, tailoring applications to the way people use them in their daily work.

With these paradigms at the heart of SAP's corporate culture, the company applies these principles and practices well beyond the development of software products. Looking at the wider enterprise, SAP uses its design practice strategically to enhance the way the organization itself interacts with its people and its ecosystem. One particular area where this type of design thinking is practiced is in the Knowledge Management Competency Center (KMCC), a SAP internal service provider, tasked with enabling knowledge work and information exchange within the company and beyond. The KMCC User Experience Consulting team asked eda.c to support a series of initiatives to redefine the way the organization works and interacts with its environment. The team has conceived and implemented solutions to drive employee productivity, facilitate social exchange, make knowledge accessible and connect actors across the enterprise ecosystem. Such work requires bridging and aligning different viewpoints to address the crosscutting concerns active in the complex Design Space between business aspirations, people's activities and technology. The following examples illustrate the variety of design themes and conceptual aspects addressed on a daily basis by the KMCC and its User Experience Consulting team.

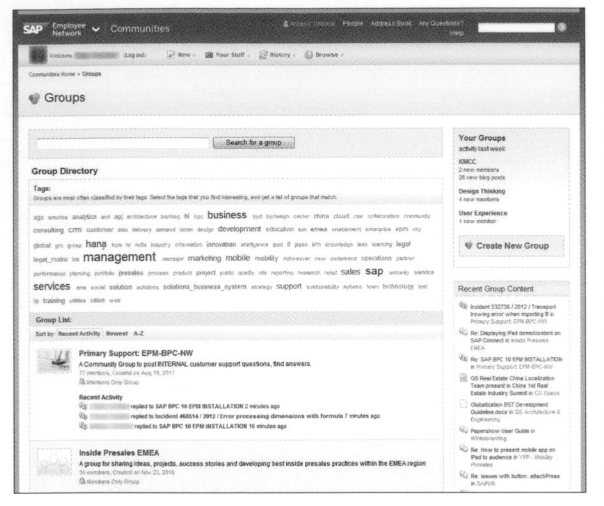

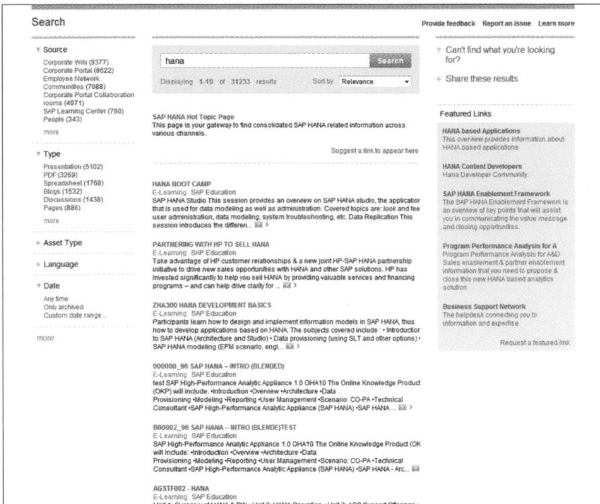

Specialized internal communication devices at SAP

_ COMMUNICATION

SAP's enterprise ecosystem is characterized by vivid exchanges between various actors, all engaged in very knowledge-dependent work. To support these interactions and to accommodate specific needs, the company has established several platforms, communities and distribution channels. One particular challenge is moving design team conversations away from telephone calls and emails and into specialized channels, thereby reducing the time and effort previously required to put messages into context and filter the relevant from the irrelevant. Such a system also means integrating communication media into other tasks, such as productivity tools or transactional applications.

Based on researching communication patterns and needs, the design team defines and enhances a coherent family of online communities, media platforms and apps, each suited for specific communication structures—including microblogging, community platforms, asynchronous collaboration and virtual meetings. As conversations are regarded as essential for any type of knowledge work at SAP, social interactions are also seen as a necessary ingredient for enhancing the effectiveness of other tools and environments regardless of the their particular purpose. Moreover, all campaigns and messages are conceived in unison with the media used to convey them, forming a unit that functions as a whole. This global approach allows them to make channels and platforms specific to the respective business context and communication modes, while also creating connections between these different nodes in the network.

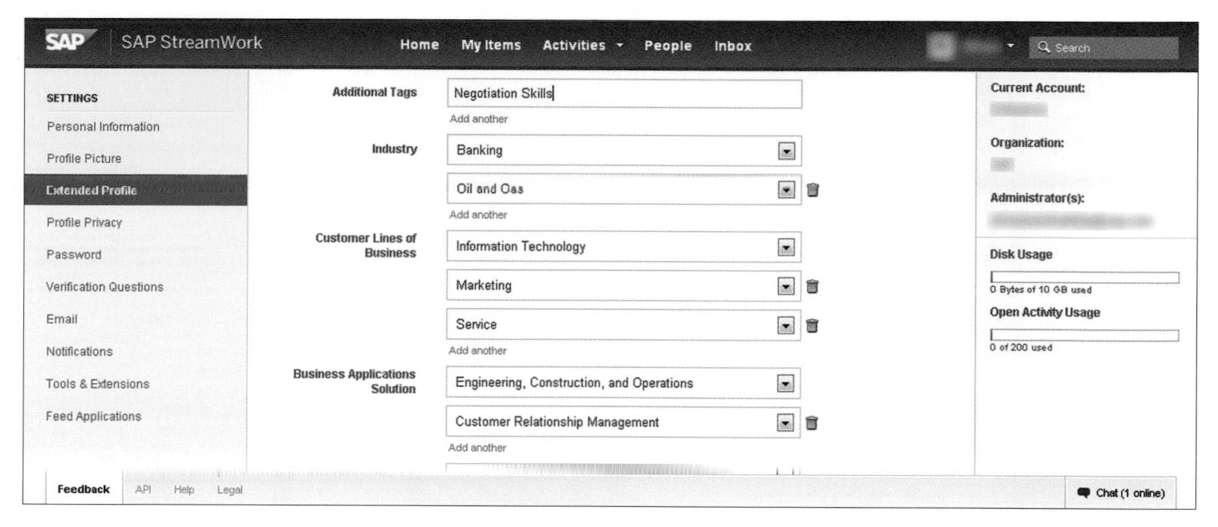

A global classification system for all content based on a corporate taxonomy

INFORMATION _

One of the special challenges SAP faces is that virtually all activities require people to access a large amount of current, high quality information to do their work. They have to deal with complex solutions and scenarios, business processes, customer needs and project settings. All design work therefore includes a strong Information component, because any outcome must effectively help people to find and use the bits of information relevant to themselves, and at the same time to make sense of large amounts of semi-structured content and operational data. In practice, this translates into the creation of classification systems, navigational structures and data architectures that make bits of information easy to find.

In order to address this issue independent of technical implementation or individual tools, Information Architecture work starts with abstract models of information structures. This work is informed by research on people's mental models and aims to create classification systems independent of particular applications. In a mix of top-down structures and social bottom-up folksonomies, these models give structure to topics and things relevant to SAP's users and business context. They provide the basis for hierarchical categorization, labeling, visualization, and metadata models. This abstract work is the basis for classifying and formatting information in a coherent fashion, accommodating to the entire range of web portals, repositories, communities, databases and search engines.

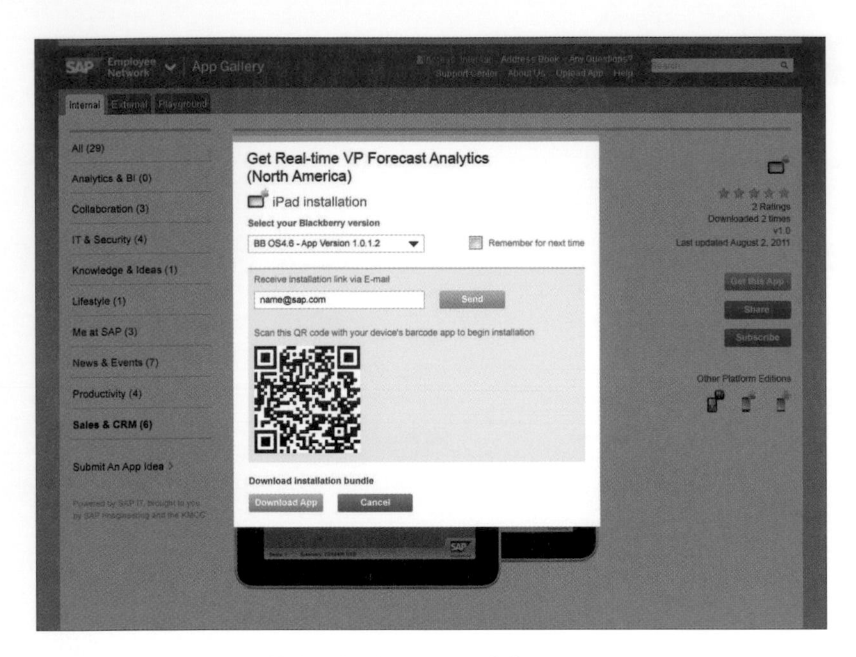

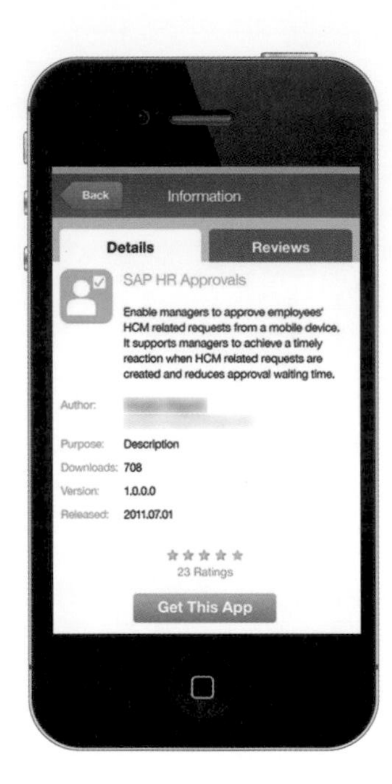

An internal repository to publish and exchange apps of all sorts

_ INTERACTION

As a software company, SAP uses digital systems to support business processes and workflows wherever possible. For conceptual design decisions, this means providing valuable tools for virtually all interactions between people and the company. By providing automated functions for all predictable operational tasks, human activities can be reduced to focusing only on the instances where people can add the most value—by making decisions, solving problems and generating knowledge, supported by tools suited to these activities.

To address the Interaction aspect, the SAP design team starts with behavioral research and task analysis to inform the creation of activity models. These models look at the wider environment of people's tasks, their business context and range of tools. In a subsequent step, the design team moves on to reshape these models and the interactions they represent in an effort to make people more efficient, remove unnecessary steps and match their behaviors to the context. Based on this modeling activity, the design team can expand conceptual work to the entire tool landscape, translate this to new user interfaces and redefine existing artifacts. In the course of eda.c's work with SAP, topics as diverse as travel management, app distribution and remote work were addressed. By looking beyond the usability or design of single applications, as in this case, Interaction Design work reshapes enterprise/people interactions as a whole.

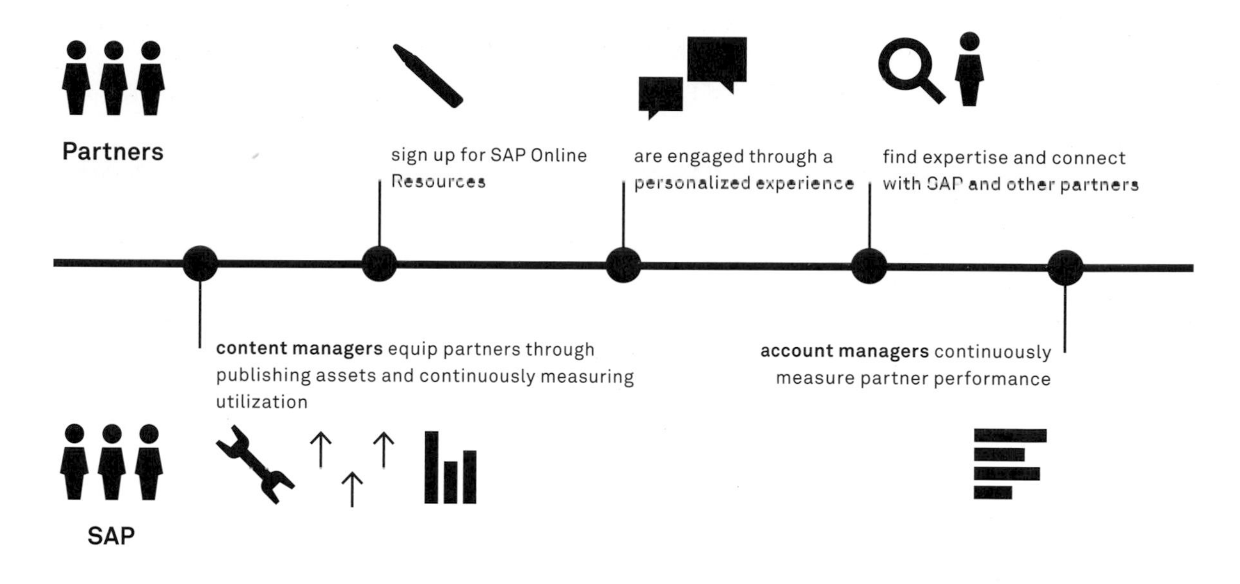

Designing the process of Partner Engagement

OPERATIONS _

SAP has an exceptional perspective on business process design and operational optimization. After all, making companies run better is its core expertise. Consequently, Operations is of particular interest for the company's strategic design initiatives, looking at the synergy between User Experience innovations and operating models to improve and accelerate business processes.

In one particular project, design teams explored operations within shared service centers. Based on research of people's activities and in close collaboration with experts in business processes, the design team reshaped work practice, procedures and tools according to new operating models in one SAP service center. In another setting, designers and researchers worked closely with the SAP Human Resources group to define a career development process to be applied across the company. In both cases, design solutions translated strategic goals into transformation projects, in a process of active collaboration with the people impacted by the redefined processes.

In projects reshaping operational design practice, a deep understanding of the work and a mapping of processes, activities and functions of the people involved enabled significant performance improvements and a better work experience. By helping employees to solve their own task problems efficiently, this design work directly contributes to SAP's objective of increased daily productivity.

We have found to be successful at creating strong user experiences, we must deeply engage with the surrounding business processes—well beyond the usual framework of the User Experience practitioner.

S. Kirsten Gay, Director User Experience Consulting at SAP

When attempting to improve enterprise efficiency and employee productivity, it is crucial to focus not only on well-engineered solutions but also on emotional value. What's fun gets done! As such, User Experience strategy, research and design are not an option; they are a necessity for developing economic value and company success.

Dirk Dobiéy, Vice President Knowledge Management at SAP

_ ORGANIZATION

All of the projects mentioned above transform the way people collaborate and do their work, which makes the Organization aspect another important point to be addressed. Any outcome of a design project that is strategically relevant comes with new ways of working together, shifting responsibilities and evolving governance models. Part of SAP's design work therefore is to engage co-workers in a large variety of roles, as well as with other actors in the company's ecosystem.

The User Experience Consulting team, for example, was tasked to explore and redefine the way the company works with SAP's partners. The partner ecosystem plays a critical role in helping customers identify, purchase and implement the solution deemed suited to their particular needs. Using an approach inspired by Organization Development practice, the design team conducted a series of workshops in collaboration with the designers behind SAP's *PartnerEdge* program, enhancing the experience of partner participation. In another program, members of the team developed a consolidated model based on the most relevant job roles, functions and skill profiles, to better tailor the different elements of the workplace supporting specific responsibilities and activities. Working on questions of responsibilities and teamwork using a design-led approach allowed turning abstract models into functional models and scenarios, making the implications of an organizational model visible.

Mobile usage
Goal: enterprise apps for my device are easy to find, install and maintain, the same way as “normal” apps
Scope: mobile apps for smartphone and tablet devices

Mixed usage
Goal: installing apps for mobile devices from the Web Store is a simple, seamless and integrated procedure

Desktop usage
Goal: I can discover, install and exchange on interesting apps for my personal/preferred technical environment
Scope: all kinds of apps and similar pieces of software for desktop, web and mobile use

Preferred solution: a **native app** that mimics the vendor's mobile app distribution channel

After selecting a mobile app for installation from a desktop PC, the mobile device is provided with a pointer to an installable file

Web Store featuring free browsing and searching, social features such as rating and commenting, idea management, developer content and support

Fall-Back solution: a **mobile web store** for unsupported devices, those without client app, and to provide additional content or social features (rating)

Dealing with different channels to enable app retrieval

TECHNOLOGY _

SAP has gained its position as a leader in the business applications industry by consistently testing and using the products it designs to drive the company's own operations. Consequently, the company is infused with SAP products and technology in every detail of its activities. These information systems designed and utilized by SAP provide the backbone of all operational processes, all stocking and processing of business information, all enabling of SAP people to work and collaborate. In short, to manage its business and to operate effectively, *SAP runs SAP*.

An important part of eda.c's design work with SAP is to combine these technical resources in a way that boosts their usefulness for people, thereby maximizing their benefits to the business activities. This involves experimenting with new solutions as they emerge and understanding a complex landscape of systems and devices. In collaboration with Research and Labs groups and the global IT organization, design teams are aligning the solution architecture, systems landscape and technical standards with the intended requirements outlined in the design processes. Beyond questions of feasibility, this involves informing design decisions with new technical possibilities. Driven by a holistic viewpoint on the enterprise, new solutions are seen as a vital source of internal innovation. Following the *SAP runs SAP* paradigm, design teams working on SAP's products are basing new market offerings on internal projects and experiments, as well as on products co-innovated with the entire ecosystem.

8 _ RENDERING

The range of aspects described in the previous chapters is meant to help a multidisciplinary team consider the intangible dimensions of strategic design work. Thoughts about suitable interactions, the organization of information, the business model or domain structures behind a certain solution, questions of brand identity or the choice of actors to address — all these elements lay the foundation for making informed design decisions and drawing a picture of the future. When working on a project meant to have strategic significance for an organization, project teams are often tempted to keep the design process on an abstract level for quite some time. The diversity of stakeholders and concerns, and the mass of questions posed and thoughts being exchanged make it easy to lose yourself in high level decisions while avoiding the concrete—this is often visible in team discussions about abstract concepts such as customer loyalty or brand values. To become reality, any design process needs to turn this abstract foundation into concrete results that have a determinate impact on people's lives.

Design is always about making, and as such it is a craft that requires us to work with materials, production techniques, and creative expression, mastering their usage and effects. Every decision made, conceptual model applied, or transformation intended that comes out of a design process must eventually result in something visible and tangible in order to come to life. Using design strategically in the enterprise does not imply abandoning this form-giving process that differentiates a design approach from a purely intellectual planning exercise. Although it starts on a conceptual level deeply grounded in strategic thinking, research activities, and systematic modeling, it is the concrete results that count.

© 2013 Elsevier Inc. All rights reserved.

The Enterprise Design framework is about defining and shaping these results as a web of connected elements rather than as isolated artifacts, and designing them to play a certain role in the enterprise as a bigger system. Conceptual aspects form the basis for deciding what to make in order to achieve what effect. The three remaining aspects of the framework portrayed in this chapter are about bringing them together and giving them a form as parts of a desired future state of the enterprise.

Visible futures _

The way we perceive the world and interact with it is largely driven by our nature as embodied physical beings. The environments that surround us and the objects we use are so closely connected to our thinking and behavior that we make use of them to arrange and structure our lives. Every morning, we move from the bedroom to the bathroom, go to work, come home again later on, and using physical environments to organize our activities. We "have a glass of good wine" together with some friends, we collect souvenirs from our vacation trips to help us remember, we associate sensations picked up from our environment with memories, places, or activities.

Physical objects or places and their configurations serve us as focal points to help make sense of the world, as symbols and contexts for our thinking and acting. We use physical metaphors such as "rising stock prices" or "going into a project phase" to describe entirely non-physical things. This universal role of physical categories is also visible in the way we apply them to activities within digital environments, where we are opening and closing application windows, moving around on the web, or "poking" someone on Facebook without leaving our computer, illustrating how the physical world rules our increasingly virtual lives. The continuing digitization of our environment makes us constantly shift between the virtual and the physical even when dealing with a single artifact.

Being an intangible conceptual construct, the enterprise comes to life only in real-life situations, as its substantial manifestation. Any intended transformation requires a translation into a set of concrete elements to be designed, produced, and introduced, turning the underlying intangible concept into perceivable signs, tangible things, and habitable places. As strategic design initiatives by definition have to start without a predefined outcome in mind, the choice of what concrete results to produce, and what shape they should assume fully rely on the ideas of the team members and the possibilities at hand. Each conceptual aspect previously described can be taken into account as an external reference to develop this set of outcomes.

One of the particular strengths of a design approach to difficult challenges is its inherent emphasis on crafting visible results that are fully controlled and envisioned by the designer who creates them. As part of this framework, we suggest three aspects to pave the way from strategic thinking to the crafted outcomes of a design process:

_ SIGNS

are carriers of messages and symbols that people are exposed to. The enterprise and its actors produce and place signs, but to make them perceivable they have to be encoded media of some kind, such as written words, visual or auditory messages. Signs are generally created to reach a specific audience with a certain effect in mind, such as making a message heard, enabling understanding through communication, conveying an identity, or supporting wayfinding.

_ THINGS

are objects people use, own, consume, take with them, or create. In the context of design, the term refers to various kinds of man-made artifacts such as goods, devices, and tools. In the enterprise context, there is a mass of things being produced, utilized, and exchanged. Designing things, thereby defining shape, characteristics, and materials, has the goal of supporting usage or marketization.

_ PLACES

are where people go, where they live or stay, meet or work. They provide the environments for activities and interactions in the enterprise, but at the same time they are also the stimuli for personal associations, memories and moods, as well as social connotations. Designing places means generating context for people by shaping their surroundings.

The aspects described in this chapter are meant to support the choice of outcomes, and employ the large variety of the arts, and classic fields of design, to make them reality. They are about giving enterprise elements a form — for example, as interiors, websites, pieces of software, print media, or signage, and making these elements work together as a system. Because of the large field of design disciplines, the practices portrayed are archetypes for individual design expertise to be called in. Depending on what best translates your vision of a future enterprise into action, you might need to involve web or mobile app designers, interior or landscape architects, or a mix of editors and illustrators, just to name a few.

The conceptual background based on the previously described aspects is the basis for working with specialist designers, envisioning the outcomes of a strategic design project. Nevertheless, it provides just the intellectual framework to guide the creative process. At this level, the myriad of strategic considerations and conceptual decisions essentially becomes a checklist for a designer to consider when coming up with a design proposal. The center of attention then shifts to a mastery of the selected creative domain and the tiny details of the design.

Bringing a strategic design initiative to a state of tangible outcomes can prove quite challenging, especially since these activities are traditionally considered to be merely tasks to be executed. They are often relegated to the lowest organizational level or carelessly left to external service providers. In reality, these outcomes are the only bridge connecting a strategic design project to the results it seeks to achieve. Neglecting their importance risks spoiling everything, and even the best conceptual approach will not help if the details of the outcomes are badly designed.

To be effective, the crafted outcomes have to embody all conceptual decisions and turn them into perceivable qualities and usable functions, to trigger a transformation of the enterprise as a system. This is exactly what distinguishes mere reactive beautification from a strategic design approach. The challenge therefore lies in turning the concrete outcomes into a rendering of a strong concept about the enterprise, and achieving the translation from thinking to creating while preserving excellence in creative work and professional craftsmanship.

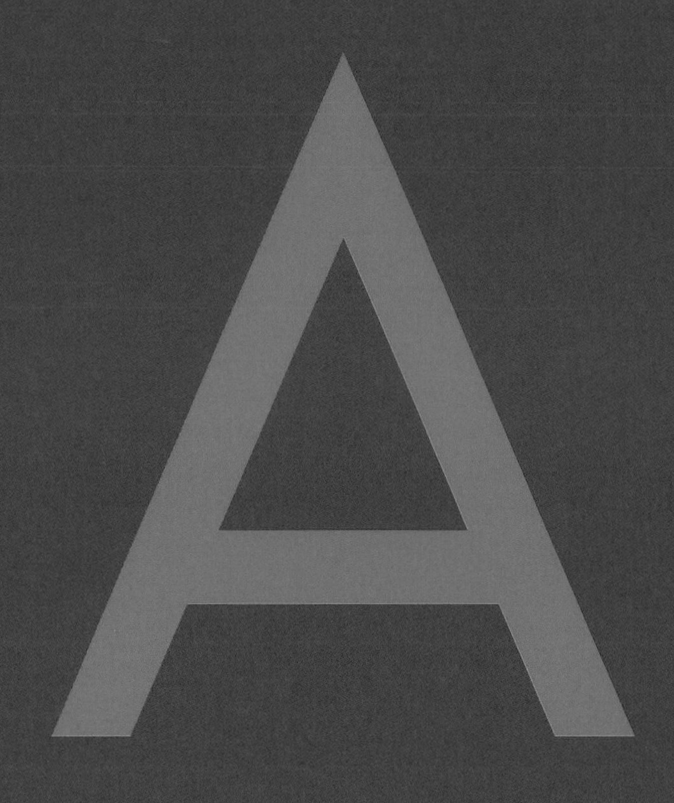

#18_SIGNS

The term *signs* in the context of this framework refers to all kinds of messages encoded in visual display devices or other types of media, created to convey a message to someone in the enterprise. Being an important part of our culture, signs and symbols permit recognition, convey information and meaning, enable wayfinding and usage of machines, trigger complex chains of thoughts, and cause profound emotional reactions. They can be used as a common communication device based on associations shared between people, making a message speak on different levels of the dialogue. Consequently, the creation and use of signs and symbols is the most straightforward way to make connections between an enterprise and its audiences.

Signs can be seen as individual interpretations of a set of impressions to our senses, for example, a pattern of colors that is picked up by the eye. The subsequent processing in our brain then enables us to recognize shapes and figures, distinguish figures from their background, and to make all of these interpretations the subject of our recognition and thinking processes, making up our minds about what a sign stands for and what it means to us.

Considering this process, although it happens quite quickly and on an unconscious level, is the basis to make use of signs in a way that they direct people's attention and have an actual impact. Anything being made in the enterprise can be seen as a contribution to a dynamic exchange of signs being created and perceived. The shapes and configurations of these signs drive the interactions with people. Because this is the perceivable part of the enterprise, it has to be addressed as one of the key aspects pertaining to the rendered results of strategic design initiatives.

Signs as a subject of strategic design _

The world is flooded with man-made systems of signs. The process of producing signs is continuously happening and accelerating, with everyone being able to participate. At the same time, there is a steadily growing number of different media transporting these signs to their audiences. Road signs share the urban space with advertisements and street art, while print and broadcast media is used in various formats to inform and educate, and tell stories in books, magazines, radio messages, and other forms of media. The Internet and interactive forms of media add to the mass, so all these channels intersect and converge and make users constantly transition between them.

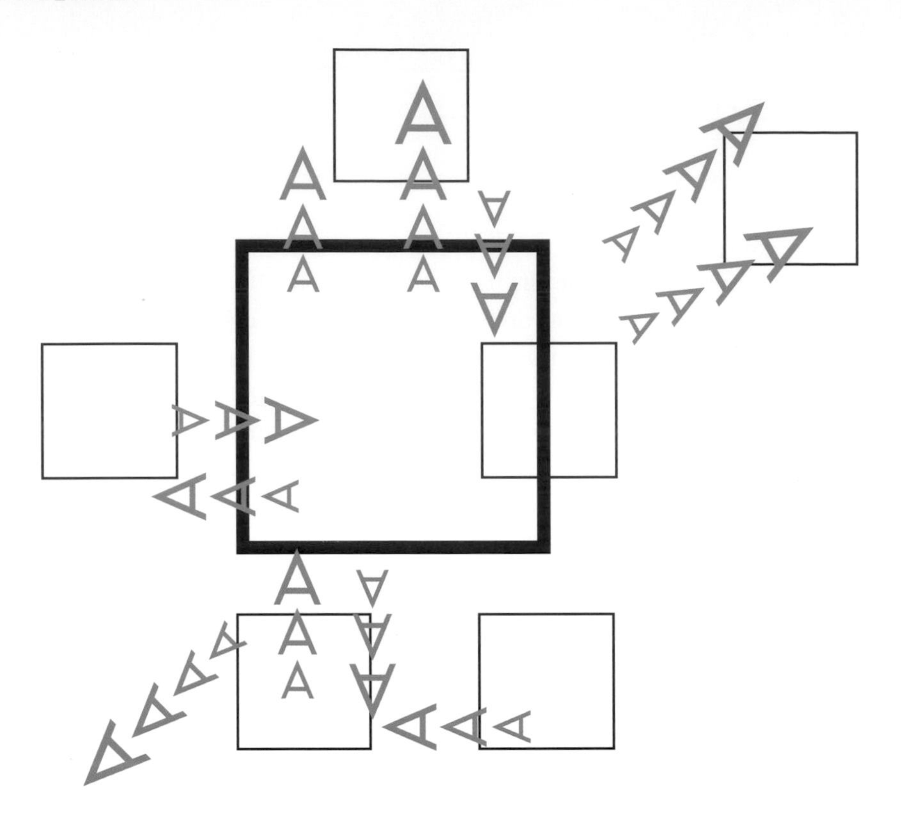

Instead of disappearing completely, legacy media often just exists alongside the new media — for extended periods of time such as newspapers, or in a niche like vinyl discs. Today, it becomes clearer that although the digital revolution has hit the media landscape with all its force, the paperless world is still far away, if ever to arrive. Paper-based media like books, magazines, and office paper still prevail. Other than shortcomings of display devices, there is something special about paper as a medium, a sensation-filled experience of haptic qualities, smell, and usage that makes it difficult to substitute.

In the context of strategic design work, this translates to the challenge of dealing with an intermingled mass of media. Virtually all organizations make extensive use of such systems to encode and deliver messages to their stakeholders, in the form of advertising, email messages, websites, packaging, stationery, or signage on their campuses. More often than not, organizations deal with signs in a limited and chaotic way, applying a set of visual brand elements just to their official mass communications for external target groups, while leaving their other roles, aspects, and audiences to chance. Such a superficial approach neglects the true significance of signs for the enterprise, as carriers for making meaningful impressions on people and to facilitate exchange and interaction.

_ EXAMPLE

As an example for signs used ubiquitously across different purposes, consider a national flag. Depending on the context of use, they indicate the country where a car is registered, an embassy in a foreign country, but may also be used by hotels to decorate their entrance in an "international" fashion. On a pragmatic level, flags are also often used to depict a language on websites or in restaurant menus, despite all the ambiguity there is in such representations. Beyond these specific purposes, flags are highly iconic symbols loaded with a wide range of interpretations, visible in the UK's Union Flag used in pop culture, or when people enthusiastically wave specific flags in street demonstrations—or burn them instead.

To be strategically relevant, systems of signs have to be designed according to the enterprise environment they seek to support—making them take up their role to get relevant messages across, thereby underpinning the enterprise. A design team has to consider all suitable media channels to shape a system of signs that makes the enterprise in all its facets visible to the actors it is addressing.

MEDIA DESIGN

Media Design in this context is in fact an umbrella term, used mainly in European countries to describe a dynamic field of specialized design disciplines, all with dealing the design and production of some sort of media. The field is also often called Communication Design or Visual Communication, all referring to slightly different aspects of the same task, which consists of giving a form to messages the to be conveyed to a targeted audience using a specific medium such as paper or the web.

Because of their shared roots, this area overlaps with the fine arts, having similar means but different purposes in mind. When talking about media and design, there is traditionally an implied emphasis on creative practice addressing visual perception, such as in graphic design or photography. Applying a broader view, the field today presents itself with a large variety disciplines and sub-disciplines with varying degrees of specialization.

The professional use of media for the purposes of strategic design depends on the collaboration of expert designers from these different fields. It might also involve other disciplines such as musicians or artists, or be completely independent of a recording technology, as in the case of a live performance for events. But more importantly, it has to be based on the larger vision of the enterprise and the way it uses media to make that vision happen, and encode its messages effectively into perceivable and significant systems of signs.

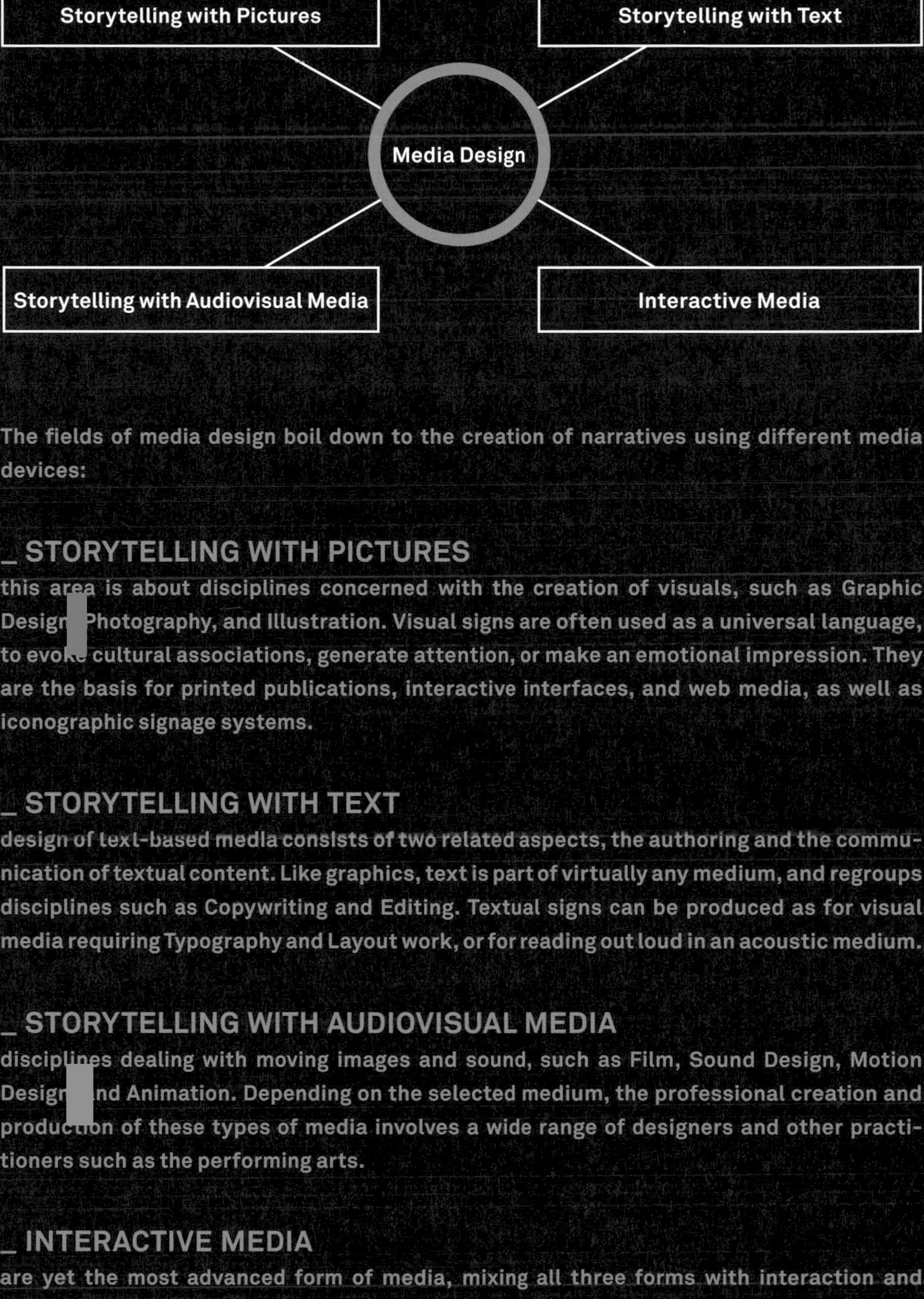

The fields of media design boil down to the creation of narratives using different media devices:

_ STORYTELLING WITH PICTURES

this area is about disciplines concerned with the creation of visuals, such as Graphic Design, Photography, and Illustration. Visual signs are often used as a universal language, to evoke cultural associations, generate attention, or make an emotional impression. They are the basis for printed publications, interactive interfaces, and web media, as well as iconographic signage systems.

_ STORYTELLING WITH TEXT

design of text-based media consists of two related aspects, the authoring and the communication of textual content. Like graphics, text is part of virtually any medium, and regroups disciplines such as Copywriting and Editing. Textual signs can be produced as for visual media requiring Typography and Layout work, or for reading out loud in an acoustic medium.

_ STORYTELLING WITH AUDIOVISUAL MEDIA

disciplines dealing with moving images and sound, such as Film, Sound Design, Motion Design, and Animation. Depending on the selected medium, the professional creation and production of these types of media involves a wide range of designers and other practitioners such as the performing arts.

_ INTERACTIVE MEDIA

are yet the most advanced form of media, mixing all three forms with interaction and behavior.

_SIGNS IN THE ENTERPRISE

_ IDENTITY

manifesting a chosen brand in signs as a perceivable identity, such as a logo, appearance, sounds, and smells.

_ ARCHITECTURE

using systems of signs that make the enterprise work, as in instructions, notifications, or news updates.

_ EXPERIENCE

making signs meaningful to people in terms of language, tone of voice, and message conveyed, like in a personal letter.

_ ACTORS

adapting signs to the nature of the relationship, engaging investors and convincing partners or customers.

_ TOUCHPOINTS

choosing the right signs for a given context, such as a mobile app to take with you, a display for wayfinding, or a movie.

_ SERVICES

making signs support and facilitate services, for example, what call-center staff members tell you about an offering.

_ CONTENT

turning content into narratives and chains of signs, like a commercial proposal or by introducing a behavioral code.

_ BUSINESS

making signs support business goals, informing decision makers about what's happening or placing up-selling offerings.

_ PEOPLE

adapting signs to the audience it is intended for, such as Braille or advertising for a specific profession.

_ FUNCTION

shaping the signs so that it fulfills its purpose, like using visualizations to enable decision making.

_ STRUCTURE

representing the subjects the sign is referring to, using consistent wording or providing accurate illustrations.

_ COMMUNICATION

using signs to convey a message and support the respective communication modes, like polls to foster participation.

_ INFORMATION

placing signs to make information accessible and support understanding, such as icons for recurring categories.

_ INTERACTION

shaping signs to trigger behaviors and engage people to act, like an audio warning signal.

_ OPERATIONS

shaping signs to support business activities and enable operations such as a roadsign or a car plate.

_ ORGANIZATION

designing signs to support team collaboration, for example, an Employee of the Month initiative.

_ TECHNOLOGY

placing signs using media technologies and techniques such as projection or mobile devices.

_ Sign systems for enterprise rhetoric

Too often, design projects are initiated with a specific target medium predefined by a commissioning party, directly addressing specialist designers such as web or motion design agencies. Applying the conceptual aspects portrayed in the previous chapters permits us to rephrase a complex enterprise-design problem as a design challenge, and to conceptualize solution approaches on a level that is initially agnostic to whatever concrete media might be used. This puts a design team in a position to question these preconceptions, and to determine suitable target media in the course of the project. Envisioning a system of signs and media in terms of its role in the enterprise makes a large part of the design decisions a logical consequence of the underlying concept.

Shaping messages into systems of signs in context of an enterprise can be seen as telling a story to its audiences. Depending on the individual purpose of the signs and the context wherein they are used, different aspects of the enterprise come into play and have to be brought together in a coherent way, such as concerns of security and operational safety, sales and Marketing, knowledge and intelligence. All of these messages are part of the same story about the enterprise, and enable its various actors to see themselves in it, define their role, commit, and participate.

In written or spoken language, the art of using words to inform, persuade, or motivate an audience is known as rhetoric. Just as speakers turn their discourse into a considerate flow of statements and figures, a system of signs arranges elements in a way that they effectively work together to convey the message. The task of design in that regard is to define these elements and rules for their application. It is the rhetoric of the enterprise, exposing some parts while hiding others, directing attention, and getting its story across.

_ EXAMPLE

Think of a corporate intranet as the primary workplace environment of employees. It is used as a medium to deliver a wide range of messages to its audiences, from product and sales information to self-service offerings and work-related tools, regulations, and critical safety information. Creating a system of signs that guides the attention of intranet users to the most relevant parts is a challenging task, also because it is often a tool that is highly disputed between different departments such as Communications, IT, and HR. Applying the conceptual aspects will help to define the key stories the intranet should convey, and turning them into a system to encode messages, present information, and guide user choices. Instead of decoration and ornamentation or running after visual trends, a thoughtful application of typography, signs, and space can visually convey its essence without distraction.

The extensive work with the conceptual aspects of the Enterprise Design framework predefines the story to be told, and how to tell it effectively. It is the basis for defining a vocabulary of signs to encode it in media, making the rules and relationships between them transparent. The logic applied to that system and its components should not be arbitrary, but motivated from the underlying rhetoric, and turned into decipherable messages both visually and by other means.

While this translation is a key step of strategic design, it is not the whole story — as good design lives off of good ideas, it is often external inspiration and the work of experienced and talented craftspeople that turns a solid design job into a great one. The goal of working in a systematic fashion is not to find a reason for everything, but to generate options that combine the rational with the inspired.

19_THINGS

We have always been cyborgs.

Michael Shanks, Stanford University

Organizations and individuals make, buy, sell, and use things as part of their activities. Such things are most commonly industrially created objects for usage and consumption. The practice of design applied to things encompasses a wide range of different objects being subjected to a creative process, from toys and household appliances, tools, and furniture made for private consumption, to industrial appliances, machines, or software applications being used as invested assets in organizations, with a great deal of overlap between those areas. Most objects we use in our daily lives are so embedded in our activities that we often don't even notice they are there. Of course watching TV requires the possession of or access to a TV set, but also more profound human needs rely heavily on objects—we sleep in beds, write with pens, read books, and call people with phones. There are numerous examples where the activity is called after the tool or device used to perform it, such as in *vacuuming* or (more recently) *googling* the web.

In the fields of archeology and anthropology, artifacts recovered from excavations are examined as objects of cultural relevance. They deliver insight into the way people lived, how they performed daily activities or individual tasks, and allow us to draw conclusions about the cultural particularities of extinct civilizations and human life in former times. These insights in turn can be used to recreate their experiences and habits—making sense of the intangible by studying the material evidence in the light of history. In his research, archeologist Michael Shanks from Stanford University makes the case that this evidence, the things we use in our lives, cannot be separated from our nature. Objects and artificial extensions to our bodies are an intrinsic part of human existence, issued from a joint evolution of the natural and the artificial—we shape our tools as much as our tools shape us.

In a similar way, contemporary artifacts being designed, produced, and used today have to be regarded in terms of their role as extensions of ourselves, applied in a certain cultural context. They are made to be a part of everyday reality, with their shape, characteristics, and functions designed to contribute to our existence in a meaningful fashion. Designing things should reflect their role as cultural artifacts for the people addressed by the enterprise.

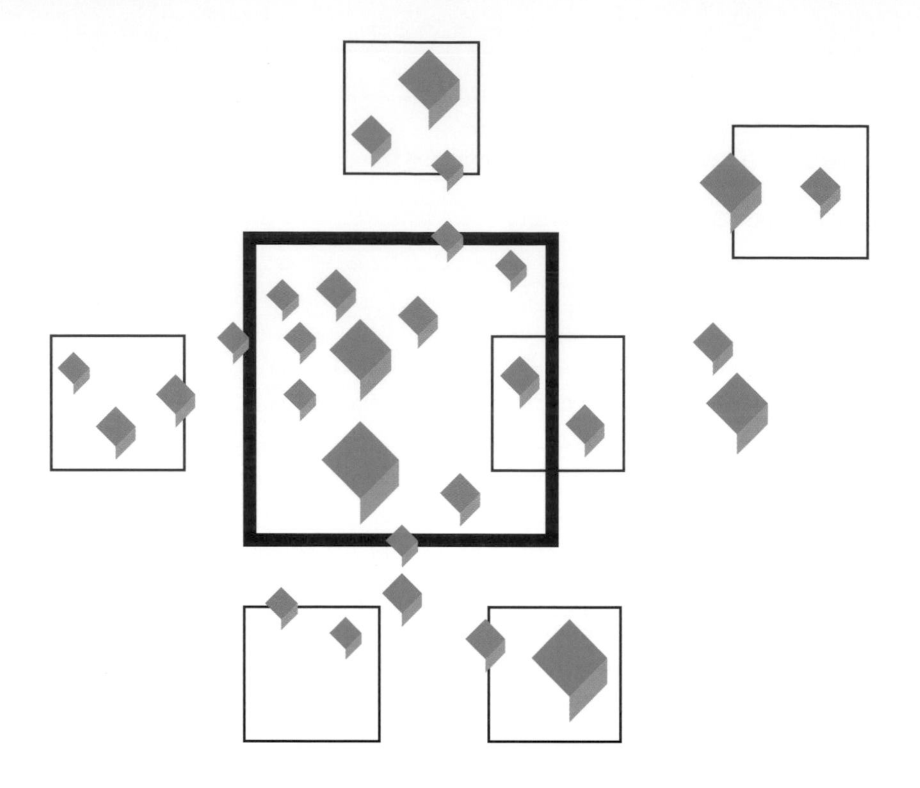

_ Designing things as enterprise artifacts

Organizations produce and use things for a large variety of reasons and purposes. A large part of the economy involves creating and selling things, such as cars or other machines, manufactured products like clothing or food. In these product-led industries, design and related disciplines are used to create the products that are offered to the market, being seen as a part of the product development cycle. As a function applied to the value chain of the organization, the role of design is to translate the aspects of use and the symbolic character of the brand into a form suitable for production.

Because of the continuing digitization of various types of products, product designers today often need to address physical materials together with software components and interfaces as part of the same product, with an increasing significance of its digital parts.

_ EXAMPLE

The task of designing a cell phone has changed significantly within the past decades. Originally a task of bringing together ergonomics and aesthetics applied to mechanical artifacts, recent generations of phones are closer to computer platforms than they resemble their predecessors. The task of the designer has shifted from defining the form of a physical thing to defining the interface and behavior of its screen contents and other channels, amplifying the need for Interaction Design and Information Architecture specialists. Moreover, the phone has become an open platform for apps (essentially virtual tools or things), enabling organizations to address specific needs in a design tailored to their audiences — including customers, employees, or others.

INDUSTRIAL DESIGN

The professional field of design applied to things for serial production is known as Industrial or Product Design. The practice is about shaping the form of a physical object with regards to its use and its meaning for people. There is a wide spectrum of branches and specialized disciplines within the field, applying the practice of design to fashion, furniture, cars, electronic devices, or industrial machines.

As in the other traditional fields of design, Industrial Design deals increasingly with things that use software components to a large degree, turning the attention of design from the outer form to the design of interface elements and their behavior. Examples of considerate Industrial Design achieve a symbiosis of the physical and the virtual elements of a product, designing for the User Experience as a whole.

The field of Industrial Design is probably the best known domain of design, being also the area of many notable designers. In professional practice, however, the task of envisioning, designing, and delivering a product is the result of interdisciplinary teamwork involving many different areas of expertise and the mastery of professional tools. This effort pays off handsomely, given the ubiquitous role tools and other things are playing in our professional and private lives. Their characteristics and shapes are a key element in any desired change.

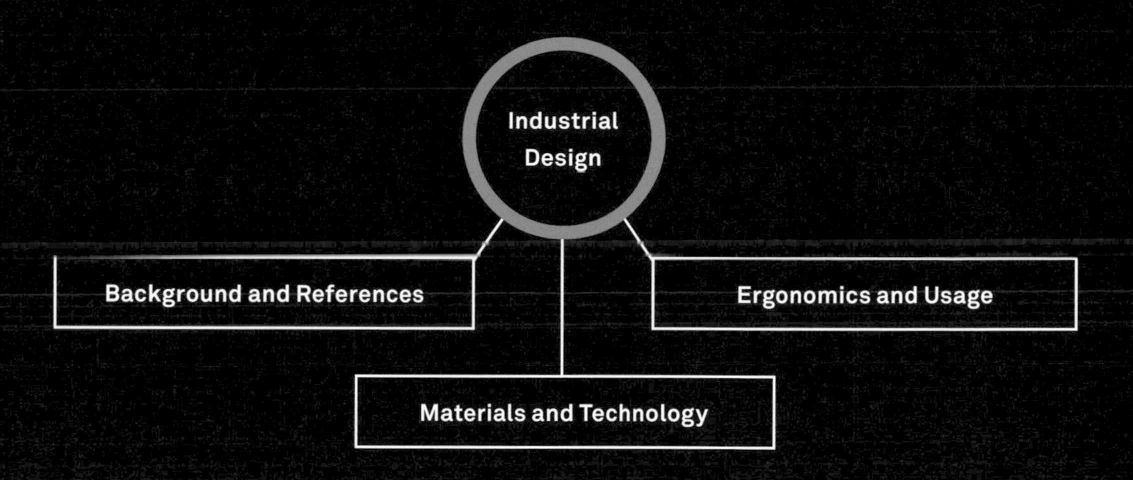

The task of coming up with a solution requires balancing different considerations to be taken into account to define a suitable form:

_ BACKGROUND AND REFERENCES

designing a product to express the goals and values behind its creation, also sometimes referred to as the styling of a product. This aspect can be defined as the embedding of references into the form. Such references can be based on the personality of a famous designer, on the brand identity of the company offering it, or the culture and context of use it is made for.

_ ERGONOMICS AND USAGE

working on the relationship between an object and its user, and adapting the form to the activities that it is supposed to support. This refers to the functional character of a product, choosing its form based on the tasks and needs of people, their characteristics and abilities. The designer might also enhance the product's functions or combine functions that go well together.

_ MATERIALS AND TECHNOLOGY

considering technology, raw materials, and engineered solutions as part of the design process is an intrinsic part of the industrial design discipline. This is especially true in technology-centric industries, where design is employed to make technical innovations accessible in a product. In the context of physical things, this involves making materials, technical components, and manufacturing considerations a part of the design process.

_THINGS IN THE ENTERPRISE

_ IDENTITY

designing things as symbols representing identities and their characteristics, such as luxury furniture in a hotel lobby.

_ ARCHITECTURE

designing artifacts adapted to their role in the enterprise, like a handheld device for an operative job.

_ EXPERIENCE

shaping artifacts according to their meaning for people, for example, a tool to help you wake up more easily.

_ ACTORS

designing things to support a specific relationship to the enterprise, like an employee access card.

_ TOUCHPOINTS

designing artifacts according to the context in which they are used, such as mobile devices suited for postal service.

_ SERVICES

designing things to make a service run, such as a kitchen bell or a self-service kiosk.

_ CONTENT

designing things to make useful content available and enable exchange, like the starter pack for a new bank account.

_ BUSINESS

designing things to match and support a business model, for example selling a cheap device with profitable consumables.

_PEOPLE

designing things to adapt to people's characteristics, like a cell phone that matches the needs of kids.

_ FUNCTION

designing things to fulfill a need or requirement according to a defined purpose, like a dentist chair.

_ STRUCTURE

designing things to enhance and fit into a given structure of related objects, such as a jet bridge connecting gate and airplane.

_ COMMUNICATION

designing things to facilitate communication processes and social exchange, like furniture for work in small groups.

_ INFORMATION

designing artifacts to convey information related to their usage, such as labels on a stereo.

_ INTERACTION

designing things to support and convey a range of defined interaction possibilities, like a touch screen or a knob.

_ OPERATIONS

designing artifacts to facilitate operations in the enterprise, such as a device for train conductors.

_ ORGANIZATION

designing things to play a certain role in organizational reality, such as the water cooler as a meeting point.

_ TECHNOLOGY

designing things to make technology accessible and usable, for example, a user interface for a complex business application.

Designing things, both in the physical and the virtual sense, is not limited to the design of salable products. Even if not directly sold, things are a part of any offering made to the market, and of all processes or services carried out in the enterprise. In any given context, things are being made or employed to carry out tasks and support interactions between people. As cultural artifacts of the enterprise, things are focal points to support and engage people, so their design should be subject to initiatives that aim at a holistic transformation. By designing things that are meaningful to people as a result of a strategic design process, we can purposefully transform the activities, habits, and practices of enterprise actors to support the underlying strategy and conceptual thinking.

_ Using things in the enterprise

Things have multiple roles and meanings in the enterprise — as cultural artifacts, as tools to support human activities, as products to be offered to the market, or as an interface for service provision or process execution. Redesigning these components of the enterprise in a way means redesigning the enterprise. The conceptual aspects deliver the background needed to inform a thoughtful and holistic Industrial Design process. They bring clarity to questions of function and usage, references and goals, as well as technical possibilities and constraints.

Designing things is at the core of the design disciplines. While the focus of the field is moving towards achieving strategic transformation through a design approach, the products resulting from a design process are the focal point for such a transformation to happen.

Nick Marsh, Design Director at the UK design consultancy Sidekick, wrote about such a product-centered approach to design in his blog. He uses the term *New Product* to describe a new generation of networked products that bring together the physical and the digital in a way that sparks great experiences for their users. These products make interactions with the web and live data exchange a part of their characteristics, while still considering the physical context of their usage as a key factor in the design process.

When people talk about the New Product they're generally describing businesses that provide a mix of content, service and experience. But the way they make that tangible is by focusing on product.

Nick Marsh (choosenick.com)

Putting the emphasis of the design process on the things to be made helps a design team to adopt the solution-oriented mindset that is critical in design work. While the objective of achieving and sustaining a lasting change that has an impact on the enterprise prevails, focusing on the making of things allows us to turn ideas into a tangible form. We can bring together strategic thinking and conceptual aspects in real objects, ready for experimentation and prototyping, validation, and iteration. This immersion into the physical use context of an object allows the design work to focus on illustrating conceptual decisions through their implications for the things to be created.

For companies that make and sell things, designing their products is obviously the most relevant use of design—but also in other industries, designing things matters. Expanding on the role of things in the enterprise, as operational tools and cultural artifacts, their design can be defined as the creation of boundary objects to trigger and sustain a transformation to a desired target state. For the enterprise as a web of things, this translates to an active consideration of product design for all actors to be addressed, and taking form factors into account when deciding what to build or buy to make things a part of commercial offerings and business operations..

#20_PLACES

The last aspect of the Enterprise Design framework is about the environments used in the enterprise to facilitate activities and interactions among actors. Organizations are designing places and use real estate for various purposes. In fact, the way physical space is organized is a critical factor in many organizations of the service sectors such as hotels and restaurants, where the facilities offered and the ambiance created are parts of the differentiation to the market.

Some businesses require a network of local stores to get near their customers, while others just work with offices and warehouses. Others again perform their activities in environments made available by other actors in the enterprise, such as market spaces or customer sites. Depending on the respective industry background and chosen business model, there are many different concerns being addressed with such structures, requiring different kinds of spaces for different purposes. Production plants are designed to support the operational flows, and strive for the most effective use of space. By contrast, corporate headquarters and environments for customer interaction are designed with representation and persuasion in mind.

Seen as infrastructure of the enterprise, spaces are created or used to bring people together. They support teams that collaborate, service staff that interact with customers, or other actors engaging in some sort of interaction. In such cases, the physical surroundings have to support the sometimes divergent needs of different actors, and provide adequate facilities for their interactions. Basic environmental factors like the floor plan, the temperature, and acoustics are mixed with questions of furniture and equipment, style, and interior design.

The ongoing shifts in the way we live and work also bring a change in the way people and organizations make use of real estate in general. While the meaning of buildings shifts towards external communication and operations of a physical nature, more and more of what is happening in an enterprise is moving to virtual spaces.

Tasks that required consumers to visit stores or markets are now performed using digital channels, such as shopping for books or opening a savings account. People are no longer required to be in an office to do their jobs, especially those performing a function dealing with information and communication. For such tasks, digital platforms and devices to access them serve as equivalents to the physical places that used to fulfill a similar purpose.

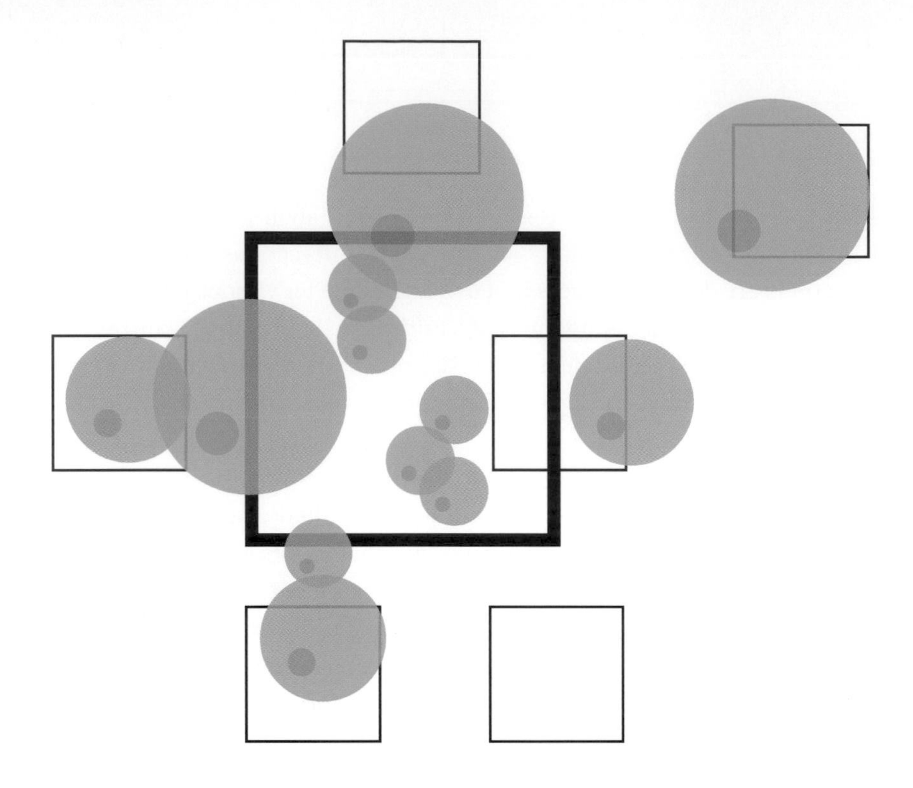

This does not mean that people are no longer leaving their homes, but that the virtual and the physical spaces overlap and form a blended spatial construct. Every virtual space can only be used when embedded into the physical space the user is accessing it from—think of tablet devices often used to surf the web in a *lean back* situation, sitting on a couch.

_ Designing where the enterprise happens

Whatever task people do in the context of an enterprise relationship, they do it in some environment—at their office desk, on a train, on the couch at home, or in a café. Each of these physical contexts imposes an environment for our activities, with certain possibilities and limitations—their effects on our lives are obvious, ranging from ideal environments such as an inspiring busy café to a loud railway station spoiling a customer call. Given the complexity of owned and shared spaces, the varying degree of influence on physical spaces and their virtual extensions, and the new freedom of taking places with you, dealing with space as a domain for design is a challenging task for organizations.

_ EXAMPLE

Many jobs today are not bound to a certain location the way they used to be. As an example, imagine the activities of a lawyer—consultations with clients to discuss cases, going into court for the actual oral argument before a judge or jury. Connected by digital media and devices, there are several tasks that can be done from anywhere in theory, such as researching law texts, drafting court papers or contracts, and even communication with clients or experts. Activities that required traveling or dealing with large amounts of documents on paper now are done by going to virtual places, which allow pulling out information into different physical contexts whenever it is required.

Designing space for the enterprise is essentially about providing suitable contexts for everything happening as part of the activities being performed in its realm. To make the aspect of space accessible for the purposes of design, it must be viewed in terms of how people use and experience it, rather than a geometrical entity or abstract infrastructure.

A key concept of how people deal with their environment is the notion of place, a subjective and personal idea of where we go and why. A place entails subjective notions of purpose, memories or emotions, and can be distinguished from the objective and rather detached concept of space. Adopted by professional architects in the last few decades, place differs from space in that it is about what a certain location and situation means to someone, and how to design spaces to become places where people go to spend their time. In their book *Pervasive Information Architecture*, Andrea Resmini and Luca Rosati describe the heuristic of Place-Making as laying the foundation for *being there*—the capability to help users *reduce disorientation, build a sense of place, and increase legibility and way-finding across digital, physical, and cross-channel environments.*

When you focus on place, you do everything differently.

The Project for Public Spaces

When thinking about our favorite places, it becomes clear that it is not about space alone. The association of memories, sensation, and behaviors makes the difference between a great place and some location we just happen to be. Applying the concept of place to the design of enterprise environments enables a design team to get a grip on the complexity of potential spaces and contexts to consider, and allows focusing on creating meaningful places for people to go and stay in when dealing with the enterprise. The basis of this work of place-making is to ground the shaping of environments in the community of actors the space is being made for, and working with local realities.

_ Enterprise place-making

Research findings and conceptual design decisions are the basis for learning about the places that are potentially relevant. Place-making is an approach with origins in the design of public spaces and urban environments, with a powerful idea at heart—designing great places requires engaging local communities to collaboratively envision and develop their future. Such an approach enables a design team to actively consider what places to create or reshape, instead of solely relying on a client briefing or statistics. Applied to the enterprise, the concept puts the question of how to make places accessible and meaningful for people at the center of the design challenge.

Places are a major part of the fabric of an enterprise, providing the infrastructure for every activity that goes on. In order to make places that make sense for people, they have to be created in a way that implements a larger conceptual vision about the enterprise. To do so, places cannot be designed in isolation, but as a system that considers their meanings, their relationships, connections and shortcuts between them, and the paths people take when traversing them as part of their activities.

ARCHITECTURE

The professional field of Architecture is about giving form to the physical environments for human existence. As for many other disciplines, the term *architecture* in this case refers not to a single practice, but to an entire range of related fields. There are many areas in the wider design and architecture industry concerned with the design of space for people to dwell in. This includes areas such as Urban Planning, Landscape Architecture, Building Architecture, and Interior Design, as well as more specific areas such as flight decks or event spaces.

While all of them are concerned with the design of space for humans to dwell in, they are working on different scales of the spectrum and have to work together in order to get the task completed. Architecture is one of the oldest design disciplines, if not the oldest one, since planning and building homes and organizing spaces must have appeared as soon as humankind settled down. As such, the different domains of Architecture enjoy a particularly long tradition and history, with schools of thought, approaches, and philosophies.

As the other design disciplines, Architecture transcends the boundaries between different levels of work on interiors, buildings, and landscapes or urban environments by working with abstract concepts. An architectural project initially works with a topological concept of space, focusing on the relationships and meanings of places to be considered rather than their locations and forms in a geometrical space. The meaning associated to places in turn has to be based on the conceptual aspects of the enterprise they are made for.

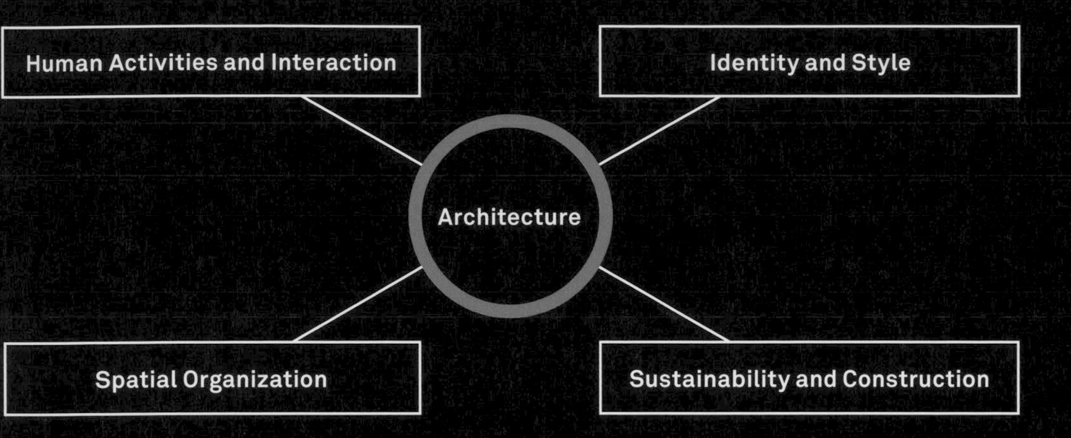

As in most design disciplines, the practice of Architecture involves addressing a large variety of considerations:

_ HUMAN ACTIVITIES AND INTERACTION

designing a space to match and support human behavior. This area involves exploring the purposes and usage scenarios of a space, and their consequences for design decisions. Considerations range from large-scale matters like staircase locations to the interior design addressing light, acoustics, signage, color, and furniture.

_ IDENTITY AND STYLE

embodying values and beliefs in the architecture, going back to the very beginnings of the discipline and visible in government buildings, churches, or other constructions. Today's company headquarters are often identity landmarks, representing the company or organization using them to customers and the public.

_ SPATIAL ORGANIZATION

dealing with the definition and separation of areas and rooms through walls, aisles, pillars, and other means of creating zones. This involves defining the relationships between topological locations based on activities and scenarios to support, and mapping these to a geometrical space.

_ SUSTAINABILITY AND CONSTRUCTION

working with the implementation aspects of Architecture, in terms of economic realities and budget, physical issues of engineering and construction, integration of technology, and meeting regulatory requirements.

Cyberspace is not a place you go to but rather a layer tightly integrated into the world around us.

Institute for the Future

Not long ago, many saw the future of how we use places in Virtual Reality, digital simulations of three-dimensional environments. The possibilities of bringing our physical world into the digital seemed endless, with the prospect of translating largely physical experiences into virtual worlds, such as shopping in a store. Except for specific purposes of games or simulations, we see the opposite is happening— we are not living in a virtual reality, but there is a virtual space expanding into our physical world through touch devices, projections, and other forms of media technologies. Today we find a hidden virtual dimension entering our physical reality, with digital media acting as our loopholes.

For a lot of people working with virtual organizations, the world of work already takes place primarily in that other dimension and only occasionally involves physically going somewhere, with collaboration tools and digital workplace technology for substituting the traditional office space. This does not mean that the physical context of work disappears, but it shifts out of control of the employer and into a private environment. Ensuring a good work environment is just one of the challenges modern enterprises face with regards to their places.

While Architecture as a discipline is clearly focused on the design of physical environments, the principles of place-making can be applied to these virtual spaces as well. Initially much like empty buildings, places occur once people associate them with activities and themes. This applies regardless of the fabric of the space to be designed—essentially because the virtual and the physical are part of the same experience. For the practice of Architecture, Interior Design, and wayfinding, designing places is essentially designing for this blended reality.

_ PLACES IN THE ENTERPRISE

_ IDENTITY

designing environments to be places where culture manifests itself, think of an office or event space.

_ ARCHITECTURE

designing spaces to support the business activities of the enterprise like production facilities.

_ EXPERIENCE

designing places for experiencing the enterprise and to make people feel comfortable, such as a café.

_ ACTORS

designing environments to match different actors and support their interactions, such as a trade show.

_ TOUCHPOINTS

designing environments to act as the physical contexts of human/enterprise interaction — think of online banking.

_ SERVICES

designing environments to support carrying out services, for example the *Sushi Circle* concept.

_ CONTENT

designing environments to support the reception and exchange of content, like the menu display in a restaurant in its interior.

_ BUSINESS

designing environments to support your business model and enable transactions, for example department stores.

_ PEOPLE

adapting environments to people, such as a university campus with dedicated spaces to study, hear a lecture, or relax.

_ FUNCTION

designing environments in a way that supports their purpose and functional context, such as calmness in a library.

_ STRUCTURE

designing environments to reflect a structure, as in production halls organized according to teams or products.

_ COMMUNICATION

designing narrative environments to enable communication and social interaction, like at a drinks reception.

_ INFORMATION

designing environments that support wayfinding and information retrieval, for example, in an underground garage.

_ INTERACTION

designing environments embedding tools and appliances to support user behaviors, such as security checkpoints at the airport.

_ OPERATIONS

designing environments to facilitate business operations, such as a canteen that supports a smooth flow of guests at lunch time.

_ ORGANIZATION

designing environments to support team collaboration and organizational structures, like team spaces in an office.

_ TECHNOLOGY

embedding technology in environments, for example, by placing power outlets.

_ RENDERING THE ENTERPRISE ACROSS CHANNELS

The enterprise comes into existence in the tangible elements that it makes available to the people it addresses. We move around in it, changing environments and contexts for certain activities, we deal with objects and artifacts, and we see and interpret various types of messages and symbols. These elements have to be treated as a web of connected elements. People dealing with this reality of hybrid systems transcend the virtual and the physical in overarching journeys. Instead of seeing signs, things, and places as isolated elements, they have to be linked together, a concept also known as *cross-channel design*.

The blending of channels and the way people use them results in an increased complexity of the design task, accounting for the increasing convergence and tight network of links between them. Design practice is shifting from a worldview of separation to an approach that brings disparate elements together. In general, traditional approaches to the design of signs, things, and places work surprisingly well if applied to the their virtual and blended counterparts, with one peculiarity: they share characteristics of all three concepts at the same time, forcing design practice to adapt and take all these aspects into account. Instead of confining design work on one single medium, this requires us to base the concrete elements on a larger conceptual vision, and to turn this into a rendering.

_ Enterprise evidencing

The term *rendering* may seem a bit misplaced when used in the context of an enterprise. This last chapter of the Enterprise Design framework describes the outcomes of a strategic design project, the generation of visible signs, things and places with the intent to have them enter the world. The term is an analogy similar to how it is used in the field of 3D graphics, where a realistic image is created from a carefully crafted model. It involves mastering a set of sophisticated techniques to bring the model to life, considering materials, textures, lights, shadows, and reflections. Rendering the enterprise works in a similar way. The conceptual work delivers a blueprint of the intangible aspects of the enterprise. The tangible outcomes of a design process have to work together as a system to implement the conceptual vision. This comes with a larger degree of complexity, so the conceptual aspects take on a more important role. Initially everything seems equally important, leading to confusion and making it hard to focus. But in our experience it is this task of balancing a proposed solution against the tradeoffs, constraints, and ideas arising from the concept that makes the task of design so interesting. The conceptual aspects have to be behind all design decisions, which is essentially what the Enterprise Design framework is about—connecting design to the reality of life and business that it seeks to reshape.

_ EXAMPLE

The field of Game Design is a good example of an emerging subfield, designing outcomes that fall in multiple categories. Games are essentially systems of signs, making a player react to certain messages or symbols on a screen or a physical game board. At the same time, they show characteristics of things — you buy a card game in a box, you can keep it on the shelf in your living room, and it consists of tangible objects. Finally, games can also be seen as places bringing people together — think of virtual multiplayer games, enabling people to spend their time together in a designed environment.

It isn't a coincidence that it's much easier to have an opinion on color or font size than it is to make a decision on product positioning.

Luke Wroblewski (lukew.com)

The term *cross-channel* is mostly used in the context of Media Design, but essentially applies to all kinds of design work. It refers to the practice of considering individual items as part of a larger experience, and designing them as a system that reaches across traditionally separate delivery channels. Every individual part is seen in terms of its role in an overall composite design. This includes physical, virtual, and blended channels, however the characterization depends on the way a device or medium is used.

_ PHYSICAL

channels are those which are part of the physical environment we live in. Interaction with these channels happens directly through our body.

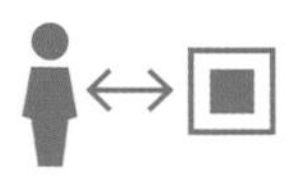

_ VIRTUAL

channels are hidden, and have to be accessed with the help of a device of some sort. They are characterized by the immersion that takes place when we use them, forgetting about the physical world for a while.

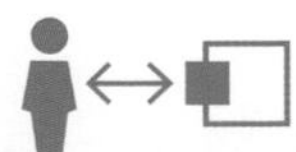

_ BLENDED

channels share characteristics of both groups, and generally pertain to usage patterns jumping between the physical and the virtual as part of the same activity.

The field of Service Design (which is subject to the services aspect described in Chapter 5) is perhaps the most conceptual and abstract professional field of design that is widely practiced today. In the research of Lynn Shostack, one of the most influential contributors to the field, the visible and tangible components resulting from a design process are referred to as *evidences*—elements that reveal to the customer an intangible service going on in the background. The pioneering Service Design consultancy live|work used this concept in their work as a means to approach the design of concrete outcomes as potential future evidences of a future service. On their website, they refer to this activity of evidencing as *archeology of the future*, or *digging out what might be.*

Applied to strategic design, evidencing is a powerful concept. Although the shift in thinking happens on an abstract level, there is a subtle difference between envisioning an artifact—say, a website—as a design proposal for today, or as tangible evidence of a future implementation of your concept. The idea of shaping signs, things, and places in the enterprise as evidence of a desirable future state is at the heart of the Enterprise Design framework. It aims at the application of conceptual work to produce a visible rendering of the future enterprise. Evidencing helps us translate a conceptual design into an actionable scope of work, to be taken up by the more traditional design practices of creative form-giving.

Embracing design practice _

The aspects considered in this chapter make the connection to the sphere of traditional design disciplines, which can be seen as a process of crafting and making artifacts of some sort for a given purpose. It is a professional field combining a long history from different roots and influences with rapidly changing trends. Public opinion sees the discipline mainly through a few star designers in the field of industrial or fashion design, while the profession is slowly recognized as a mainstream discipline that can help in every aspect of life and business. In that endeavor, essential design philosophies such as the teachings of the Bauhaus from 1919 are still applicable and valid in today's practice.

The large variety of subjects that could be taken on as part of a design process has led to the emergence of so many design fields and specializations that it is sometimes difficult to determine the discipline best suited for a problem, even before struggling to find the right expert designer for the job. On the other hand, the designer's role is one of integration, of coming up with something concrete grounded in the abstract thinking which is was least initially neutral of a particular medium or field..

The key issue in managing the design process is creating the right relationship between design and all other areas of the corporation.

Donald E. Petersen, Former Ford CEO

Working with the diversity of the field is as much a challenge for its practitioners as it is for their clients commissioning design work. For a person with a strong focus on a classic field like Graphic or Product Design, other creative areas often are a blind spot. Mastery of their main medium, through talent, experience, and skills, does not easily translate to another one. There is a need for overlap and collaboration, such as designers and architects working together to put together an exhibition or a signage system, but it is hard for them to break out of their respective boxes. As mentioned before, the emergence of a blended reality of interconnected artifacts that combine the characteristics of signs, things, and places blurs the boundaries. It sparks the need both for generalists and for experts in certain creative subfields.

For strategic design initiatives in the enterprise context, there is a need to bring together generalists and specialists, talent and collaborators, as well as rational conceptual thinking and opportunistic external inspiration. Design professionals rely on the ability to walk through the world with open eyes, an open mind, and the possibility to transfer things from here to there. The best ideas and approaches in design often come unexpectedly and from somewhere outside the thinking about the design challenge itself, or from dialogues between experts in seemingly unrelated fields. The tension between these different approaches makes the case for actively managing the way design is practiced in an organization, and adapting it to the specific situation and challenges of the enterprise..

DESIGN MANAGEMENT

The professional area of Design Management is about making design work as a competency for an organization. It is concerned with putting into place the formal structures that such leverage requires, but also sparking and sustaining a culture of creative experimentation in all aspects of the business. On a more concrete level, Design Managers need to establish systems of governance and asset management, manage internal design resources and external partners commissioned with design work, and facilitate the integration of this work into the functions and business areas of the organization.

Although many organizations realize the potential and the need to employ design as part of their capability set to achieve innovation and market differentiation, the aspect of formally managing design is often neglected or overlooked — instead, the practice is vaguely seen in the area of creativity, often assumed to be rather unmanageable both in terms of processes and people. This is reflected in the teachings of some business schools, putting out dogmas like "never hire a designer" for the reason of not to stifle their creativity by a nine to five job.

To advance to the next level of integration beyond a superficial and therefore ineffective use of design, it has to be managed systematically as a strategic competency much as any other. This involves orchestrating a complex piece of teamwork, aligned with a long-term vision where to go. It requires putting into place design programs, strategy, governance processes and Change Management to embed it into the regular business practice of the organization. In the next part of *Intersection*, we will share our experience on how to make a strategic design initiative happen as a program and capability.

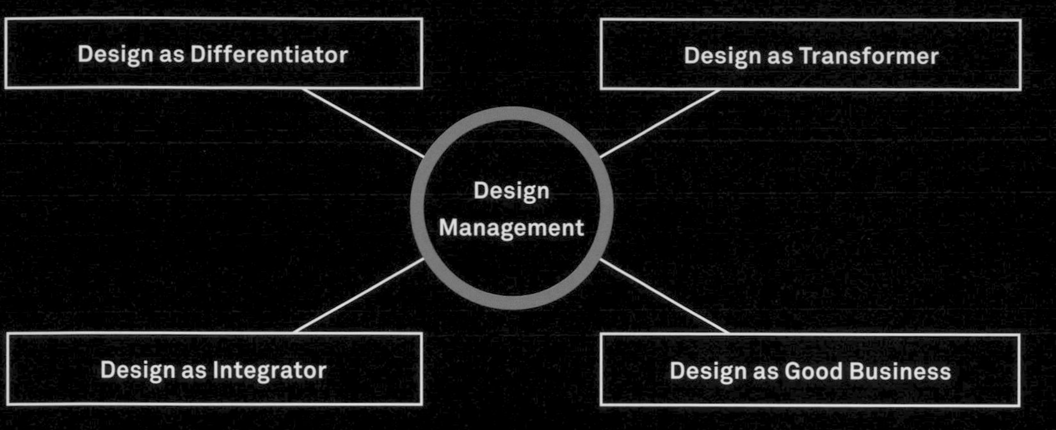

The Four Powers of Design: A Value Model in Design Management, Design Management Review, Vol. 17(2) 2006, Pages 44-53

French scientist Brigitte Borja de Mozota developed a scientific model on the role of design in organizations. The *Four Powers of Design* differentiate levels of integration:

_ DESIGN AS DIFFERENTIATOR

Establishing a design competency for Marketing purposes, as a means to differentiate products and influence customer behavior. It uses Communication Design to position brands in advertising, product packaging and visual brand identity, while the products and services themselves are untouched.

_ DESIGN AS INTEGRATOR

The design competency is used for its creative input for the enterprise, as a management tool for innovation. The integration happens on the level of idea generation and exploitation, and focusing on human-centric approaches. It employs product-oriented disciplines such as Industrial and Interaction Design.

_ DESIGN AS TRANSFORMER

The design competency is seen as a strategic enabler for all sorts of problems, supported by senior management and firmly grounded in the organizational culture. It focuses on creating business opportunities, and is therefore linked with Business Design and Architecture and Organizational Design.

_ DESIGN AS GOOD BUSINESS

The third level of design integration into an organization is both broad and deep, touching virtually every area and function. Design is part of the DNA of the company and therefore an intrinsic part of every endeavor.

AT A GLANCE

To deliver on its promise to draw a picture of the future, any strategic design initiative needs to result in visible and tangible outcomes as evidence of an evolved enterprise. This requires working with talented designers of relevant fields, embodying abstract concepts, and generating ideas for a hybrid system that spans the virtual and the physical reality of the enterprise.

Recommendations

_ Design systems of signs that effectively encode the overall story of the enterprise and its key messages into media, and apply them consistently

_ Produce things according to their roles both as useful tools and artifacts of the enterprise as a cultural space, adapting them to their use context and causing behavioral change

_ Provide personal, meaningful places as environments to support the activities of the enterprise, and enable people to find their ways and inhabit places important to them

_ Use the Rendering aspects to map and create paths and links between all these designed elements as part of an overarching Design Management, accounting for their role as triggers for the intended change to happen

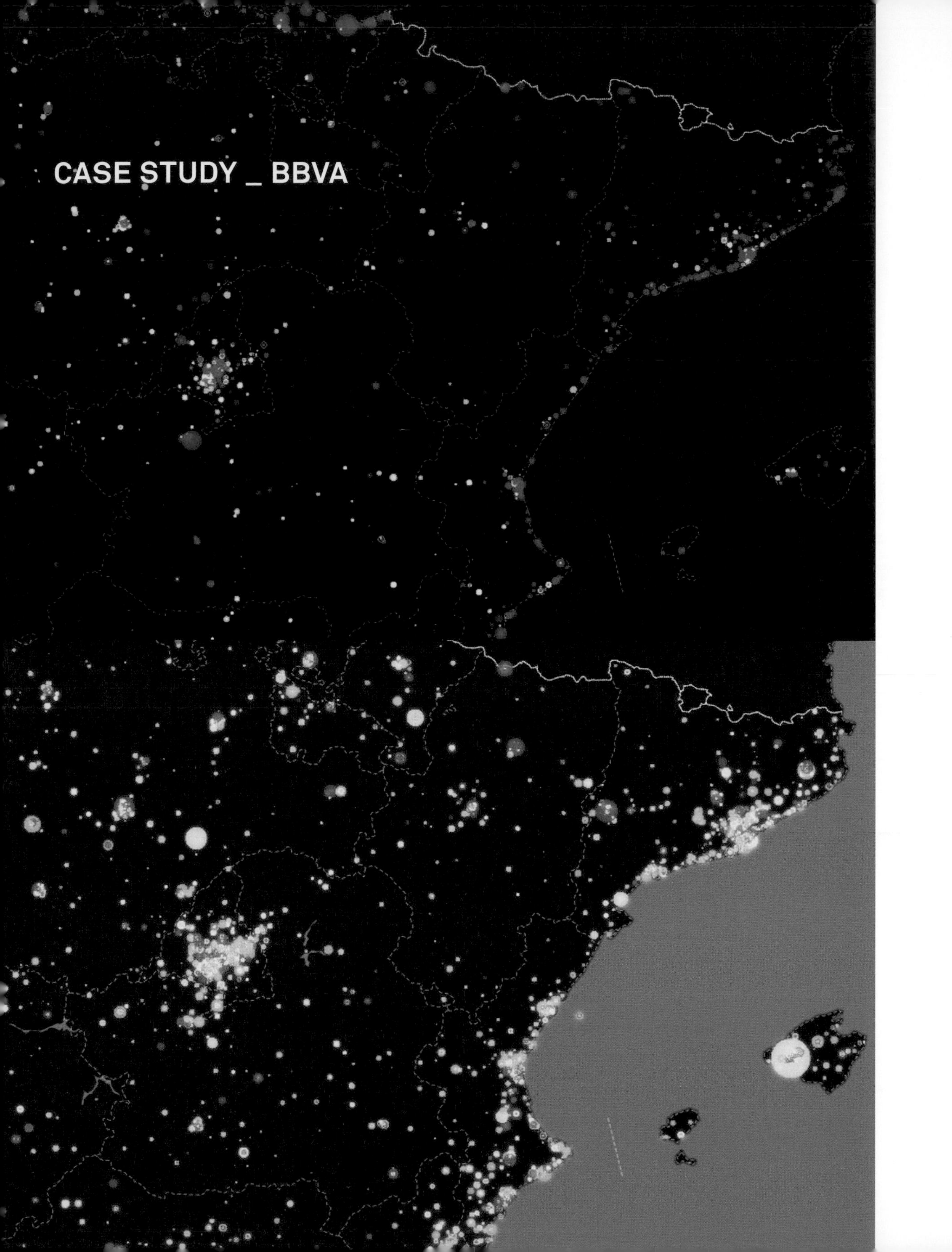
CASE STUDY _ BBVA

BBVA'S ENTERPRISE

Simplifying banking and delighting customers, to become the people's favorite bank - working towards a better future for people.

As one of the biggest financial groups in the world, Spanish Grupo BBVA today serves over 50 million clients in 32 countries, with various offerings in banking, insurance, and related services, and looks back on over 150 years of history. A strong market segment of the company is the retail business, where the bank is present in many markets with a network of branch offices and a wide range of services for personal and small business banking.

Driven by a management paradigm that focuses on value creation for people, the group has launched the BBVA Innovation Center. Its initiatives reach across all business units, functional departments, and services. Although the banking sector is one of the oldest in the world, management believed that there is still a lot of innovation possible in the area of dealing with people. The company strives to make people's lives easier by offering transparent and simple solutions, integrated into daily activities. This innovation initiative is approached like an industrial process, creating value for people in a sustainable way.

As part of these activities, BBVA launched a strategic design program with a strong focus on cross-area innovation. In a series of projects, BBVA addressed a variety of design themes, using a design-driven approach to produce remarkable outcomes as part of the Rendering of their enterprise. The group has worked with some of the most renowned design consultancies such as IDEO, Saffron, Continuum, and Fjord. All design work is embedded in a network of academic partners and technology vendors. Achieving such outcomes would be impossible without the underlying conceptual decisions, bringing together enthusiasm, strategic vision, and attention to detail. Although BBVA's design program is about producing tangible results, it is their role for people in touch with the enterprise that makes them meaningful.

The future of innovation can only have one purpose: to make people's lives easier. It is useful, practical, focused on actual solutions, meaningful, user-centered, functional and ergonomic. It must add value, to close gaps.

Ignacio Villoch

_ SIGNS

As in all large companies, BBVA group is in constant exchange with a large and diverse enterprise environment. As part of the innovation initiative, the company looked for new ways to engage its entire range of stakeholders and drive the communication process. As a result of the design process, the company overhauled the media elements of its brand identity, and applied it across all its messages, communication platforms and other sign-based systems facilitating this exchange. Elements such as a defined visual language and tone of voice are the basis for a coherent presentation across all these Rendering elements.

Based on the paradigm of open communication, BBVA moved away from the traditional ways of corporate communications such as publishing polished reports. Instead, it concentrates on formats such as online media, videos, and microblogging, developing new channels to distribute its messages and facilitate a two-way conversation. Driven by the goal of a better future, the group addresses topics of sustainable and inclusive banking, openly co-creating new solutions with their environment. These formats enable BBVA to tell the story of their enterprise beyond the one-way messaging of the past. This ranges from visualizations based on the group's huge base of transactional data, to sponsoring Spain's national soccer league, connecting to people by placing signs in their reality.

The same fundamental sign systems are used to shape the Customer Experience of BBVA. The conceptual design decisions behind BBVA's offerings are based on radical simplification, equally represented in the visual clarity of signs used for Marketing materials or tangible elements of a product offering like different credit card options, designed to communicate the difference instantaneously and unambiguously. Another task of sign systems is structuring BBVA's digital systems across different services and devices, offering a system of simple, elegant, and reduced tools to the client—but also to internal systems for employees, reporting for investors, and communications to the public. The conceptual design choices about the way people interact and communicate are the basis for a systematic use and distribution of signs within BBVA's enterprise environment.

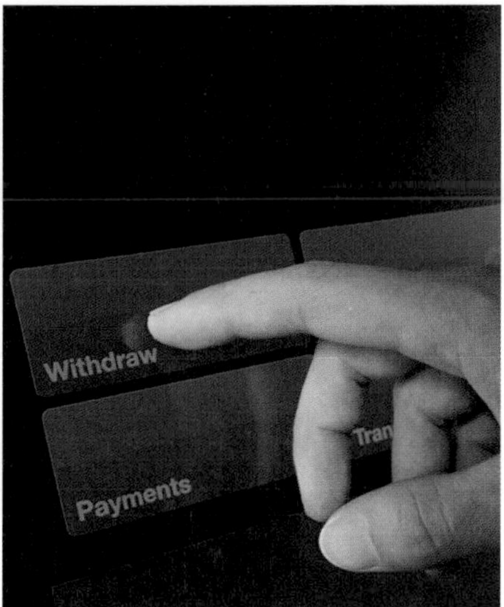

THINGS _

BBVA's innovation initiative turns the design themes identified into vision projects, always with the goal of translating ideas into concrete outcomes. In such open-ended settings, BBVA is indifferent whether such an outcome would reside in Marketing, internal operations, products and services, or other functions, or span multiple areas. Any activity by a stakeholder group that could be supported is a potential case for a tool, a physical or virtual artifact created to address a concrete need.

One of the most famous projects of the BBVA Innovation Center is ABIL, its next-generation vision of the ATM — a field where not much has changed for decades. Initially looking at ways to bring more offerings into the self-service channel, BBVA quickly discovered that many of the current functions of their kiosk systems were not used. Apart from cash withdrawal, most customers preferred waiting in queues for a human operator, even though the functionality was available in the self-service environment. So instead of adding new functions, the design team realized it had to make the existing ones much more accessible, simpler, and better adapted to the customer.

In a four-year project, the BBVA Innovation Center adopted a design-led approach together with the American consultancy IDEO, starting with extensive research and conceptual design to bring together human activities and the corresponding interactions and transactions. Based on a collaboration of over 30 professionals from many different fields of design and related areas as well as finance and technology professionals, the design team conceived, prototyped, tested and iteratively defined the ATM of the future, which is currently being rolled out in branches. It addresses issues of privacy, security, transparency, and efficiency with an elegant and dynamic interface design, making interactions much more fluid and humane than any of its predecessors.

In similar projects, BBVA redesigned its online services, produced apps for mobile banking and investment, and physical evidences such as credit cards — all considered as part of a larger customer journey. Moreover, BBVA produces tools for internal or external use with the same approaches. BBVA makes heavy use of custom iPad apps for staff activities, aiming to make their workforce more productive in any situation, including interactions with customers.

PLACES _

As a company active in the retail banking sector, BBVA is at the forefront of the shift from physical to virtual environments where their business is happening, resulting in a set of highly disruptive developments. Many customers interact exclusively via online banking, mobile apps, and other distance services. Employees are empowered to work from home, working with a cloud-based intranet and remote tools. Even face-to-face meetings in the branches are facilitated using these tools, bringing in experts from other locations as needed using video calls. Finally, conversations move to entirely different virtual places, including social media such as YouTube, Twitter, or Facebook, both with customers and other stakeholders.

BBVA has recognized that this new reality has to be addressed by making places that fit their context of use, bringing together elements of physical, virtual, and hybrid spaces. These places are the basis for interaction and conversation that drive business processes, customer interactions, stakeholder communication, and other essential activities of the enterprise. Led by the BBVA Innovation Center and based on a large conceptual body of work, the group introduced places on the web dedicated to various themes, target groups, and activities. They all are part of one larger, overarching concept, providing paths from one to another and guiding people on their individual journeys.

The blog platform Bancaparatodos (a bank for all) is an example of an owned place dedicated to an open exchange with the whole stakeholder community—including everyone interested in participating. Focusing on social topics such as financial inclusion and service availability to everyone, responsible investing, and financial literacy, it is a place to discuss and share ideas. It gives the innovation center a place to co-create the future of banking with others.

In any case, the design of the spaces strives to make places that relate to the purpose and activities to be addressed. Physical facilities such as branch offices and interiors, and virtual or hybrid places, are made to drive interactions and conversations. They combine connections on a social level with working touchpoints for the actual interactions—for example, the bank's Twitter team is ready to discuss the future of banking with you, or help you with a problem in your online service. With their place-making strategy, BBVA attempts to be where people are—a strategy also reflected in sponsoring Spain's LigaBBVA.

PART 3

Enterprise Design Approach

The final part of *Intersection* is an introduction into a use of the Enterprise Design framework as part of real world design programs. It is based on extensive experience, applying the different aspects for our work at eda.c for organizations from the very big players to garage-like startups, from various industries, and on a wide range of challenges.

The framework with its twenty aspects is meant to be a guide helping us navigate through a complex space of intermingled concerns. They go beyond the views of traditional design disciplines, considering also things that touch the areas of business decision-makers, engineers, and social scientists. Considering the non-tangible and conceptual aspects makes a design practice strategically relevant, and allows us to make the enterprise's future visible.

When working on a given problem setting, you will find that some aspects are more applicable then others depending on the challenges we wish to address. The framework is not an instruction manual, so its application has to be adapted to the situation at hand. This part features two chapters. Chapter 9 describes the application of the framework in a design project, detailing each step with some detail. Chapter 10 is about launching a strategic design program in an organization, and making it part of work practice.

9 _ DESIGN PROCESS

The Enterprise Design framework is meant to be applied in design processes, both as a vocabulary and a way of thinking about complex problem settings. This chapter proposes a generic process that is applicable to strategic design projects of all kinds, going from an initial preparation phase, through stages of research, concept and ideation all the way to bringing the ideas to life in implementation. Each step describes the key elements to take into account, and maps the activities to the twenty aspects of the Enterprise Design framework. This is meant to guide your own individual approach, and help you select professionals to involve during the process.

The process we propose has seven phases: Prepare, Discover, Define, Ideate, Validate, Implement, and Deliver. This includes the five phases which are a part of almost all generic process models for design work, plus a preparation phase at the beginning and a change or transformation phase at the end, to account for the specifics of the enterprise context. The process also incorporates two different cycles of iteration, a small one inside a project, and a big one as a roadmap of sequential projects. While lean or agile approaches that focus on delivering results quickly work in many small iterations, formal methodologies applied to large projects put an emphasis on a clear up-front definition of the intended outcomes. Whatever the chosen approach, design projects usually go through all these stages regardless of what process they employ, but with different degrees of speed and attention to detail.

In reality, such a process does not exist. There is no clear point in time when a project switches from research and analysis to concept development and ideation, and then to implementation and change, so a formal separation between these phases is unsuitable to reflect the reality of design work. Likewise, a design project, especially a courageous and daring one, requires adapting the approach to the specifics of the design challenge. Every enterprise is different, and projects can start from different points and work towards different objectives.

© 2013 Elsevier Inc. All rights reserved.

Properly gaining control of the design process tends to feel like one is losing control of the design process.

Matthew Frederick in *101 Things I Learned in Architecture School*

Redesigning an enterprise is difficult—the sheer number of generic aspects described in the previous part of this book and their interrelationships is a good indicator of how complex and challenging such an endeavor can be. Often, it seems much easier to approach a project as though it were separate from the wider environment of the enterprise.

Our experience tells us that it is worth taking on the challenge of an overarching enterprise-wide approach. This opens up the design process to questions you might not be aware are relevant at the beginning, but later on prove to be critical. Spending time understanding how the different aspects influence your project enables you to make informed conceptual decisions, aligning those elements to a common goal.

By doing this based on a holistic model, organizations can ensure the viability, desirability, and feasibility of an intended transformation. It allows them to make the consequences of a transformation visible, and iteratively adapt to the reality of their enterprise.

Like other design process models, this model puts an emphasis on the early phases of a project, where the design practice makes its main contribution of defining a potential future of the enterprise. In subsequent phases of implementation and delivery, design switches to a supporting role. At that point, questions of engineering, governance, and change take center stage. The design team takes on tasks of executing and producing everything needed to make the vision reality, managing quality and transformation processes, and launching further projects for missing pieces or a subsequent iteration.

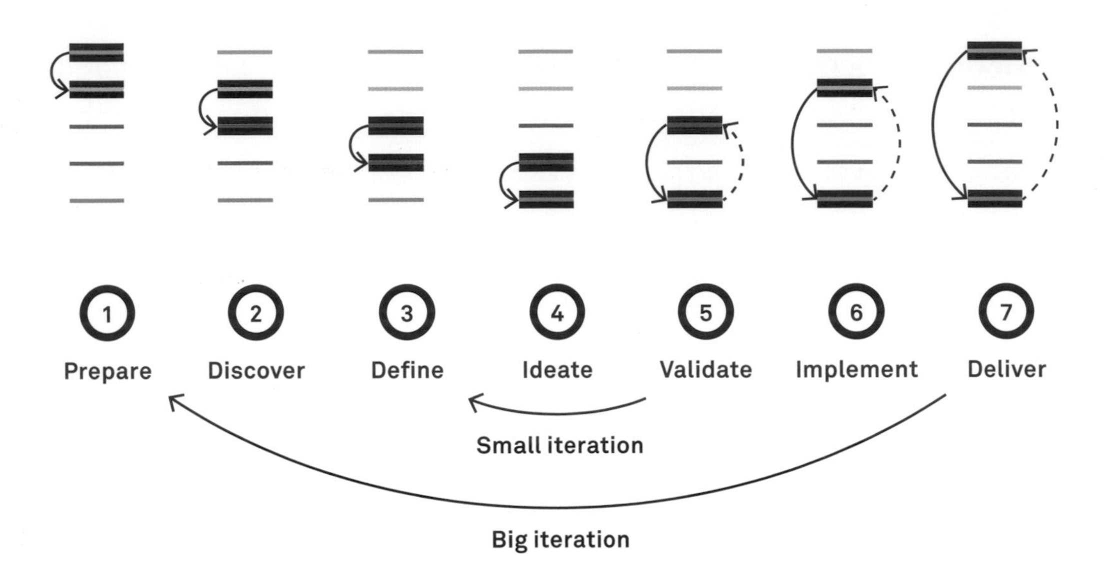

_ 1 PREPARE: GETTING STARTED
BIG PICTURE / ANATOMY

_ 2 DISCOVER: EXPLORING THE PROBLEM SPACE
ANATOMY / FRAMES

_ 3 DEFINE: DEVELOPING A SOLUTION APPROACH
FRAMES / DESIGN SPACE

_ 4 IDEATE: GIVING A FORM TO THE SOLUTION COMPONENTS
DESIGN SPACE / RENDERING

_ 5 VALIDATE: PROTOTYPING, SIMULATION AND TESTING
RENDERING / FRAMES

_ 6 IMPLEMENT: PRODUCTION AND DEVELOPMENT
RENDERING / ANATOMY

_ 7 DELIVER: DEPLOYMENT AND TRANSFORMATION
RENDERING / BIG PICTURE

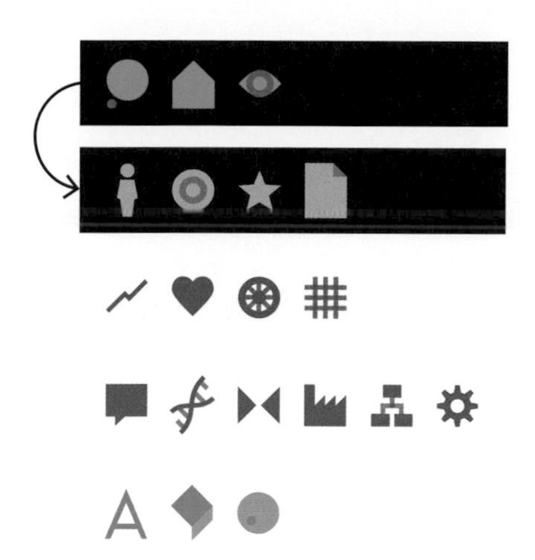

PREPARE _

Obtain an initial briefing and perform a high-level scan to develop a Big Picture view of the enterprise environment, and understand how it addresses the three universal qualities. Develop a first understanding about its anatomy in stakeholder interviews and workshops.

Activities _

_ Obtain an initial client briefing from the party commissioning a strategic design project
_ Set project objectives, overall planning, and timetables
_ Identify the most relevant aspects and establish a project team mixing people with multiple relevant backgrounds and design expertise
_ Conduct interviews and stakeholder workshops to develop a high-level view on the enterprise
_ Conduct initial desk research to find out about the history, structure, and positioning of the enterprise, and benchmark the offerings and assets relevant to the project
_ Establish a basic vocabulary of terms and concepts
_ Set the scope of the intended transformation
_ Make an initial plan and organize the people involved in the project, assemble a diverse team, and get started

Challenges _

_ Getting access to all relevant stakeholders
_ Finding a viable compromise between high aspirations and a realistic scope
_ Avoiding falling in love with initial ideas, which leads to a narrow and predetermined project scope

Typical techniques _

_ Stakeholder Workshops
_ Interviews
_ Briefing Document
_ Project Scoping

_ BIG PICTURE

The focus of the initial project phase is to develop a high-level view of the enterprise. It starts by engaging in an intense dialogue with the client or party commissioning the design work, understanding what the enterprise is about and where people see its future. Using the interrelated Big Picture aspects of identity, architecture, and experience as the guiding topics of this exchange helps to focus on the universal qualities that make the enterprise, drawing a holistic picture of the design challenge.

_ IDENTITY

elaborate with project stakeholders how people think about the enterprise, what it is about, and what it stands for.

_ ARCHITECTURE

evaluate the structures of the enterprise, how it works or is intended to work, and the portfolio of formalized systems applied.

_ EXPERIENCE

examine how and where people are experiencing the enterprise, and define what role it seeks to play in their lives.

ANATOMY _

The Anatomy aspects are very useful in this phase for developing a first rough blueprint of the enterprise to be transformed, and identifying the particular elements that are relevant for a specific project. The goal is to develop a comprehensive view of actors and their journeys across touchpoints, as well as services being provided and content being exchanged. This is the prerequisite for the following phases or research and analysis to inform the project, since it provides the list of enterprise elements potentially subject to design work.

_ ACTORS

determine the key stakeholders the project owner intends to address and map their relationships with the enterprise.

_ TOUCHPOINTS

find out about the journeys actors are experiencing across key interactions with the enterprise, and the channels used to facilitate them.

_ SERVICES

define and describe activities the enterprise carries out to generate benefits for the actors addressed in terms of services.

_ CONTENT

list relevant content elements made available in the enterprise as part of its activities.

_ DISCOVER 2

Perform research and analysis of the Anatomy of the enterprise, to gain an overview of the elements to consider. Frame the problem space by a deep exploration from different Perspectives with the help of external input, to develop conceptual models of the enterprise and generate ideas for a transformation.

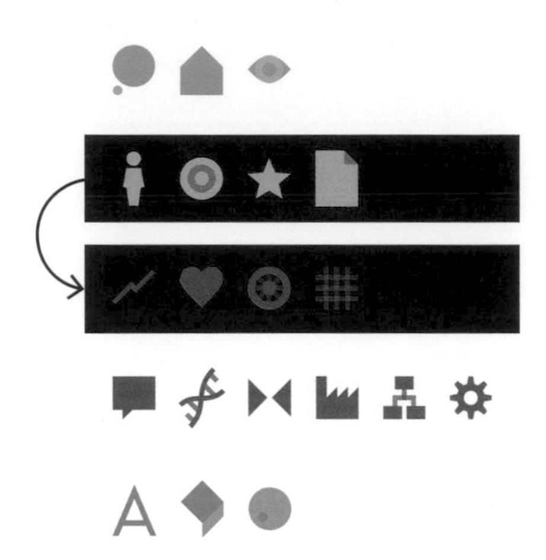

_ Activities

- _ Perform a detailed investigation of the Anatomy of the enterprise, if possible backed by quantitative data on the way it currently functions
- _ Conduct qualitative research working with clients or business owners, prospective target groups of the design process, technology and domain experts
- _ Model the current situation from different perspectives, capturing the way things are now, and envisioning a potential transformation as well as approaches to achieve it
- _ Perform a detailed analysis of the research results, applying the four Frames to filter relevant bits out of the findings
- _ Map the touchpoints, services, content elements, and actors and how they interrelate and interact in the enterprise space,
- _ Formulate a set of strategic recommendations based on the research findings as input for the Definition phase

_ Challenges

- _ Avoiding getting lost in the details of research and analysis activities and losing the connection with the design challenge to be solved
- _ Avoiding commiting to ideas coming out of initial research from just one perspective and failing to account for the remaining aspects
- _ Finding a good balance between effort put into research activities and a transition to the next phase
- _ Translating research results into actionable insights for Definition and Ideation
- _ Gathering suitable data to perform quantitative analysis

Typical techniques _

_ Data Extraction and Analysis
_ Ethnographic Research
_ Market Research
_ Reverse Engineering
_ Interviews
_ Inventories
_ Mappings

ANATOMY _

As the second phase puts an emphasis on deeply comprehending the enterprise as a problem space, the Anatomy aspects are used to evaluate the enterprise and the way it works (or is supposed to work, for a startup), as the environment for the design work. This auditing activity has the goal of analyzing the enterprise elements so that they can be mapped and prioritized based on the objectives of a design project. Ideally, this is informed by conducting quantitative research, such as gathering data from information systems. This is then used to create a mapping of the current or planned enterprise elements and how they are related.

_ ACTORS

validate and complete the initial group of actors identified based on a mapping of stakeholders, and analyze how these groups are addressed.

_ TOUCHPOINTS

analyze the contacts actors have with the enterprise, and identify the journey they are taking across them along with key situations to consider in the design.

_ SERVICES

develop a comprehensive view of activities being carried out in the enterprise, including the steps they involve, their interdependencies, and benefits they create for actors.

_ CONTENT

inventory content assets being exchanged and made available in the enterprise to support its range of activities, mapping content elements to their usage.

_ FRAMES

The different Frames come into play to ground and define the design challenge, establish a good direction to go, and inform the design decisions to be made. They are the basis for conducting deep qualitative research with people involved into the project in different roles, and seeking input from a variety of sources. Findings from the research activities are used to develop simplified conceptual models describing the enterprise based on this input. Applying the four perspectives ensures that the research and analysis activities take a balanced approach with regards to the different perspectives to consider.

_ BUSINESS

explore and understand the markets the enterprise addresses, formulate the design challenge in business terms, and learn about the business rationale and business models driving the enterprise.

_ PEOPLE

conduct human-centric research with people to achieve an in-depth understanding about their characteristics, goals, and social and physical context, and develop empathy to inform the design process.

_ FUNCTION

understand the functional context of goals and processes to be supported, and envision potential functional components to support this context.

_ STRUCTURE

develop a model mapping the structural context of the enterprise, and capturing key objects and relationships to be reflected in the design process and envisioning a transformation.

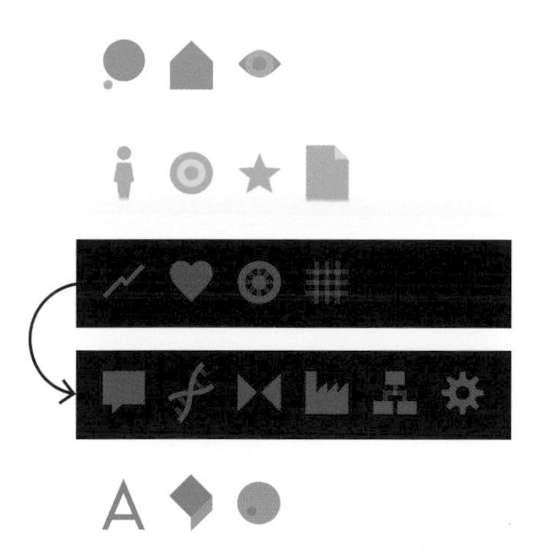

DEFINE _

Use the research results to envision a possible future state of the enterprise applying different Frames, and to inform the development of a conceptual Design Space that outlines how the Enterprise Design could work.

Steps _

_ Use the four Frames to identify key challenges to be tackled and formulate Design Themes guiding all further activities
_ Make sense of the insights gained in the preceding phase to develop an informed approach to solve the problems identified
_ Use these themes to envision a potential future state of the enterprise and ways to make it happen
_ Work with your multi-disciplinary design team to collaboratively explore different possibilities how the results of thc project might transform the enterprise
_ Apply the four Frames to model the future of the enterprise generating new business models, imagining its impact on people's lives, and modeling the structures and functions it puts in place
_ Use the six conceptual aspects to bridge these viewpoints and map the Design Space of decisions and ideas
_ Define how the future enterprise should is supposed to work in conceptual models, aligning divergent concerns and plotting a common course toward a coherent vision
_ Generate initial ideas on visible and tangible outcomes to implement that vision in real life

Challenges _

_ Making choices in a jungle of insights, requirements, ideas, feedback from peers, opinions of important influencers, and personal experiences
_ Boiling down the design challenge to a set of key issues to be tackled, taking into account all inputs from external influencers and a multidisciplinary design team
_ Keeping a focus on the Design Themes identified and avoiding the temptation to tackle everything at once
_ Dealing with conflicting goals that arise from applying different Frames and stakeholder perspectives

_ Typical techniques

- _ Persona Development
- _ Business Model Design
- _ Domain Modeling
- _ Requirements Documentation
- _ Modeling Techniques
- _ Alignment Diagrams
- _ Scenarios

_ FRAMES

The thorough research activities executed in the preceding phase result in an extensive body of findings, as well as key themes to guide the definition of potential solutions. At this point the focus of the design process switches from inquiry to the definition of an intended transformation. In order to translate these findings into actionable insights that open the path to a solution definition, the Frames are used to identify a set of fundamental Design Themes to guide all further activities, and to model a potential future state of the enterprise. Depending on your individual design challenge and enterprise context, you might approach this work in a traditional *design studio* setting, or working together with members of the target group, business owners, and experts from across the organization in a co-creative approach.

_ BUSINESS

brainstorm multiple evolved or new business models based on the exploration of the current business, experimenting with bold ideas to achieve market differentiation and better ways of working.

_ PEOPLE

imagine different ways to improve people's lives based on the findings from a human-centric perspective, and make the outcomes fit into people's lives, making them valuable and engaging.

_ FUNCTION

gather a collection of requirements for the intended outcomes to fulfill, in order to support goals and processes in the evolved enterprise, taking on their role as functional components.

_ STRUCTURE

work on evolved structural models, taking into account the insights about the domain, the outcomes of the design process, and the required transformation of other parts of the overall system.

DESIGN SPACE _

The six aspects of the Design Space come into play in this phase, guiding all conceptual design decisions that define what desired target state the outcomes should be designed to support, and how these pieces fit together as a whole. This involves thinking about the conceptual aspects as the synthesis of different visions of a potential future state. The process expands to defining the details of the intended transformation and enabling us to envision the impact on the way people interact and communicate, clarifying how it helps the enterprise do what it has been created for.

_ COMMUNICATION

envision how your design will support people engaging in communication processes, and establishing the exchange of information, as part of the business activities. Formulate key messages and model communication processes, to reach audiences and support and sustain the dialogue.

_ INFORMATION

envision how enterprise information will be handled in your design. Define classification systems and delivery channels that proactively provide information to their audience where it makes sense, but enable people to actively search and find what they are looking for in all other cases.

_ INTERACTION

define the interactions and behaviors that you intend to support with your design, and envision suitable tools to turn these functions into a dialogue with people. Model and define the behaviors these tools should exhibit towards their users, keeping the larger context of people and organizational behavior in mind.

_ OPERATION

define how the evolved enterprise should work by defining its operating model to support the functions it carries out. The operation aspect is about defining future work practices, considering the nature of tasks, their interdependencies, and balancing defined procedures against the need for flexibility.

_ ORGANIZATION

define roles needed for the evolved enterprise to implement its business model, and how to make people work together as a team. Define organizational structures of formal and informal elements to sustain this change and engage the people involved.

_ TECHNOLOGY

define ways to leverage and deploy technology as means to automate functions and structures of the evolved enterprise. Define technical options and capture their strengths and shortcomings with regards to the intended outcomes of the design process.

IDEATE _

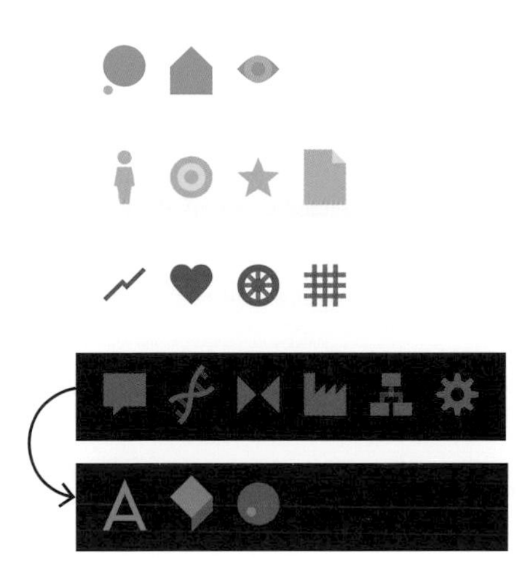

Develop ideas about the potential outcomes of the design process, implementing the conceptual basis across the different aspects of the Design Space, and turn them into sketches or other forms of illustration that make the Rendering of the enterprise visible.

Activities _

_ Use the Design Space aspects as starting points to develop ideas for innovation, inspired by research results and models defined in the preceding phase
_ Combine conceptual decisions with external sources of inspiration to open up the design space for unconventional approaches and surprising ideas
_ Engage in an open exchange among the design team and other parties involved, generating ideas and taking time to discuss them in depth
_ Combine visual thinking and rapid sketching to give these ideas a form, and communicate them as pictures of the future to team members and stakeholders
_ Produces sketches, more sketches, and even more sketches, knowing that most of them will not be taken further
_ Turn your ideas into concrete outcomes based on the three Rendering aspects, and envision what these elements of your design could be like
_ Work with expert designers to develop perceivable high-fidelity renderings of the outcomes, serving as evidences of the future enterprise and in turn influencing the conceptual decisions
_ Make Rendering elements high fidelity in terms of how well they represent the overall experience, not an isolated element
_ Explore, develop, and refine ideas collaboratively, continuously adding more detail and testing out different potential directions

_ Challenges

- _ Encouraging team members and stakeholders to come up with bold ideas, and to actually sketch them out visually
- _ Overcoming assumptions and predeterminations resulting from research results and conceptual work, questioning and informing design decisions instead of just implementing them
- _ Evaluating design ideas based on how well they work, not the personal taste of team members or stakeholders
- _ Not marginalizing the work of specialist designers, reducing their contribution to mere decoration

_ Typical techniques

- _ Personal Design Journals
- _ Brainstorming
- _ Sketching
- _ Storytelling
- _ Storyboards
- _ Visual Thinking

_ DESIGN SPACE

By now you should have an idea of the intended outcomes of the design process, and a vision of the transformation that they should support. The conceptual models defined in the preceding phase provide a solid foundation for envisioning how the future enterprise will work and how the design should support it. The goal is still the transformation itself, which will only come to life in the tangible results of the design.

Therefore, the conceptual vision has to inform the creation of concrete Renderings of potential outcomes, based on creative thinking and a much greater level of detail. Key to this is working with sketches of some sort. Sketching allows us to capture this projected reality in illustrations of how the future might be like. In this stage of the design process, the team should keep the focus on a definition of what might be, not exactly the way it should look or feel.

_ COMMUNICATION

envision and sketch out communication devices and media to convey key messages and facilitate the dialogue with the audience, supporting the communication processes.

_ INFORMATION

envision ways to make information available to people, and produce sketches on how to make it accessible via navigation and wayfinding systems, as well as suitable representations of information.

_ INTERACTION

envision digitally supported systems to facilitate key interactions, and generate ideas to implement the defined behaviors by sketching out interfaces adapted to their respective use context.

_ OPERATION

envision procedures and workflows to be executed to implement the operating model, and how they are supported with resources and artifacts such as machines, documents, or facilities.

_ ORGANIZATION

envision how the organization design comes to life in daily practice, defining job and skill profiles and potential candidates, and sketching work environments and tools to support key tasks.

_ TECHNOLOGY

envision how technical systems fulfill their part as components to automate the enterprise. Sketch their role in terms of the visible effects they produce as devices, products, or other outcomes that are made available to people.

_ RENDERING

Thinking conceptually enables the design team to plot a path to where you want to be, and how the design should support the intended transformation. The ideation phase is about turning this vision into actionable options to pursue, translating the thinking behind into concrete outcomes of the design process. The phase relies as much on insights and conceptual models as it does on creative thinking and good ideas. Decisions made and models produced based on the six aspects of the Design Space deliver the blueprints that define the Rendering elements. Their design needs to bring together these different aspects into logical arrangements and compositions, to shape a coherent whole. This involves working closely with specialist designers who have mastered the craft of form-giving in a certain medium, material or other design domain, to come up with strong ideas about what to produce in order to make a transformation happen.

_ SIGNS

turn the enterprise into stories to be told, and design systems of signs to reach the audiences to be addressed. Envision media to drive this communication process, crossing the boundaries of channels and human senses.

_ THINGS

envision things as components of the enterprise and help facilitate its activities. Design tools, devices, software systems, or other objects as artifacts to embody both their utilitarian purposes together with their cultural meanings.

_ PLACES

envision the space for the activities in the enterprise to happen. Design spaces to bring people in together in physical, virtual, and blended environments, making and connecting places people can go and relate to.

5 VALIDATE _

Build prototypes that simulate how the Rendering of the enterprise would work, and validate these designs, looking at the outcomes using the four Frames. Work with external people representing different perspectives, concerns, and knowledge areas relevant to the project. Iteratively develop and refine the concept based on validation results.

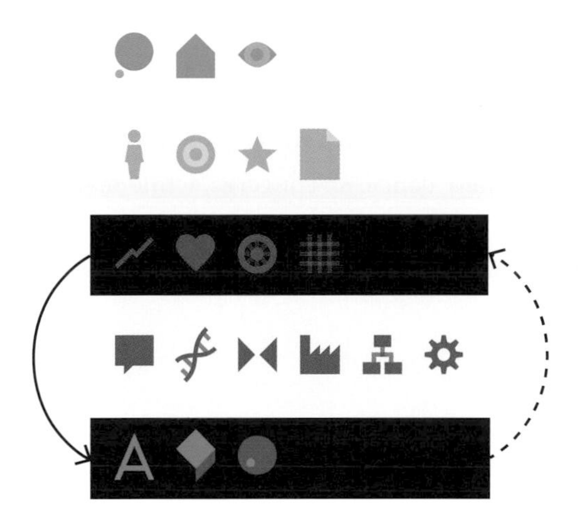

Activities _

_ Design and develop prototypes that simulate the outcomes of the design project, simulating their role as Rendering elements of the future enterprise

_ Use these prototypes as a basis to discuss the potentials and shortcomings of Rendering elements within the design team engage external stakeholders and work with experts

_ Put your models into action in the larger context of the project, using the four Frames as a guide to validate the concerns of viability, meaningfulness, feasibility, and structural fit

_ Use the Design Themes to identify the concerns to address, and develop success criteria to decide out how to incorporate them into a prototype

_ Turn these themes into stories that illustrate how the outcomes work in real life and fulfill their role in the enterprise, and use them to guide the definition of states the prototypes support

_ Make use of whatever materials, techniques, or methods work best to create your prototypes, and use paper drawings or spreadsheets to create high-fidelity simulations

_ Make different prototypes adjusted to the audiences they are made for, and include areas with great uncertainty

_ Produce multiple prototypes to validate and revise them regularly, opening a dialogue with everyone involved and iteratively refining the solution

_ Challenges

- _ Developing prototypes with just the right degree of fidelity that makes validation possible
- _ Finding a suitable way to represent a design challenge in a prototype of some sort
- _ Having a clear objective in mind when making a prototype, as a basis to decide how to build it
- _ Combining different aspects in a single prototype where concerns converge, while making distinct ones for separate issues

_ Typical techniques

- _ Prototyping
- _ Simulation
- _ Feedback Sessions
- _ Usability Testing
- _ Roleplay
- _ Surveys

_ RENDERING

Validation of potential outcomes relies on the design and development of prototypes to be used as focal points to validate ideas, concepts, and design decisions. This phase of a design project is about creating representative models that enable the team and others involved to evaluate the design in action. Such models are essentially "fakes" of the actual results of a design project. The most promising outcomes of the preceding phase, which define the Rendering of the future enterprise, are the focal point of these efforts. While there are many potential techniques that might be used, on different scales and with varying degrees of fidelity, prototyping is best approached in an opportunistic manner. This involves using whatever materials and tools prove useful to build something that fits the purposes of validation and communication, and benefits greatly from working with experienced practitioners in a relevant field.

_ SIGNS

create prototypes that show and tell the messages you need to get across to the audiences in the enterprise, encoded in media and signs such as graphic, text, or sound elements.

_ THINGS

create prototypes of the things being part of the design, shaped to be useful to people addressed, and simulating their functional and symbolic meanings.

_ PLACES

create models of environments to be made available in the enterprise, as places for people to dwell in, prototyping them to reflect activities, movements, communication, and ambiance.

FRAMES _

In the context of design work, a prototype is primarily a communication device. The details of its technical implementation do not matter that much, but its capability to communicate the intent of a certain design to different audiences. Its essential purpose is to adequately represent different concerns to be addressed. The goal of prototyping is to create something that speaks to the people looking at it, to illustrate the different states of a designed outcome. It can be used to tell a story, try it out, gather feedback, provoke reactions, and explore alternatives.

The Frames of the Enterprise Design framework are used to look at prototypes from different perspectives, evaluating how well they work. They allow us to measure them against the priorities identified in earlier phases, as expressed in the fundamental Design Themes to be addressed. To do so, they have to be turned into stories or scenarios, which drive the creation of prototypes and their usage in validation sessions. Depending on the qualities to be measured, techniques applied may include test settings, reviews, co-creation, or even an acting exercise—anything to make the scenario come to life and simulate the effects of the actual design is a valid approach.

_ BUSINESS

assess the viability, working with business owners and key stakeholders to simulate how a solution would work from a business point of view. Simulate the business model in terms of revenue streams and value generation, and refine the underlying strategy.

_ PEOPLE

work with people to assess the usability and desirability of the design, using working models that simulate the visible characteristics and qualities of the intended outcomes. Human-centric validation techniques help to develop new views and determine the next steps.

_ FUNCTION

design prototypes to incorporate the outcomes' functions and behaviors, and measure them against defined requirements. Work with implementation experts to refine their definition, decide on ways to implement potential solutions, and confirm their feasibility.

_ STRUCTURE

work with domain experts to validate the accurate representation of the problem domain in your design, and the fit of intended outcomes into the structural context. Refine the to-be domain model to make structural design elements relevant to the enterprise.

_ IMPLEMENT 6

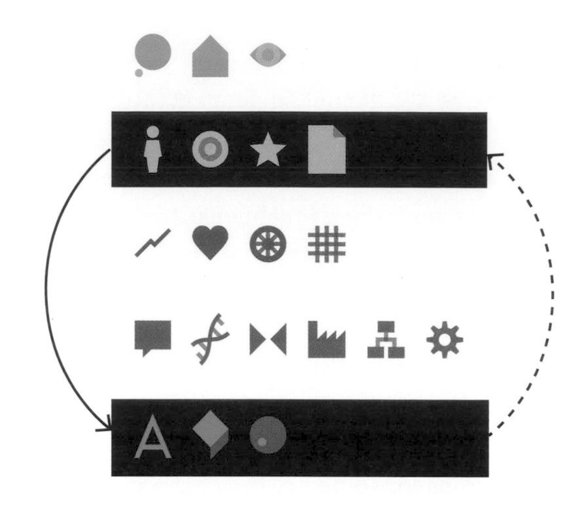

Produce the outcomes of the design project, turn them into documentation and specification to be shared with a wider audience of people involved in their implementation, and create a plan to actually build the components of the redefined enterprise.

_ Activities

_ Finalize the design of all outcomes, working with specialist designers and implementation experts
_ Develop a plan to implement and distribute all outcomes of the design project as components of the redefined enterprise
_ Specify activities and milestones as well as a longer roadmap to follow, and plan the business side of implementation
_ Align with the general project portfolio planning and change management initiatives in the organization
_ Plan subsequent projects to make the next steps of the larger roadmap happen after launch
_ Plan the migration of all technical or automation components where the design foresees a transformation
_ Prototype, document, and specify the outcomes to share them with everyone involved in their implementation
_ Produce the tangible artifacts, media, objects, and devices needed for launching
_ Organize people involved in the activities, launch a communication campaign engaging in a dialogue with them, and conduct training sessions
_ Develop an initial plan to introduce the project outcomes into the enterprise and transition to the new state
_ Prepare to rapidly adapt the design to real-world developments after implementation and launch

_ Challenges

_ Adjusting to last-minute changes and changing requirements
_ Maintaining the integrity of the design over the process of implementation and the compromises that have to be made
_ Determining the right moment to transition from validation and refinement to implementation

_ Avoiding just throwing documents and deliverables over the fence, engaging with the people charged with implementation tasks to ensure a shared understanding
_ Getting design results out of the door and avoiding skyrocketing effort trying to make it perfect
_ Managing and mitigating risk of implementation activities going not exactly as planned

Typical techniques _

_ Prototyping
_ Work Breakdown Structure
_ Blueprinting
_ Architecture Drawings
_ Specifications and Design Manuals
_ Issue Tracking Tools

RENDERING _

The implementation phase is about making an intended transformation happen, translating planned outcomes into reality. Rendering elements are the tangible outcomes of a design project, so this involves planning, construction, and production following the way they were specified in the preceding phases. For a design team, this translates into the challenge of documenting and sharing the design with others involved, and adapting flexibly and pragmatically to changes that emerge during implementation. Achieving unambiguity in terms of specification involves a lot of work, and results often in quite detailed and extensive documents. In many cases, we have successfully engaged with engineers and others involved in the realization of a design very early in a project, collaborating closely from thereon. Such an approach makes it easier to concentrate on the deliverables that really move the project forward and avoid waste, as well as to achieve a sense of shared project ownership.

_ SIGNS

create the final contents, layouts, artworks, and media, and oversee their production and distribution.

_ THINGS

finalize the design of products and other artifacts that are part of the solution and move into manufacturing activities.

_ PLACES

plan the construction and development of places to be created, to move into construction and production activities.

_ ANATOMY

In order to account for the enterprise as the larger system where the outcomes are applied, non-tangible aspects have to be addressed as part of the implementation. The Anatomy aspects come into play as elements to be created, changed, or replaced as part of introducing the design, and build the basis for planning this transformation process. Key to this is the technique of blueprinting. Originally used in different engineering domains, a blueprint is a visual mapping revealing the scope of a system, and the dependencies between its components across different domains. In strategic design projects, blueprinting can be used as a technique to map out the Anatomy of the evolved enterprise, and create detailed plans to place Rendering outcomes according to their role in that wider system. Such a blueprint does not need to be a drawing or visual mapping. Any format will work as long as it serves as an overarching plan of the elements envisioned and the outcomes to be put in place to inform planning and organization activities.

_ ACTORS

plan how actors are involved in the implementation of the design in the enterprise, and address people associated with these roles with hiring, communication, training, or other measures.

_ TOUCHPOINTS

plan how the outcomes drive an evolved system of touchpoints, and re-orchestrate these touchpoints to support different journeys people take as part of their interactions with the enterprise.

_ SERVICES

plan what services are to be provided as part of the evolved enterprise, and how outcomes facilitate service provision. Organize resources, responsibilities, and activities to make services available.

_ CONTENT

plan the content to be made available across the enterprise as part of the project outcomes, and coordinate its creation, production, and distribution to all audiences.

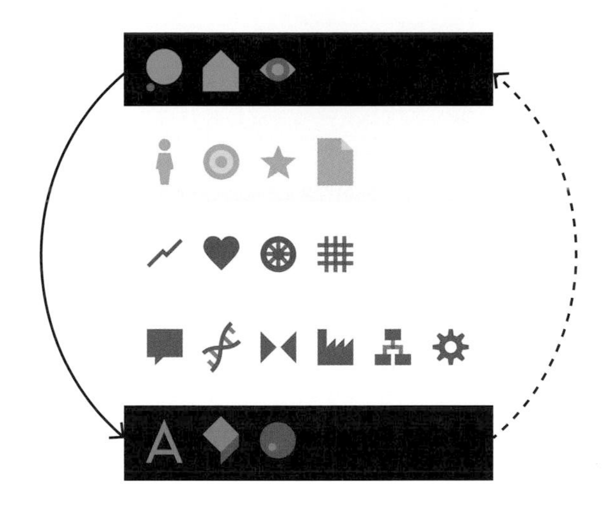

7 DELIVER _

Make the finalized outcomes available to the enterprise audiences you designed it for, manage the change processes involved in transitioning to the new state, and set up governance and continuous improvement processes to monitor performance as well as evaluate and refine the design.

Activities _

- _ Roll-out the tangible outcomes of the design process, making signs and media, places and things available to external stakeholders and staff members
- _ Deploy all technical components and systems needed to put the design in place
- _ Communicate the changes to the organization and the wider enterprise to all audiences, choosing tools and formats adapted to the nature of the project and its environment
- _ Assess the actual usage of the outcomes, address shortcomings identified, and refine the design to fit reality
- _ Perform long-term evaluations of the effects of the results on the Big Picture qualities addressed with the design initiatives, leading to further refinements
- _ Perform scans of the enterprise as a wider environment to identify upcoming changes, trends or themes to be addressed in a strategic design initiative
- _ Identify the design challenge and preliminary scope for follow-up projects

Challenges _

- _ Making the task of rethinking the enterprise everyone's duty, leveraging ideas and feedback from people all over the enterprise
- _ Being flexible in the way outcomes are managed and interpreted once the results are out the door, but at the same time protecting the essential ideas and underlying concepts
- _ Keeping an open mind for other ways to do things in the enterprise and constantly questioning the decisions made in the design process
- _ Sustaining the design by protecting the assets and outcomes developed as results of the process against weakening support or undesired changes

_ Typical techniques

_ Newsletters or Magazines
_ Events and Seminars
_ Books or Other Publications
_ Scorecards
_ Regular Audits
_ Surveys

_ RENDERING

The last phase of the design process is about introducing the results of a design process into the enterprise, as triggers for change and drivers of the transformation towards the intended target state. In the enterprise context, big bang success stories are rare — much more often, successfully designing, implementing, and delivering transformation involves an initial step followed by a longer phase of transitioning and adapting to the new reality.

Therefore, redesigns and project launches are just the beginning of the journey. Strategic design requires continuously improving the Rendering elements after their introduction, based on a measurement of actual usage, adoption, and problems that occur. Informed by collecting quantitative data and performing qualitative observations, minor adjustments can lead to dramatic improvements of the overall performance. It also reveals larger issues or opportunities to be addressed in new projects.

_ SIGNS

assess how people navigate through systems of signs, consume media and content, and make improvements to support reception and understanding.

_ THINGS

assess how things are actually used, support emerging usage patterns and learn for subsequent versions of the objects made available in the enterprise.

_ PLACES

assess how places across the enterprise are frequented, how people move through them and what they use them for, and shift them based on these insights.

BIG PICTURE _

Returning to the top layer of the Enterprise Design framework, the Deliver phase is about managing the transformation process which is the goal of the design process. Design projects can be called strategically relevant only if they have a significant impact on the organization and its enterprise. Therefore, outcomes have to be tied to the Big Picture aspects of the enterprise. This translates into a two-fold challenge.

To have an impact, changes have to reach everyone involved in the transformation process, and impacted by its results. Potential activities to be planned and executed include communication and launch activities, aimed at getting people engaged and committed to the new situation, as well as explaining how things work now. This also requires visible backing from sponsors, executives and other key stakeholders.

The other part of this challenge involves maintenance and governance, long-term change management and performance measurement, making sure the results of a design initiative are sustained in the long run, and continuously refining the design further and launching new projects as needed. In a world of constant change, this might happen sooner than expected.

_ IDENTITY

set the stage for new or redesigned brand identities and their introduction into the enterprise ecosystem, and make long-term assessments of its impact on culture, behaviors, and perceptions.

_ ARCHITECTURE

manage the transition to new ways of working in the enterprise, new formal structures and systems to be used, and measure them against performance and quality objectives.

_ EXPERIENCE

make the rendered outcomes of the design project available to people, and perform long-term studies to evaluate the impact they have for people and their interactions with the enterprise.

I think it's good to have to do something quickly and not to have time to spoil it.

David Gentleman, Painter

_ Getting the design process right

The topic of design processes has been the subject of academic research and vivid discussions among practitioners for quite some time. Countless models have been invented to describe how designers work, attempting to turn the professional practice of design into a manageable approach for organizations.

One of the core problems of putting design into such a framework is that it hardly ever fits the reality of design work. Attempts to impose on such work a rational, deterministic model like those used in large engineering projects have largely failed. Through empiric studies it has become clear that most designers do not actually work that way. In order to deal with the underdetermined nature of design work, design processes often are messy and opportunistic. Designers iteratively inquire into the problem space, to make moves in a design space that is constantly being questioned and redefined. While working with a large number of techniques and methods, even for the most methodical designers no single design project is the same.

This is no different for the process discussed in this chapter. Although most strategic design projects go through all the phases and address all the aspects mentioned in one way or another, the way they do that is far from a linear and planned process, and much closer to the non-directed way an artist comes up with an oeuvre, or a college student learns about complex field of study. There seem to be no clear boundaries between the phases, and activities which are at opposite ends of the model in reality might have significant overlap.

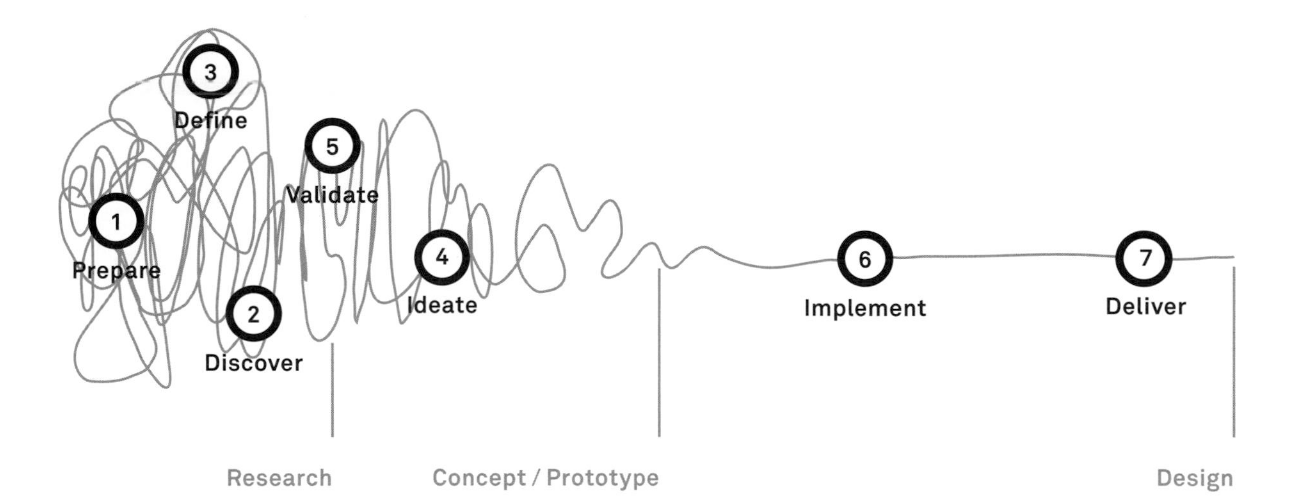

The Squiggle, made famous by Damien Newman: starting in a space of great uncertainty and jumping between insights and ideas, conceptual decisions and iterative inquiry. In later phases the work switches from opportunistic exploration to defining and refining a desired target state identified.

Understanding the problem and inventing a solution both happen in parallel, and ideas for potential outcomes can occur in any phase of the process. One project might make you move slowly from research to definition to ideation, developing a thorough understanding of a complex problem before deciding how to proceed. Another may require developing a first shot in a matter of days and co-creatively reshaping it with peers, looping through the process steps in short cycles.

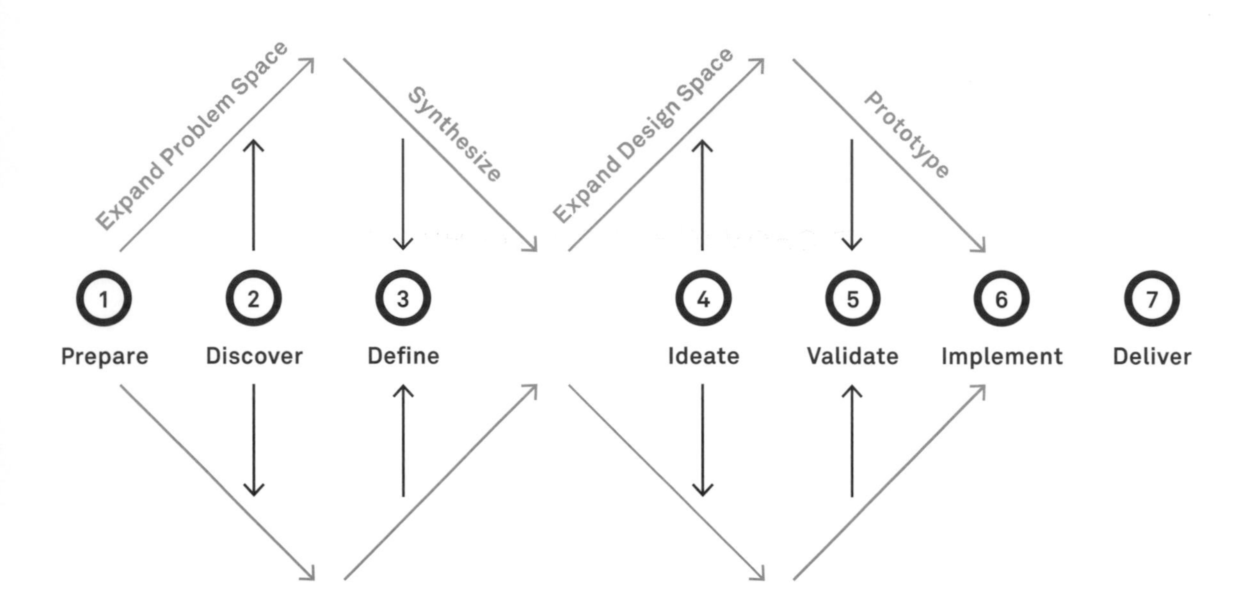

The Double Diamond model from the UK Design Council illustrates how a design process first opens a space to explore the problem, and then consolidates to synthesize potential solutions. In a second diamond the space is opened again to creatively explore ideas, which are turned into prototypes that bring together multiple ideas.

In the enterprise context, figuring out what to do next is often as hard as the actual design challenge. Just as the outcomes have to fit into all sorts of environmental variables, so does the design process itself. Key to this is developing a coherent design strategy, making individual projects a part of a larger program aiming to transform the enterprise.

AT A GLANCE

A design process needs to touch all relevant aspects and bring a project from early ideation to delivery. Key to a successful design project is the constant adaptation of the chosen approach, balancing an opportunistic and dynamic selection of methods and approaches with the constant quest towards a well-defined objective.

Recommendations

_ Start with an exploration of the context of the design project, consulting with stakeholders, performing research, and developing framed models of the problem space

_ Formulate the objectives, the principles, and the success conditions that you want to achieve, and envision a desirable future state in concepts and sketches

_ Create prototypes of the envisioned future enterprise to be validated, refined, and iterated, and turn the most promising results into outcomes and specifications

_ Consider the delivery of results and the transformation process part of the design initiative, adapt and protect the outcomes, and plan for the next steps

CASE STUDY_INSTAGRAM

INSTAGRAM'S ENTERPRISE

Providing the best possible experience in mobile photography for sharing your life with your friends and peers.

The story of Instagram, a Silicon Valley startup that was purchased by Facebook for $1 billion in cash and stock, has left many transfixed and inspired. The company created the most successful app for mobile photography, optimizing the entire product for the experience of taking and sharing photos on the go — quickly, effortlessly, and conveniently. Released in October 2010 after just 8 weeks of development, it was used by 24 000 people on its first day. Over the course of a year the number of users rose higher and higher. By the time the company was sold in 2012, the Instagram app was being used by over 30 million registered users.

While this is clearly a tale of overnight success, the way it has been picked up by the media often neglects the fact that the development and release of the software was just one part of a much longer story. The two founders had complementary skills and aspirations, and a shared passion for the design theme they chose to work on — providing the best possible Customer Experience in mobile photography.

Instagram's success is the result of a holistic design process applied to an enterprise. Instead of putting their effort into new technologies or developing new markets, the team focused on an existing field where they saw a deficit. The idea of photo-sharing was hardly new, and players like Flickr were already successful in that market. Led by Mike Krieger, a technologist co-founder with a User Experience background, they started at the activities of people, designing their product and the company around the User Experience. Unlike the development itself, this process lasted over a year — possibly longer if looking at all activities of preparation and problem definition. It demonstrates well how bumpy and unpredictable such a process can be — the team sprinted through all the stages described in this chapter, applying strategic design in a non-linear, adaptive, and iterative way.

We decided that if we were going to build a company, we wanted to focus on being really good at one thing.

Kevin Systrom

_ 1 PREPARE

The Instagram story started with Burbn, an app developed by Kevin Systrom, with a much broader scope. When Mike Krieger joined, the two founders realized that this multitude of features actually made their product too complex, and that they wanted to concentrate on doing just one thing, and that one really well. This resulted in a fundamental change of direction, although the team first continued to develop the original app.

But at the same time, the idea of mobile photo sharing became the story of the startup, so that the two founders started defining their vision of the enterprise. They developed ideas on the focus area, and performed scans of the ecosystem to determine other actors to address, such as Facebook and other social networks becoming external touchpoints for their offering. They developed approaches specific to the content of photography, and services tailored to sharing pictures with friends. This eventually led to a new vision for the experience the company wanted to create for their users. The new name, Instagram, reflected the evolved identity, and a revised architecture supported the new strategic direction based on existing assets from the previous project.

2 DISCOVER _

Instagram's offering is based on an insight into the way people make and share pictures with their handheld devices. Being a startup business, the team gathered this knowledge in a rather informal way, by experimenting with the product and listening to the feedback of their users. This personal but subjective knowledge was backed by usage data, which proved quite revealing when the team discovered that the photo-sharing feature of Burbn was the app's most used functionality. This made them question the initial idea, and be courageous in changing their plans.

Through these two ways of gaining detailed knowledge, Instagram was able to discover the enterprise elements that they needed to incorporate into their design. They determined the profiles and behaviors of users, and dived deep into the domain of mobile photography to discover the structural elements that mattered to them. They were able to determine a set of functions that matched the context of people taking photos on the go and sharing them.

The Discovery activities were clearly driven by applying a People frame, while questions of the business model were considered less pressing at that point — a prioritization that happens in any design project. Instagram concentrated on researching people's habits, to make an offering that was meaningful them.

3 DEFINE _

The initial design work in a startup is characterized by a large degree of uncertainty, for the whole time that the founders and their team are struggling to define a suitable direction. At Instagram, this changed when they set mobile photography as the new design theme, and radically reduced the scope to just those elements connected to that theme.

Once the basic goal was defined, the team was able to further elaborate their intended solution in terms of the different perspectives. While the primary driver was the people perspective and the User Experience considerations, Instagram defined a business model based on advertising revenue, and defined both the functional scope and the structural elements to execute on the shared vision of their enterprise.

I think the scariest part of a startup is that you don't have your product defined, you don't know what you're working on.

Mike Krieger

This enabled them to work towards a well-defined future state, and make a series of fundamental conceptual design decisions aligned with that vision. The decisions to keep the app free for everyone to install and use, and to make shared pictures public unless the user explicitly decided otherwise, provided the basis for the rapid adoption of Instagram. Their consequent focus on providing a fast and easy sharing mechanism let to the decisions to make classification optional, and remove as many steps and barriers as possible from the interaction.

_ 4 IDEATE

Instagram was created in a lean way, by way of experimentation rather than a traditional software development cycle. This emphasis on trying out things was also leading to some of the ideas that make a solid User Experience a remarkable one.

When the team invented the sharing mechanism as the primary communication mode for users, they streamlined the process to adapt it as much as possible to the mobile use context, not bothering users with any details or additional options. This thinking about the context also led to one of the more progressive ideas.

Instead of just enabling people to share photos, Instagram wanted to help them share the greatest moments of their lives. This was in contrast to the bad quality of mobile phone cameras at the time, and also the fact that most users shared snapshots rather than high-quality photos. The solution was the introduction of a set of image filters and effects—hardly a new idea, but a suitable solution for the exact problem to be solved.

Every Monday we would say: what's a problem that we think we can solve this week around taking mobile photos? By Friday we had a prototype, by Saturday we had it in the hands of users.

Mike Krieger

In putting the conceptual Design Space into outcomes, the Instagram team proved to be obsessive about design details, inspired by years of experience in the field of photography and online communication. This allowed them to bring together all relevant aspects in a coherent Rendering of the app as a service interface and place for the community. Being quite passionate about photography, Kevin Systrom designed every component of their offering with great attention to detail.

5 VALIDATE _

From the beginning, the work at Instagram was driven by a culture of purposeful experimentation. Approaching every design challenge as a small research project, the team started with an experiential concern to address, putting their ideas into code to be tested with people within one week. From the basic set of functionality to the selection of which filters to include or how to design the interaction, every detail is the result of a hypothesis that got designed, implemented, and put before users, including different variants subject to A/B test setups.

This enabled them to constantly validate their design in terms of adoption and fit with the functional context. They kept carefully adding structural elements, iteratively refining the vision, and bringing their product closer to it.

With this approach, they were able to drive spectacular user adoption, giving confidence to their investors that they were on the right track and validating their efforts from a business perspective. Although there was no detailed business model defined, they raised $7 million of venture capital, buying them more time to continue their experimentation. The quality of the product and the increasing user adoption eventually led to the Facebook takeover even before any substantial revenues were generated.

_ 6 IMPLEMENT

The founding team of Instagram brought together both talent and expertise in technology, Experience Design and photography—the domain of their project. Instead of addressing concerns of product design, technical implementation, marketing or service in isolation, they bridged the gaps between all those aspects and produced a Rendering that executed on a holistic strategy. This strong vision led to important technical decisions, for example, the uploading and processing the data in the background to the sharing interaction exceptionally fast.

All elements were approached as the Anatomy of a greater whole, including the app and its functions, the web site, the background operations, and the integration with external platforms. To be able to implement and test quickly, the team first developed any given component in a generic HTML5 version, iterating and refining any design before making it a part of the native apps for Android and Apple iOS devices. These early prototypes then became the blueprints for the real application, already tested for feasibility and quality.

The experimentation mindset continued well into implementation, flexibly adapting the approach to new situations. The initial release of the Instagram app was a repurposed version of the preceding Burbn project, flexibly leveraging existing technology and assets to follow a new vision for the enterprise.

In the long run you have to build a company. That is way harder than coding a few lines and building the design out. I did all the design for our app. But I had to stop doing that and say to myself: we need an office, we need to recruit people, we need to think about vision, we need to think about where we stand in the ecosystem—and that's building a company.

Kevin Systrom

7 DELIVER _

The holistic and strategic design approach at Instagram led to an active consideration of their enterprise ecosystem. This includes the founding team's valuable network of Bay Area contacts. Instagram reached out to more external actors in their enterprise ecosystem, creating new touchpoints and mindfully extending their services on new platforms such as Android. Although there was strictly no business model defined, there was a great customer acquisition strategy executed in their design.

This vision encompassed not just their product, but also all other parts of the company. They leveraged Apple's App Store, Facebook, Foursquare, and other platforms as channels to reach more people and make Instagram a part of a larger experience.

They hired a community manager dedicated to driving the conversations with their audience across all communication channels when the team was still quite small, although it must have been tempting to just hire more developers to reduce the coding time. Instead of adding more features, they consciously threw away functions that didn't contribute to the experience they wanted to support. As adoption increased, their main effort shifted to refactoring their operational and systems architecture to support that growth, expanding their hardware, and optimizing performance. That way, Instagram continuously adapted the Rendering of their enterprise to an evolving Big Picture vision.

10_ DESIGN PROGRAM

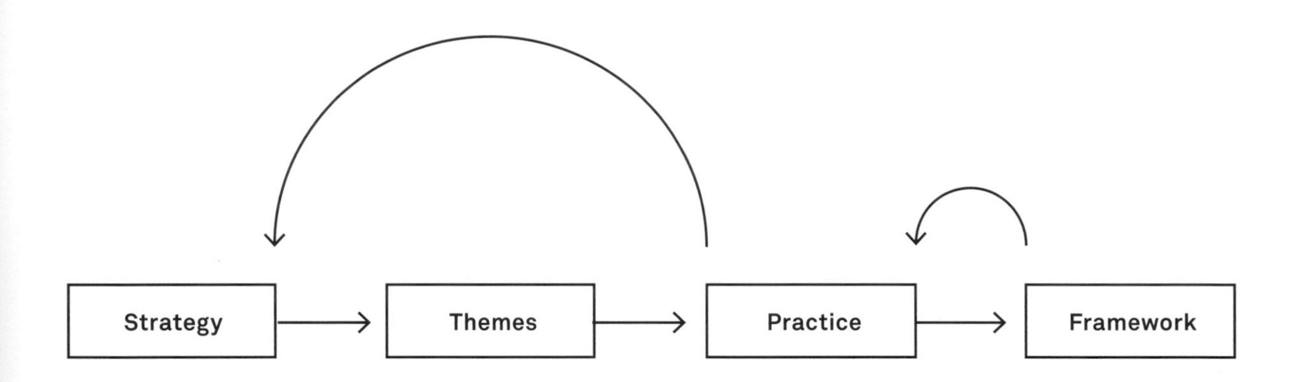

Given the ambiguity and complexity of the design process at the enterprise level, it is clear that integrating design practice into organizational life is no easy task. Assembling a skilled and diverse team, determining a suitable approach, and integrating stakeholders are the right ways to go for individual design projects with a limited scope. In order to address enterprise-wide design challenges however, the problems get too large and unmanageable to be tackled all at once. While the concept of project work is deeply embedded in all design activities, in isolation they fall short of addressing long-term strategic matters.

This chapter is about embedding the practice of design in the enterprise context. Using the twenty aspects of the Enterprise Design framework as a guide, it deals with the strategies, themes, and practical issues of delivering a strategic design program in the enterprise.

Putting design into actual initiatives and organizational structures, integrating various groups and departments, defining themes and philosophies, encoding design into guidelines and frameworks, or simply deciding what to tackle next— the challenges you might be facing are manifold and too diverse to discuss them here at length. Instead, this chapter is a collection of thoughts and hints, based on years of professional experience that gave us an idea of what actually works.

© 2013 Elsevier Inc. All rights reserved.

It is not enough to have a talented designer; the management must be inspired too. The creative process is very disorganized; the production process has to be very rational.

Bernard Arnault, LVMH

STRATEGY _

This book talks a lot about strategic design, and we see the Enterprise Design framework as a way to make design as a practice strategically relevant to an organization. Applied in reality, this means connecting design programs and strategic management practice in a way that they assume a joint responsibility to shape or re-shape the enterprise, and to bring a strategy to life.

Bringing together strategy and design results in a symmetrical relationship. From one perspective, a key task of design initiatives is to illustrate how a certain strategic choice might turn out once it is executed. Vice versa, the outcomes of a design process serve the purposes of informing and guiding strategic thinking, illustrating desirable futures. Once defined in concrete scenarios, the choice of which future to pursue becomes much easier.

Seen from a design perspective, the topics of strategy boil down to the question how the organization chooses to appear in the lives of people it addresses and those involved in running the enterprise, and the value it creates for them. A vital prerequisite for such a strategic design is grounding this work in an enterprise-wide view, and connecting design practice as applied to concrete outcomes to the conceptual thinking behind it.

This section is based on a model from Michael Treacy and Fred Wiersema. It describes three generic value disciplines for a company to excel in, while not leaving behind the remaining two. We applied them to envision specific options to pursue, defining both a design strategy and a strategic vision.

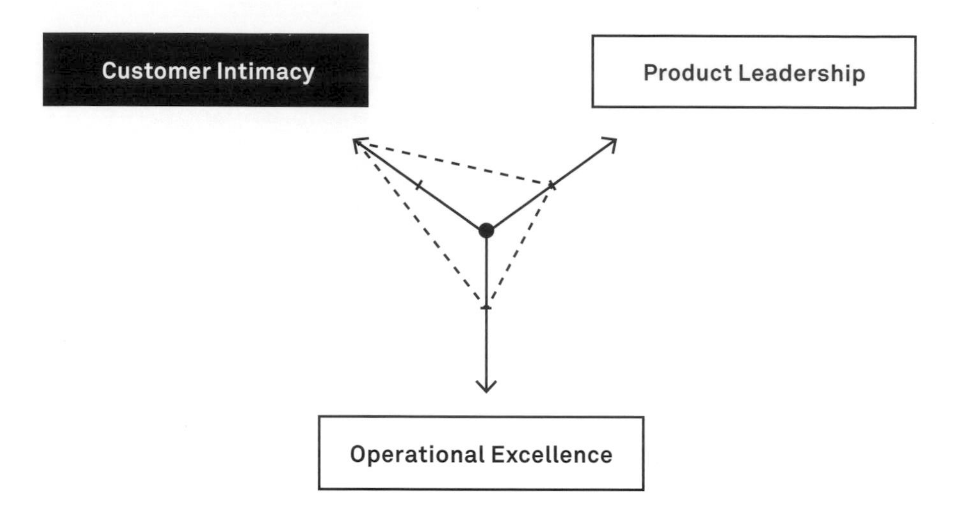

_ Customer intimacy

This area of focus involves envisioning a strategy that centers on designing the relationship with individual customers, and what they experience when interacting with the enterprise. As the label suggests, Customer Intimacy strategies are about turning that relationship into a genuine love affair.

There are few brands that can claim to engage customers in such a deep and profound way, but all that do have something in common: a deep understanding of those they are addressing with their products or services, and tailoring their offerings almost to the individual customer. They excel in constantly meeting and exceeding customer expectations, handling interactions on a personal level, and managing a strong brand identity that has a meaningful place in their lives. The role of design is to shape the enterprise in a way that it supports creating this kind of close and trusted relationships.

_ *Design Strategy:* Design all parts of the enterprise to delight and surprise customers, and to empower all other actors to deal with individual customer needs and situations

_ *Strategic Vision:* Envision what such an intimate relationship can be like in day-to-day practice, and picture meaningful roles for the enterprise to play in people's lives

The strategy, energy and resources of customer-centered organizations are aimed at processes enhancing knowledge of an engagement with their customers. And prioritizing these over keeping conventional competitive barriers.

Peter J. Bogaards (bogieland.com)

_ EXPERIENCE

designing the enterprise to serve the customer experience, making offerings and services valuable and meaningful to individual clients.

_ ACTORS

designing for the wider group of actors involved in delivering the customer experience.

_ PEOPLE

understanding addressees on a personal, empathetic level and learning about their lives.

_ INTERACTION

designing for customer interactions that reflect personal relationships and individual behaviors.

_ ORGANIZATION

empowering front-line staff to make informed ad hoc decisions when an individual need arises.

_ PLACES

making places to bring together customers with other actors, and foster an active exchange.

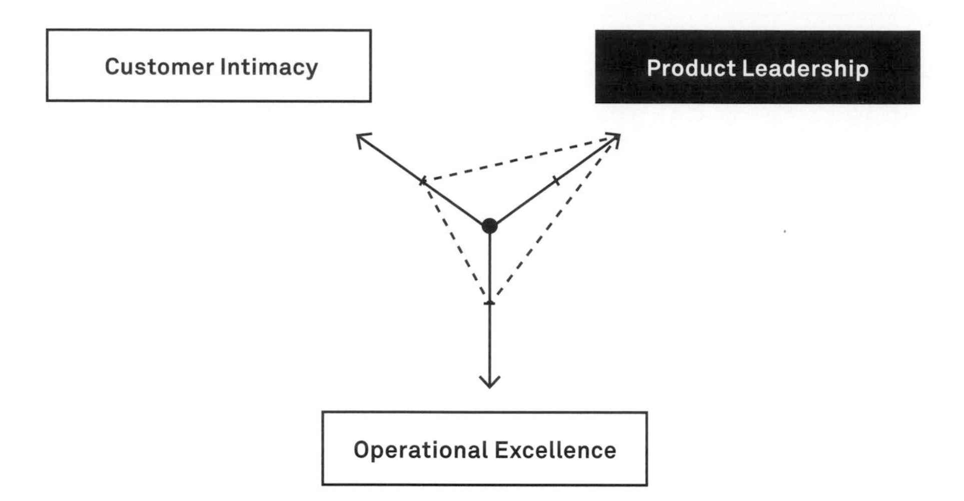

_ **Product leadership**

This type of strategy is for organizations focusing on making innovative and disruptive offerings to the market. Achieving excellence through Product Leadership entails a strong focus on market trends and research, to identify, design and implement new products and services in a rather short time frame. The idea behind such an approach is to stimulate the enterprise as the organization, its wider ecosystem and its market environment to deliver the next big thing.

The role of design in such an environment is to bring together the dimensions of creating a strong brand identity and a winning product design, as well as the people, processes, and assets behind their delivery.

The enterprise of a product leader consists of managing a dynamic environment of actors involved in the identification and creation of new opportunities, and leading this dialogue with research probes, ideation, and vision prototyping. For design practice, this translates to opportunistically co-creating the future with those at the forefront of its discovery and definition.

_ *Design Strategy:* Focus on continuous inquiry into a dynamic enterprise and market environment, generating ideas and turning them quickly into *concept cars*

_ *Strategic Vision:* Anticipate future offerings both near-term and long-term, and designing the enterprise as a system of services and internal assets to deliver them

_IDENTITY

making the brands of the enterprise synonymous with meaningful, cutting-edge products and services leading the market.

_ACTORS

engaging an ecosystem of partners, lead users, and customers to generate ideas and deliver on them.

_STRUCTURE

mapping the structures of the enterprise, its domains and people's mental models, to find a suitable niche for future products and services.

_COMMUNICATION

designing modes and channels to facilitate open communication, announce new developments, and gather expectations.

_TECHNOLOGY

leveraging technical possibilities to create novel products and services.

_THINGS

producing artifacts that inherit the trends, ideas, and concepts envisioned and make them accessible.

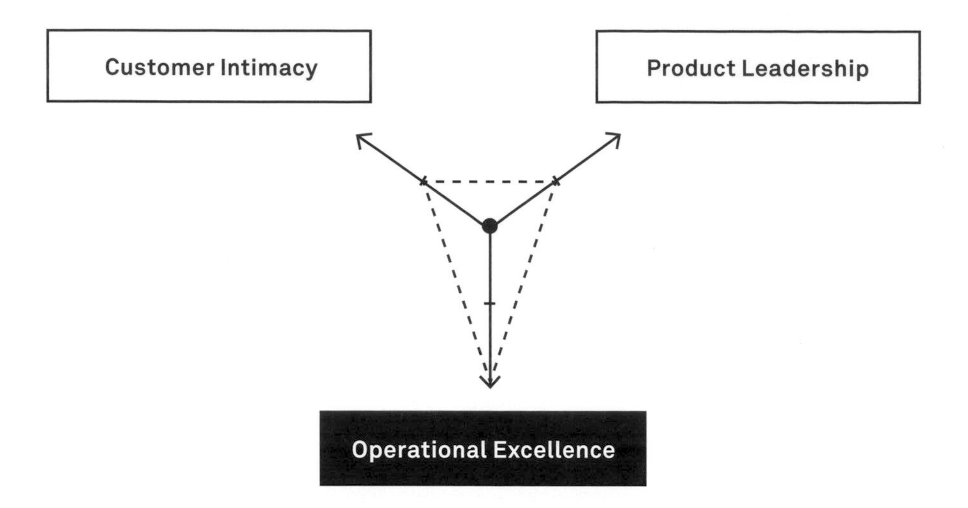

_ Operational excellence

This strategic focus area is about optimizing the way the enterprise works, using automation and intelligent operations to minimize efforts and cost, in order to outperform the competition. It means designing the enterprise as though it were a well-oiled machine, striving to provide more value to customers for less cost and effort. Organizations pursuing this type of strategy rely on achieving a good price and high volumes in order to benefit from economies of scale, while maintaining the expected quality and value propositions.

Design for Operational Excellence involves a continuous search for improvements, often refining rather than redesigning the enterprise. It contributes to this strategic focus by designing all kinds of systems that are involved in delivering on that promise. Capabilities to understand actual needs for a large customer base allow us to gain insights into what elements of an offering are considered essential, and what can be removed or downgraded to improve the competitive positioning. This definition in turn informs the design of processes supporting their execution and automation.

_ *Design Strategy:* Design the enterprise to support process execution, striving for a smooth flow of communication and work between people, and mindfully automating activities

_ *Strategic Vision:* Develop illustrations of systems such as products and tools, procedures and services that support running the enterprise and connect actors to underpin processes and activities.

_ ARCHITECTURE

making the enterprise a system working in an efficient and lean fashion, supporting the flow by streamlining structures and processes.

TOUCHPOINTS

Orchestrating, connecting, and designing touchpoints to make business processes accessible and directly involve customers and other actors in their execution

_ FUNCTION

designing functions as capabilities to better run the enterprise, serve customer goals, and make it fulfill a clearly defined purpose.

_ INTERACTION

designing interactions with automated components and self-services to support all typical transactions and activities as well as exceptions and errors.

_ OPERATION

defining a well-tuned operating model to support human work and automated tasks and measure performance.

_ SIGNS

developing systems of signs to guide people navigating the enterprise and support handling a high volume of cases.

_THEMES

This section describes design themes, topics to be addressed with a design program. Themes usually represent fundamental aspirations of key stakeholders. They guide both the choice of challenges to tackle and the chosen approach. To a design team, well-defined themes often take the form of a few statements or short stories, and provide both a rich source of inspiration and a basis for making conceptual decisions. In the enterprise context, themes for high-level design programs tend to cycle between innovation and consolidation as guiding trends.

_ *Innovation:* Limiting the scope of the design program to a well-defined theme, often following an external trend, new technology, or other kind of development. The design task is to define a way to creatively respond by seizing opportunities as they emerge and laying out a bold vision of the future.

_ *Consolidation:* Mapping the existing assets and components of the enterprise to better align them with each other and benefit from global platforms and common resources. This task of "silo-busting" is one of the core competencies of design, bridging different viewpoints by delivering a unified vision.

Themes often depend on global decisions or events such as an acquisition, a re-branding, or simply a new CEO. This section portrays a set of topics which are archetypes for individual themes and narratives for design programs in the enterprise context to start from.

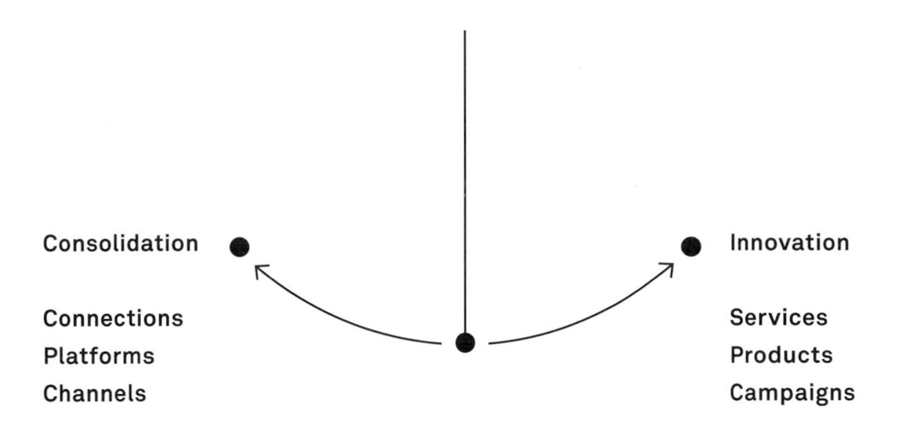

Customer experience _

Taking the Customer Experience as the guiding theme defines the objectives of a design program in terms of its impact on customers, and recognizes them as the most important stakeholders of the organization. It means designing the overall shape of the enterprise as it manifests itself in customer interactions, from initial awareness to conversion and sustaining customer relationships.

Connecting a business to the customers it addresses is a typical task for design practice, working in conjunction with Marketing, Product Design, and Customer Service. In today's organizations, there often seems to be a mismatch in design practice as it is applied to communication, to the offerings themselves and to the services provided at the different stages of the customer journey.

Instead of a coherent whole, customers have to deal with isolated and disjointed parts. Designing for an end-to-end experience is a way to leverage design in order to stand out in the market.

A remarkable Customer Experience can come from many factors, but it is key to define the theme in alignment with at the chosen strategy—any of the three focus areas previously described can provide the leverage to make it happen:

_ Capture concisely and in a few words what the experience you envision should be like, and the particular benefits it brings to the customer.

_ Use this vision to guide the re-arrangement of enterprise elements, bringing them together to reshape the customer journey.

_ EXPERIENCE
how is the desired Customer Experience like?

_ TOUCHPOINTS
what touchpoints are used to facilitate interactions?

_ PEOPLE
who are the customers we are designing for?

_INTERACTION
how do we interact with customers?

_ORGANIZATION
how to allocate responsibilities to support the customer?

_PLACES
what places are driving the Customer Experience?

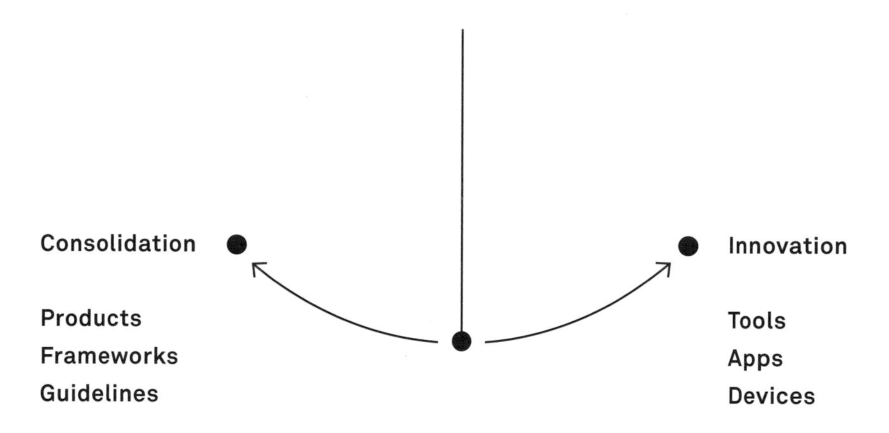

Digital workplace _

A key topic to be addressed in the enterprise is how people contribute to its endeavors, as employees, partners, or in other roles. The emerging concept of work technologies and environments, now often labeled the *Digital Workplace*, is about designing the enterprise in a way that it supports people connecting, collaborating, making decisions, and solving tasks.

The addition of *digital* is essential, because it is the digitalization of technologies and structures that is transforming the world of work at an astonishing pace. It allows a large part of the working population to be liberated from fixed locations, potentially working with people all over the world. The communication channels, tools, and systems running today's organizations are at the forefront of an ongoing shift to a more dispersed, flexible, and dynamic economy. In today's work environments, tools and technologies remain neffective and hard to use, often preventing rather than supporting productivity and exchange.

As a theme for a design program, the objective of reshaping work translates to the creation of an interconnected set of useful tools and social environments:

_ Capture in a short narrative how you envision connecting people to information, conversations, and business processes that drive the enterprise.

_ Describe ways to enable both the enterprise and individuals to realize the potential of working digitally

_ IDENTITY

how to support the brand identity and culture?

_ SERVICES

what services are driving the Digital Workplace?

_ FUNCTION

what purposes does it serve, what processes does it enable?

_ COMMUNICATION

how does it support collaboration and communication?

_ INFORMATION

how does it make information accessible and findable?

_ THINGS

what devices are used to access it?

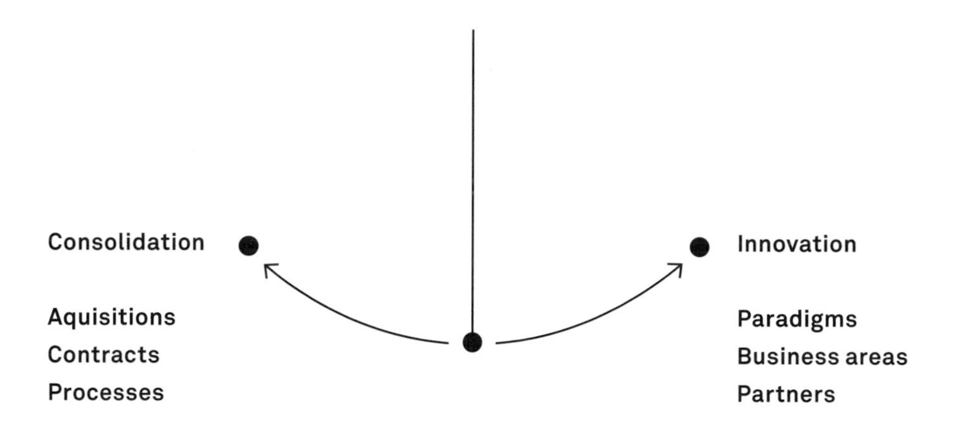

Business platforms _

The idea of designing enterprises around Business Platforms out of the emerging business models using the Internet as the main driver of the enterprise. Inspired by companies like Amazon, Apple, or eBay, they attempt to turn themselves into a digitally supported platform driving customer interactions, marketplace collaboration, and internal operations, thereby expanding their business significantly into new domains and segments.

As a theme for strategic design initiatives, creating a business platform for the enterprise is about designing a central system that serves a large number of actors in the enterprise ecosystem at once. It is about supporting continuous business development by integrating new suppliers and partners and expanding the scope of offerings. Platform companies excel at redefining and extending their business, leveraging existing assets to reach new target groups, and gathering data to better serve customers and identify opportunities.

_ Bring together operational processes and workplace technologies to drive the Customer Experience with e-business offerings, and identify opportunities for collaboration open to the wider marketplace

_ Determine ways to create flourishing ecosystems around your market offerings, supported by a digital infrastructure of owned and third-party platforms

_ Capture in a design theme how that ecosystem could be structured, and how the platform to be created could support its development over time

_ ARCHITECTURE

how does your platform work?

_ ACTORS

who accesses your platform?

_ CONTENT

what is being echanged?

_ BUSINESS

how will the revenue stream work?

_ OPERATIONS

what is needed to operate it?

_ TECHNOLOGY

what are its technical components?

_ SIGNS

how to navigate across the platform?

PRACTICE _

There is a multitude of potential options for integrating design in an organization, considering not only ways to make the practice part of organizational structures and operational processes, but also aspects of working culture and stakeholder engagement. The success of a design program depends on finding the right approach, balancing both the hard and soft factors of creative work in a complex environment. Taking design practice to the enterprise level also requires granting practitioners the space and resources needed to have an impact on the larger ecosystem. Confined to the common silos of Marketing, Product Development, or Corporate Communications, the results it produces are likely to be as limited as the selected scope predicts. But even when seen as a strategic resource for developing the enterprise, design often seems to be at odds with management practice and the mindset of many business people.

This section is about addressing these practical issues of embedding design in daily work with a strategic impact in mind. While this without doubt depends on a variety of factors, in our experience it boils down to the creation of a supportive culture across the organization backed by executives and stakeholders, and assembling great design teams to run the projects. We strongly believe that consciously addressing these factors tremendously improves the chances of the design program being a success.

Culture _

Many organizations struggle when implementing strategic design practice in real life and making it part of the way they work, even when it is approached from an enterprise-wide view and with top-level support. A common challenge is figuring out a way that sustains bold ideas coming out of design projects, taking them all the way from discovery to execution.

This applies particularly to larger organizations, where decisions and activities touch many parties. They often face issues of making design part of the whole organization instead of just a single department or team, and moving beyond isolated projects to its application to the big picture. But even in smaller settings, there is often a collective mindset antagonistic to the messy process and the flexible way designers work, which prevents design practice from creating and sustaining the stunning results it is expected to deliver.

Groupthink and consensus, or focusing on decisions or feasibility too early are poisonous to design practice. Ironically, attempts to prevent such developments by planning the steps of the process up-front tend to further stifle innovations. Therefore, a key element in this regard is the working culture — the way teams and their leaders deal with design activities, collaborate across disciplines, and approach problems and ideas. We found three key values to be particularly important when developing a strong design culture, to encourage an attitude of creative discovery in the collective mindset.

_ *Championing:* A strong design practice needs charismatic leaders who advocate a relentless quest for the best possible outcome, keeping the process agile and flexible, and who commit to achieving excellence in all key aspects.

_ *Experimentation:* We have seen great enterprise-wide and strategically significant results coming out of rather limited initial pilots. A prerequisite is that they act as role models for many follow-up projects that together spark a large-scale change.

_ *Courage:* A design culture requires being bold in many ways. It means investing in research that many would see as a waste of time, trusting experts but also bold ideas that strike you with their elegance and simplicity, and allowing people to make mistakes even if this takes the project back to the drawing board.

Management _

Embedding a design approach in an organization requires turning it into structures and processes, to make it formally part of the activities being carried out. This applies to different levels of management and work practice, from decisions made on a strategic level down to operational management of design-related resources and assets, as well as managing design professionals and working with external partners.

The idea behind the Enterprise Design framework and strategic design initiatives is to bring together aspects often seen as separate matters. However, putting design practice into departments, defined methods, and governed procedures comes at the risk of losing the very characteristics the business aims to profit from. A particular challenge is to make sure design practice is deeply linked to all key parties and everything going on in the enterprise, but at the same time keep design teams independent and agile.

This also means being quite selective about who to involve, and positioning the design approach alongside other valid paradigms such as quality management and continuous improvement, making them co-exist and contribute to a greater whole.

Many projects fall short of their goals because they have a restricted scope, or aim too closely at today's reality, which stifles their potential even before work begins. Conversely, too broad or far-reaching goals can make it hard to set a direction, and make design teams involve too many parties, leading to a mass of conflicting interests and opinions. We have seen most success stories use a balanced approach.

_ *Executive support:* Create a steering group to prioritize and set themes to be tackled by design initiatives. Strategic design requires strong backing by senior executives, on the CEO level or very close to it. Depending on the selected theme, a particular function or line could take a lead role, but a global and relevant theme often makes it hard to pick one.

_ *Two types of projects:* Plan a portfolio of strategic design initiatives (lead projects), involving external expertise, and corresponding to the chosen themes, to lead the transformation process. These projects aim to swiftly produce vision prototypes of potential futures. Translate selected outcomes into larger realization projects to define the details, build components and drive change processes.

_ *Mindful integration:* The theme should guide the selection of members of the core teams, external parties and contributors, stakeholders taking part in the steering group, and participants in research and validation.

_ *Governance structures:* Bringing visions to life requires a proactive management of design activities, separating global strategy and constraints from individual projects, and using selected themes to make them contribute to a larger vision. A governance team should manage the project portfolio, and create a framework that covers their shared base of global assets and guidelines.

_ Teams

Designing in the enterprise context requires building teams which are diverse in many ways. They need to bring together people representing a variety of disciplines depending on the theme to be tackled and the environment where a transformation is supposed to happen. Moreover, they need the ability to engage in a dialogue about strategic priorities and scenarios with people from an even greater variety of backgrounds, achieving a shared understanding with executives, technologists, domain experts, and most important, people addressed by their designs.

The best design teams we have experienced combine several traits and characteristics. They keep a high level of motivation and dedication even in times of great ambiguity and uncertainty. They share ideas and thoughts without switching into a mode of compromise or committee-like decision making. Experts lead the work in their respective fields, without losing sight of or interest in what others are doing, constantly making an effort to align different viewpoints.

Most of these characteristics are related to social dynamics and human behavior, elements which are not easy to change or influence. Therefore, organizations have to bring together the right people and provide suitable conditions for design to thrive. In most cases, this means working with external design professionals who can bring in the necessary expertise and guide the approach, and also contribute an external and unbiased view on the themes and challenges to be tackled

_ *Design leadership:* Find people combining design expertise with a strong enterprise vision to lead the projects, identify key aspects, and generate visions of the future enterprise. In addition to experience in creative processes, design leads need two capabilities—getting their team to becreative and excited to contribute to something great, and at the same time being demanding and reluctant to compromise on the overall vision in order to sustain it through to implementation.

_ *Mixed teams:* The diversity of teams is vital to achieving the open minds needed to come up with bold ideas. Working with aspects of the Enterprise Design framework enables us to identify disciplines and backgrounds to involve in a project. A good design team mixes generalist and specialist profiles, junior and senior people, as well as team members from different functions or units in your own organization with external professionals.

_ *Building bridges:* Use research professionals and design generalists (Hybrid Thinkers) as connectors, bridging lenses applied by members of the design team, and reaching out to the project environment. Such profiles help when developing the understanding needed to inform design work, as well as for engaging with stakeholders to mix in concerns and priorities.

Engagement _

The role and strategy of a strategic design program depends largely on the enterprise environment it is embedded in. In business practice, the term *enterprise* is often equaled with large organizations, and integrating strategic design into a large corporate environment is a key scenario for the Enterprise Design framework. But also in other cases, the multitude of different things to consider makes its twenty aspects highly applicable to guide any design program.

At eda.c, we applied the framework in different cases, with clients of varying size, from diverse industries and following different strategic objectives. Depending on the environment, there are different strategies to engage a client or organization to define and face a design challenge.

_ *Large organizations:* this applies to work with large companies and international groups, but also public bodies or non-profit organizations. Guided by the aspects to be considered, the scale of design projects tend to grow quickly, learning about a multitude of interests and parallel projects already in the discovery phase. Therefore, it is important to keep the project in a fluid state for some time, allowing for research and alignment activities and questioning the original briefing. Nevertheless, it is critical to quickly generate and communicate a vision of a potential future, for example, by producing a comic or video to be shared. Instead of attempting to boil the ocean, the initial goal is to create the engagement and visibility needed to go a first step. Even if the proposed initial outcome is rather limited in scope, it helps to align people on a more significant transformation in a later stage.

_ *Small and medium organizations:* Compared to large corporations, working with small organizations is somewhat different. While the exposure to the organization and involvement of executives can be achieved much more easily, it often entails achieving significant change within limited time and tight budget restrictions. When applying the Enterprise Design framework to such an environment, it is vital to early select a limited set of aspects considered initially relevant. This enables a rigorous scoping of the early project phases to get from research to vision work in a matter of weeks, guiding subsequent larger implementation-oriented projects.

_ *Startups:* Working with (technology) startups is particularly exciting, because it entails both the opportunity and the challenge to invent everything from scratch. The aspects of the Enterprise Design framework are a good guide to selecting the challenges to tackle and defining the overarching design theme. The often chaotic way startups work requires working virtually as part of the startup team of entrepreneurs, technologists, and often also shareholders. The process needs to focus on rapidly co-creating working prototypes, constantly expanding the scope to tackle the next most urgent design challenge.

FRAMEWORK _

Project work embedded in a strategic design program is characterized by the fact that the outcomes of design work are only meaningful as a means to gradually transform the whole enterprise toward a desired target state. In order to be effective, individual initiatives have to effectively contribute to a shared goal. The underlying strategy and design themes are suitable ways of expressing that goal across multiple teams and projects. But the challenge of achieving a coherent design of the enterprise requires us to make its components work together as a system.

In many ways, a strategic design program applied to an enterprise ecosystem is fundamentally different from a single design project. While the scope of such a program is very broad in comparison, it has a much lower level of control on individual outcomes. Instead of designing a finished object, artifact ,or other type of outcome, the goal of a design program has to be a transformation of the enterprise. Yet, any such change has to happen at a small scale of transactions, experiences, and communication flows.

Therefore, an important part of any such program is the elaboration of an overarching framework, put in place to support design work on individual challenges, and holding the different pieces and projects together. The main goal behind a framework is ensuring alignment and fit of outcomes into the enterprise context. Just as the Enterprise Design framework described in Part II, a design framework for a specific enterprise is about a set of aspects that is relevant for design teams to consider, aspects the enterprise strives to address in a coherent and meaningful fashion. Consequently, one of its key roles is that of a list of questions to be addressed in the course of the design process. In practice, a design framework serves several interrelated purposes of design work:

_ *Principles:* A list of answers to recurring questions in the form of principles, guiding conceptual decisions, and reflecting values and strategic priorities

_ *Guidelines:* A set of guidelines describing the bigger picture, how elements are related, and how to use them within common platforms and infrastructure

_ *Library:* A collection of resources, reusable elements, and solutions for common problems, which fit into the greater structure as interconnected modules

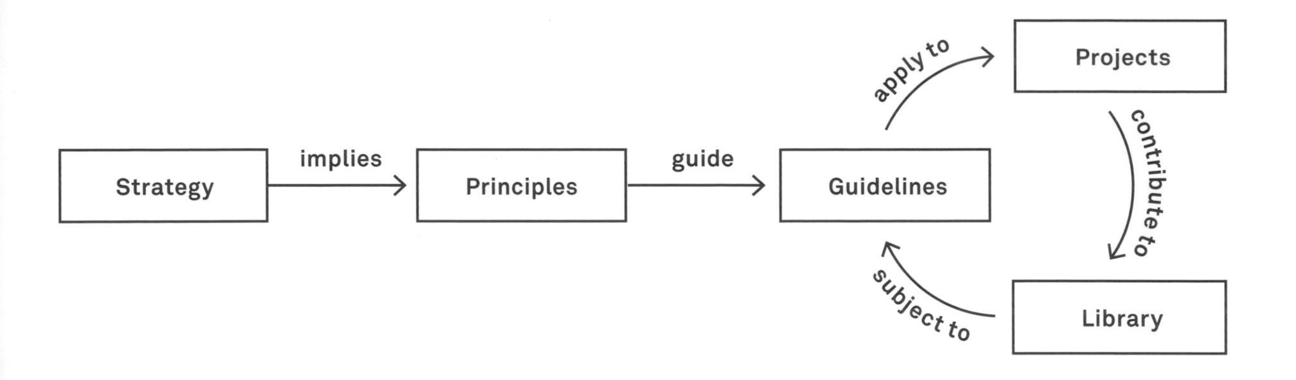

It is therefore essential to treat framework as a genuine design project, choosing adequate ways to describe the elements and convey them to the relevant audiences. The most difficult challenge of a design program in an enterprise environment is not its creation—with a strong underlying strategic vision made visible by lead projects, the direction to go becomes much easier to choose. Making all parts of the enterprise contribute to this idea, and putting it into action step by step can prove to be the harder part. The following elements are the basis for documenting a shared vision into online and offline media, and facilitating the conversations that engage all parts of the enterprise.

_ Principles

In design practice, principles are high-level statements that guide both design decisions and craft. For strategic design, such principles play a key role. They articulate the shared idea of the enterprise, its identity, values and beliefs. At the same time, they are derived from strategic priorities, experiential goals of key actors, and external conditions. The number of influencing factors is great, and therefore the creation of principles no trivial task.

Unlike principles as found in science or law, design principles do not represent universal truths of nature or binding regulations, but merely expressions of an attitude behind the creative process. They do not contain any direct instructions, requirements, or descriptions.

They should be formulated in a way that is actionable for teams, helping them set priorities and decide on what to address with the outcomes of a design process. But principles should also be generic enough to be applied across different contexts, being the traceable roots for individual requirements or design decisions.

Design is attitude.

Helmut Schmid

A set of well-defined design principles can express the thinking, motivations and mission of the enterprise, thereby giving design work a direction to pursue. Such principles should articulate the enterprise's strategic choices, design themes, and core ideas.

Guidelines _

Anyone already involved in traditional design work is probably quite familiar with the set of guidelines companies use to align the effort. Typically called *design manuals* or *style guides*, design guidelines specify global standards, components, and processes that apply to design projects, imposing design constraints and describing shared assets for reuse.

Most classic approaches of standardizing design work fall short of providing such a comprehensive foundation at the enterprise level, which results in projects and their outcomes drifting apart. Brand guidelines describe a set of visual elements, but ignore the behavioral aspects of identity. An architecture blueprint maps the way information systems are connected, but without grounding this in the Customer Experience served by those systems. For a design team, it is hard to develop a comprehensive view of the environment since many questions are not addressed. On the other hand, many practitioners feel quite constrained in their creative work by overly rigid instructions.

We found the best guidelines to adopt a systematic approach, by addressing each concern separately while rooting them in a fundamental set of principles. The Enterprise Design framework provides a good basis to do this, and allows us to develop our fundamental thinking and standards about the various aspects of design work. This includes the way an enterprise appears, communicates, behaves, operates, and structures its business. These guidelines then support our use of the library, the collection of assets to be reused in projects.

_ Library

In our own work at eda.c, we are asserting a paradigm shift from monolithic solutions and detailed deliverables to flexible systems. Instead of making websites, we are designing for apps, information feeds, and mashups. Instead of documents describing just one design, we are designing templates, standards, and patterns. This development is visible across all domains touched on in *Intersection*, and means that the focus of design work changes from looking at individual solutions to creating modular structures.

A design library is a catalogue of different types of modular elements to be leveraged, reassembled, and reused in design projects. It contains all elements that are considered part of the design framework, and provides design teams with the rationale, description, and documentation to incorporate them into their work.

_ *Patterns:* Documented solutions to recurring design challenges, to be adapted to their context of use

_ *Components:* Functional or structural assets ready for direct reuse in a project

_ *Platforms:* Assets used as infrastructure elements providing a foundation to base on

Again, the aspects described in the Enterprise Design framework are a good basis for defining the elements to include in a design library according to different concerns of design. The different elements managed here are then used as building blocks to speed up the design and implementation of new and enhanced outcomes, leveraging infrastructure and assets. These in turn add reusable outcomes as new elements of the library.

AT A GLANCE

A strategic design program aims to make an enterprise benefit from design practice, incorporating design into all aspects of its activities and endeavors. This requires alignment with strategic decisions, integrating design work into the organization, and ensuring that everything contributes to one greater vision of the enterprise's future.

Recommendations

_ Engage in a dialogue between strategy and design, translating the strategic vision into suitable design strategies, and in turn informing strategic decision making by turning the vision into tangible illustrations

_ Express the strategic intent behind all design work as a set of a few connected themes, capturing in a short statement how the future enterprise should appear in people's lives

_ Integrate design work into the organization, combining a formal organizational and operational model with by leadership, cultural change initiatives, and stakeholder engagement

_ Elaborate a design framework to align different design initiatives, leverage shared assets and reusable outcomes, and convey the common ground that brings everything together

OUTLOOK

Intersection is the result of a personal journey. Jumping from one engagement to the next, first as freelance designers and then with eda.c, we found the same challenges again and again. They were all related to the perceived scope of design work, the alignment among different professionals and departments, and the difficulty of giving the project a strategic dimension. These experiences resulted in a constant expansion of the aspects that we considered relevant to making a design initiative successful beyond ornamentation.

There is a lot to be done. In every project or program we are involved in, we find elements that don't fit together, and conditions that prevent good relationships from developing. When relationships fail, it is rarely a single issue that can be blamed. It is the interplay of all the parts which, together with the circumstances at hand, lead to a complex picture of problems and shortcomings. Turning this statement around, it is also that interplay that needs to be transformed to have an impact on the enterprise.

We strongly believe that applying design strategically to enterprise can make relationships much more humane, and enterprises less awkward or alienating. To do so, the practice itself will have to reflect and evolve, further expand its scope, and develop new approaches suited to this new dimension of design work.

Without any doubt, our house will gradually become more and more humane. The closer the machine gets to perfection, the better it hides behind its role.

Antoine de Saint-Exupéry

Instead of solutions for problems
programs for solutions.

Karl Gerstner

THE ENTERPRISE AS A PROGRAM _

In design history, there have been many proponents of a modular, systematic approach to creative work. They favor methodologies that result in systems of elements, in rule-based programs, grids, and modular forms that can be recombined to create new configurations, uses and renderings. Instead of working on just one artwork, object, or solution, the designer would develop a program that generates the outcomes by applying the rules as defined.

For design in enterprise ecosystems, such approaches become more and more relevant. Regardless of whether we are working on buildings, brand identity, web portals, or other types of design artifacts, the rising complexity demands the creation of programs rather than individual solutions. In light of digitization, social networks, and the shift from perception to participation, the role of a designer shifts to creating frameworks for others to configure and adapt. It means giving up control, favoring the flexible over the finished, and providing a space for others to fill with life.

We believe that strategic design work will necessarily have to adopt such a generative approach. It involves designing the enterprise as a program of framework elements, key components, and rules connecting these bits to form a coherent whole. Applied, it moves design work to the place where it can provide the most value, and takes into account the reality of a social age.

_THE SOCIAL ENTERPRISE

Most of the content in *Intersection* is linked to the world of business. Questions of strategy, operationalization, or customer focus are the typical themes of commercial organizations. Looking at other areas, we find the same issues and opportunities also in non-profits, public bodies, educational institutions, and other forms of enterprises.

Just like a company, any kind of organization has to develop a vision of where they want to be and follow a strategy to make it reality. It has to embed it as a shared idea with the stakeholders in its ecosystem—all of this is the essence of what an enterprise pursues. Moreover, we find new companies emerging which, while clearly making profits, have chosen a wicked social problem as their enterprise.

In most cases, it is not the a question of commercial orientation that makes a difference. In a talk at TED Madrid, designer and entrepreneur Harry West even made the case that the commercial aspect can act as the most important lever that gives an initiative the impact it needs to drive cultural change, by basing a solution on scalable and repeatable options and leveraging market mechanics. More than that, it is the choice of design challenge, researched in a design process and addressed with its subsequent problem-solving activities, which can give an enterprise a social dimension.

If man is to survive, he will have learned to take a delight in the essential differences between men and between cultures. He will learn that differences in ideas and attitudes are a delight, part of life's exciting variety, not something to fear.

Gene Roddenberry, the creator of *Star Trek*

ENTERPRISE — THE NEXT GENERATION _

In my youth, I watched a lot of science fiction series and movies. One that I enjoyed in particular was *Star Trek: The Next Generation*, where the crew of Captain Jean-Luc Picard embarked to explore the galaxy, keep peace, and have adventures. The future envisioned by Gene Roddenberry and his co-creators has many elements of a utopia. On its mission, the crew of the USS Enterprise acts based on altruistic values, and solves problems on an interstellar scale.

Today it strikes me how close we are to this vision—working and communicating anywhere in real time, handing each other tablet devices, travelling faster (although not yet instantaneously), and talking to our computers. But also how far away we still are from the humanitarian side of the vision—eliminating poverty and misery, saving our planet, and living together peacefully.

It is unclear who designed our universe, or our planet, along with all the things that surround us. But as far as the human race is concerned, a large portion of our reality is being continuously designed by us. I hope this book has given you the thinking, vocabulary, and approaches to make a part of this change your enterprise—to tackle the world's most challenging problems, and to redesign our reality for the better.

_ INDEX

_ REFERENCES

Erik Roscam Abbing (2010). *Brand-driven Innovation*. Lausanne: AVA Publishing
Douglas Adams (1979). *The Hitchhiker's Guide to the Galaxy*. London: Pan Books
Marcia J. Bates (2002). *Toward an Integrated Model of Information Seeking and Searching*, in New Review of Information Behaviour Research, Vol. 3, pp 1-15
Jörg Beringer, Karen Holtzblatt (2006). *Designing Composite Applications*. Bonn: SAP Press
Joy Beatty, Anthony Chen (2012). *Visual Models for Software Requirements*. Sebastopol: Microsoft Press
Richard J. Boland Jr., Fred Collopy (2004). *Design Matters for Management*, in Managing as Designing, pp 3-18. Stanford: Standford University Press
Brigitte Borja de Mozota (1990). *Design Management*. New York: Allworth Press
Brigitte Borja de Mozota (2006). *The Four Powers of Design*, in Design Management Review, Vol. 17.2. Boston: DMI
Tilmann Buddensieg, Henning Rogge (1993). *Industriekultur: Peter Behrens und die AEG 1907-1914*. 4th Edition. Berlin: Mann
Richard Buchanan (2001). *Design Research and the New Learning*, in Design Issues, Vol. 17, pp 3-23, MIT Press
Bill Buxton (2007). *Sketching User Experiences*. San Francisco: Morgan Kaufmann Publishers
John C Camillus (2008). *Strategy as a Wicked Problem*, in Harvard Business Review, May 2008 Edition, pp. 98-106
Alan Cooper (2004). *The Inmates Are Running the Asylum*. Indianapolis: Sams
Alan Cooper, Robert Reimann, David Cronin (2007). *About Face 3*. Indianapolis: Wiley
Eric Evans (2003). *Domain-driven Design*. Boston: Addison-Wesley
Roger Evernden, Elaine Evernden (2003). *Information First*. Burlington: Butterworth-Heinemann
Antoine Fenoglio, Frédéric Lecourt, Claire Fayolle (2009). *L'objet du design*. Paris: Cité du Design
Edward R. Freeman (1984). *Strategic Management: A Stakeholder Approach*. Boston: Pitman
Matthew Frederick (2007). *101 Things I Learned in Architecture School*. Cambridge: MIT Press
Nicholas Gall, David Newman, Philip Allega, Anne Lapkin, Robert A. Handler (2010). *Introducing Hybrid Thinking for Transformation, Innovation and Strategy*. Gartner Research
Jesse James Garrett (2003). *The Elements of User Experience*. Berkeley: New Riders
Karl Gerstner (2007). *Programme entwerfen*. Baden: Lars Müller Publishers (Original Edition from 1973: Programme entwerfen. Basel: Niggli)
Tom Graves (2009). *The service-oriented enterprise*. Colchester: Tetradian Books
Tom Graves (2012). *The enterprise as story*. Colchester: Tetradian Books

Kristina Halvorson, Melissa Rach (2009). *Content Strategy for the Web.* Berkeley: New Riders
William Hudson (2001). *Toward Unified Models in User-Centered and Object-Oriented Design,* in Object Modeling and User Interface Design, pp. 313-362. Boston: Addison-Wesley
Mary Jo Hatch, Majken Schultz (2008). *Taking Brand Initiative.* San Francisco: Jossey-Bass
Karen Holtzblatt, Jessamyn Burns Wendell, Shelley Wood (2005). *Rapid Contextual Design.* San Francisco: Morgan Kaufmann
Robert Johansen (1988). *Groupware.* New York: Free Press
James Kalbach, Paul Kahn (2011). *Locating Value with Alignment Diagrams,* in Parsons Journal of Information Mapping, Vol. 3.2, pp. 1-11
Wolfgang Keller (2007). *IT-Unternehmensarchitektur.* Heidelberg: dpunkt.verlag
W. Chan Kim, Renée Mauborgne (2005). *Blue Ocean Strategy.* Boston: Harvard Business Press
John Kolko (2007). *Thoughts on Interaction Design.* Savannah: Brown Bear
John Kolko (2011). *Exposing the Magic of Design.* New York: Oxford University Press
Gene Leganza (2010). *Topic Overview Information Architecture.* Forrester Research
Löwgren, Jonas (2008): *Interaction Design.* In: Encyclopedia of Human-Computer Interaction. Aarhus: The Interaction-Design.org Foundation
Roger Martin (2009). *The Opposable Mind.* Boston: Harvard Business Press
Peter Merholz, Brandon Schauer, David Verba, Todd Wilkens (2008). *Subject to Change.* Sebastopol: O'Reilly Media
Bill Moggridge (2006). *Designing Interaction.* Cambridge: MIT Press
Bill Moggridge (2010). *Designing Media.* Cambridge: MIT Press
Roger L. Martin (2009). *The Design of Business.* Boston: Harvard Business Press
Clement Mok (1996). *Designing Business.* Berkeley: Adobe Press
Peter Morville, Louis Rosenfeld (2006). *Information Architecture for the World Wide Web,* 3rd Edition. Sebastopol: O'Reilly
Marty Neumeier (2005). *The Brand Gap.* Berkeley: New Riders
Marty Neumeier (2006). *Zag.* Berkeley: New Riders
Marty Neumeier (2009). *The Designful Company.* Berkeley: New Riders
Don Norman (2002). *The Design of Everyday Things.* New York: Basic Books (Original Edition from 1988: The Psychology of Everyday Things. New York: Basic Books)
Don Norman (2004). *Emotional Design.* New York: Basic Books
Wally Olins (2008). *The Brand Handbook.* London: Thames & Hudson

Martin Op 't Land, Erik Proper, Maarten Waage, Jeroen Cloo, Claudia Steguis (2009). *Enterprise Architecture*. Berlin: Springer

Alexander Osterwalder, Yves Pigneur, 470 practitioners from 45 countries (2010). *Business Model Generation*. Second printing, self published

Chris Potts (2010). *RecrEAtion*. Bradley Beach: Technics Publications

Richard Poulin (2011). *The Language of Graphic Design*. Minneapolis: Rockport Publishers

Andrea Resmini, Luca Rosati (2011). *Pervasive Information Architecture*. San Francisco: Morgan Kaufmann Publishers

Horst Rittel, Melvin Webber (1973). *Dilemmas in a General Theory of Planning*, in Policy Sciences, Vol. 4. Amsterdam: Elsevier Scientific Publishing

Victor Malsy, Philipp Teufel, Fjodor Gejko (2002). *Helmut Schmid*. Berlin: Birkhäuser

Herbert A. Simon (1997). *The future of information systems*, in Annals of Operations Research, Vol. 17, pp 3-14. Baltzer

Phil Simon (2011). *The Age of the Platform*. Henderson: Motion Publishing

Michael Shanks (2011). *We have always been cyborgs*, in Archaeology and Things, a special issue of Science, Technology & Human Values, SAGE Journals

Lynn Shostack. *Designing Services that Deliver*. Harvard Business Review, vol. 62, no. 1 January - February 1984, pp. 133–139

Marc Stickdorn, Jakob Schneider (2010). *This is Service Design Thinking*. Amsterdam: BIS Publishers

Roberto Verganti (2009). *Design-Driven Innovation*. Boston: Harvard Business Press

Fred Wiersema, Michael Treacy (1993). *Customer Intimacy and Other Value Disciplines*, in Harvard Business Review, Vol 71.1, pp. 84-93

David Weinberger, Rick Levine, Christopher Locke, Doc Searls (1999). *The Cluetrain Manifesto*. New York: Perseus Books

Indi Young (2008). *Mental Models*. New York: Rosenfeld Media

John A. Zachman (1987). *A Framework for Information Systems Architecture*, in IBM Systems Journal, Vol. 26, pp. 276-292

Design Council (2006). *A Study of the Design Process*. London: The Design Council

Institute for the Future (2009). *Blended Reality Report*. Technology Horizons Program

OECD (2001). *The Internet and Business Performance*. OECD Digital Economy Papers, No. 57. OECD Publishing

IMAGE CREDITS _

19_left © Tristan3D / Fotolia.com, middle © dimedrol68 / Fotolia.com; 32_AEG Archiv / Stiftung Deutsches Technikmuseum; 34_AEG Archiv / Stiftung Deutsches Technikmuseum; 35_AEG Archiv / Stiftung Deutsches Technikmuseum; 38_sierragoddess (Darla Hueske) / Flickr.com; 44_Paul Butler / Facebook.com; 58_La 27e Région; 60_La 27e Région; 68_Dmitry Fironov; 80_Randy Le'Moine Photography (Randy Lemoine) / Flickr.com; 84_goodrob13 (Rob DiCaterino) / Flickr.com; 91_left © Ilan Amith / Fotolia.com, middle © Jan Kranendonk / Fotolia.com, right © mema / Fotolia.com; 96_NASA Earth Observatory; 106_photohome_uk (steve gibson) / Flickr.com; 108_CERN / Cern.ch; 124 _left © Leonardo Franko / Fotolia.com, middle © famwa / Fotolia.com, right © Studio DER / Fotolia.com; 128_used with the permission of Inter IKEA Systems B.V.; 131_used with the permission of Inter IKEA Systems B.V.; 132_used with the permission of Inter IKEA Systems B.V.; 135_left slasher-fun (Mathieu Marquer) / Flickr.com, right © Benicce / Fotolia.com; 144_SML See-ming Lee / Flickr.com; 157_left © Rohit Seth / Fotolia.com, right © atlasphoto / Fotolia.com; 176_© THesIMPLIFY / Fotolia.com; 178_Verband der Agenturen (VdA); 179_Verband der Agenturen (VdA); 181_Verband der Agenturen (VdA); 194_Business Model Foundry / Businessmodelgeneration.com; 232_top Mike Atherton, bottom BBC Food / bbc.co.uk - © 2012 BBC; 240_Jeppesen Sanderson, Inc.; 244_Jeppesen Sanderson, Inc.; 246_Jeppesen Sanderson, Inc.; 247_Jeppesen Sanderson, Inc.; 268_kozumel (Camilo Rueda López) / Flickr.com; 291_© Miredi / Fotolia.com; 309_© Borna_Mir / Fotolia.com; 323_Indi Young; 326_SAP AG; 328_SAP AG; 329_SAP AG; 330_SAP AG; 331_SAP AG; 333_SAP AG; 341_Uwe Hermann / Flickr.com; 351_top © japolia / Fotolia.com, bottom © goodluz / Fotolia.com; 361_left © hacohob / Fotolia.com, right FutUndBeidl / Flickr.com; 376_SENSEable City Laboratory / MIT / BBVA; 379_IDEO / BBVA; 380_BBVA; 411_Damien Newman; 412_UK Design Council; 414_Edited with Instagram; 420_Screenshot Instagram; 424_Based on a model by Michael Treacy / Fred Wiersema; 426_Based on a model by Michael Treacy / Fred Wiersema; 428_Based on a model by Michael Treacy / Fred Wiersema

Every effort has been made to contact copyright owners and obtain permission. The author and publisher apologize for any omission or mistake and request that the copyright owners contact us for corrections in future editions.

_ THE TEAM

_ Author

Milan Guenther is a founding partner of the enterprise design associates—eda.c, a European strategic design consultancy created in 2009. Milan has a rather mixed background, graduated in Communication Design at the Fachhochschule Düsseldorf (Germany), and holding an MBA specializing in Business Information Systems from the Institut Supérieur du Commerce in Paris. Having worked as a Freelancer most of his professional life, he touched the many areas related to Experience Design, Branding and Enterprise Architecture, and has been a popular speaker on international conferences on these subjects. He is living in Paris, or wherever his project work might take him.

_ Design Team

Benjamin Falke
Sarah Matzke
Dennis Middeke
Eva Pika
Andrea Santema

_ Copyeditor

Philip Hellyer

_ ACKNOWLEDGEMENTS

This book is the result of a long thinking process, sparked in 2007 during a conference session in Nancy, France (see The story behind this book). It would not exist without the help, contributions and critique of the people mentioned below, who are listed in random order.

_ My reviewers

Sally Bean, Paula Thornton, Katharina Weber, Eva Pika, and Philip Hellyer

_ My partners at eda.c

Benjamin Falke and Dennis Middeke

_ The team at MK

especially Robyn Day, Andrea Dierna, Lawrence Shanmugaraj, Mohanambal Natarajan, and Steve Elliot. Also the editors that helped me through the process of making a book proposal: Mary James, Rachel Roumeliotis, Jenifer Niles, and David Bevans

The subject of *Intersection* was also the theme of my master's thesis at the Institut Supérieur du Commerce in Paris, so I would like to thank everyone who supported this work: Gilles Enguehard, Stephan Schillerwein, Jean-René Ruault, Matthew West, Bryan Minihan, and Bob Goodman.

I also would like to thank everyone contributing to the 9 case studies: S. Kirsten Gay, Melissa Visintin, Marian Gunkel and Dirk Dobiéy from SAP, Jens Schiefele, Thorsten Wiesemann and Cindy Dorfmann from Jeppesen, Michael Zirlewagen from °visualcosmos, Ignacio Villoch from BBVA, Dirk Fehrecke and Bernhard Hoestermann from VdA, Stéphane Vincent and Romain Thévenet from La 27ème Région, and Cecilia Emanuelson and Susanne Rolf from IKEA.

Thanks for many inspiring conversations _

Sally Bean, Paula Thornton, Katharina Weber, Sylvie Daumal, Peter Bogaards, Louis Rosenfeld, Sylvain Cottong, James Kalbach, Nigel Green, Paul Teeuwen, Andrea Resmini, Luca Rosati, Roger Evernden, Tom Graves, Nick Gall, Jason Hobbs, Julian Masuhr, Jan Jursa, Mark McElhaw, Hervé Mischler, Alec Sharp, Ansgar Sporkmann, Chris Potts, Paul Miller, Arnaud Bonhomme, Fadi Dayoub, Melissa Visintin, Johann Sarmiento, Marian Gunkel, Tom Hirt, Olivier Ageron, Christina Wodtke, Christina Eddiks, Mark Morell, Benoît Drouillat, Markus Weber, Thomas Immich, Thomas Küber, James Robertson, John Zachman, Janna DeVylder, Wiesław Kotecki, Hubert Anyzewski, Mike Atherton, Philipp Teufel, Victor Malsy, Gene Leganza, John Gøtze, Joe Lamantia, Frank Rausch, Timm Kekeritz, Giuseppe Attoma Pepe, Paul Kahn, Louise Kennedy, Søren Muus, Stephan Schillerwein, Nick Marsh, Nathan Shedroff, Brigitte Borja de Mozota, Marty Neumeier, Marc Stickdorn, Jakob Schneider, Clement Mok, Peter Bogaards, Claudia Alsdorf, Enno Lüthje, Cennydd Bowles, Silvia Calvet, Lee Clemmer, Hervé Mischler, Florence Gailledreau, Natacha Hennocq, William Hudson, Stefan Krüger, Joe Lamantia, Petra Meyer, Tim Ostler, Stephan Schmotz, Pauline Thomas, Martin Janssen, and Mike Clark. My apologies for everyone I have forgotten to mention here.